Molybdenum (Mo) deficiencies in field-grown plants were first
recorded in Australia more than 55 years ago, and this book
condenses all the information currently available on the subject of Mo
as it relates to soils, crops, and livestock.

The book reviews our knowledge of the chemistry and mineralogy
of Mo, the extraction of available Mo from various soils, the various
analytical methods of determining Mo content in soils and plants,
the biochemical role of Mo in crop production, the technology and
application of Mo fertilizers to crops, the responses to Mo of various
temperate and tropical crops, Mo deficiency and toxicity in various
plant species, the interaction of Mo with other plant nutrients, and the
distribution of Mo within the plant. Factors affecting the availability of
soil Mo to plants and Mo status in the semiarid and subhumid tropics
are also discussed.

The book will be a worthwhile reference tool to assist agricultural
researchers, professors, and extension personnel.

MOLYBDENUM IN AGRICULTURE

MOLYBDENUM IN AGRICULTURE

Edited by
UMESH C. GUPTA
Agriculture and Agri-Food Canada
Research Centre, Charlottetown

CAMBRIDGE
UNIVERSITY PRESS

CAMBRIDGE UNIVERSITY PRESS
Cambridge, New York, Melbourne, Madrid, Cape Town, Singapore, São Paulo

Cambridge University Press
The Edinburgh Building, Cambridge CB2 8RU, UK

Published in the United States of America by Cambridge University Press, New York

www.cambridge.org
Information on this title: www.cambridge.org/9780521571210

© Cambridge University Press 1997

First published 1997
This digitally printed version 2007

A catalogue record for this publication is available from the British Library

Library of Congress Cataloguing in Publication data
Molybdenum in agriculture / edited by Umesh C. Gupta.
p.   cm.
Includes index.
ISBN 0-521-57121-9 (hc)
1. Molybdenum in agriculture.   I. Gupta, Umesh C.
S587.5.M64M65   1997
631.8′1 – dc20   96-14069
                                                    CIP

ISBN 978-0-521-57121-0 hardback
ISBN 978-0-521-03722-8 paperback

Contents

Contributors

James F. Adams, Ph.D.
Department of Agronomy
   and Soils
Auburn University
Auburn, Alabama 36849-5412

Laurie S. Balistrieri, M.Sc.
U.S. Geological Survey
School of Oceanography WB10
University of Washington
Seattle, Washington 98195

C. Chatterjee, Ph.D.
Botany Department
Lucknow University
Lucknow 226 007, U.P.
India

Frieda Eivazi, Ph.D.
Division of Food and Agricultural
   Sciences Cooperative Research
Lincoln University
Jefferson City, Missouri
   65102-0029

Umesh C. Gupta, Ph.D.
Agriculture and Agri-Food
   Canada
Research Centre
P.O. Box 1210
Charlottetown
P.E.I. C1A 7M8
Canada

Chris Johansen, Ph.D.
ICRISAT
Patancheru
Andhra Pradesh 502 324
India

Peter C. Kerridge, Ph.D.
CIAT
Apartado Aero 6713
Cali, Colombia

J. A. MacLeod, Ph.D.
Agriculture and Agri-Food
   Canada
Research Centre
P.O. Box 1210
Charlottetown
P.E.I. C1A 7M8
Canada

John J. Mortvedt, Ph.D.
6420 Compton Road
Fort Collins, Colorado 80523

Larry C. Munn, Ph.D.
Department of Plant, Soil, and
   Insect Sciences
P.O. Box 3354
University of Wyoming
Laramie, Wyoming 82071-3354

V. K. Nayyar, Ph.D.
Department of Soil Science
Punjab Agricultural University
Ludhiana 141 004
Punjab, India

N. S. Pasricha, Ph.D.
Department of Soil Science
Punjab Agricultural University
Ludhiana 141 004
Punjab, India

Katta J. Reddy, Ph.D.
Wyoming Water Resources
   Center
University of Wyoming
P.O. Box 3067
Laramie, Wyoming 82071-3067

R. C. Severson, Ph.D.
U.S. Geological Survey
Box 25046, M.S. 973
Denver, Colorado 80225-0046

C. P. Sharma, Ph.D.
Botany Department
Lucknow University
Lucknow 226 007, U.P.
India

John L. Sims, Ph.D.
Department of Agronomy
N-122 Agricultural Science Bldg.
   North
University of Kentucky
Lexington, Kentucky 40546-0091

R. Singh, Ph.D.
Indo Gulf Fertilizers and
   Chemical Corporation Limited
312-A World Trade Centre
Barakhamba Lane
New Delhi 110 001
India

Kathleen S. Smith, Ph.D.
U.S. Geological Survey
Box 25046, M.S. 973
Denver, Colorado 80225-0046

Steven M. Smith, M.Sc
U.S. Geological Survey
Box 25046, M.S. 973
Denver, Colorado 80225-0046

P. C. Srivastava, Ph.D.
Department of Soil Science
G.B. Pant University of
   Agriculture and Technology
Pantnagar 263 145, U.P.
India

Barrie Stanfield, M.S.A.
Agriculture and Agri-Food
   Canada
Research Centre
P.O. Box 1210
Charlottetown
P.E.I. C1A 7M8
Canada

Adib Sultana, Ph.D.
ICRISAT
Patancheru
Andhra Pradesh 502 324
India

Liyuan Wang, Ph.D.
Mineral and Chemical
   Division
J.R. Simplot Company
P.O. Box 912
Pocatello, Idaho 83204

Preface

The chief purpose of preparing this book was to condense all the information available on the subject of molybdenum (Mo) as it relates to soils, crops, and livestock. Because the problems related to the requirements for Mo in soil and in crop production differ considerably from one part of the world to another, I attempted to solicit the assistance of experts from the different regions of the world who were best suited to write about the topics of the various chapters. These contributions by authors from different geographical areas have helped to provide a broader viewpoint of the subject matter than would have been the case if only a single author had prepared the book in its entirety.

Molybdenum deficiencies in field-grown plants were first recorded in Australia more than 55 years ago. This book contains a chapter authored by two Australian scientists, Drs. Chris Johansen and Peter Kerridge, who have advanced our understanding of the responses of agricultural plants to Mo in Australia and elsewhere, particularly in tropical regions. Currently, they are senior research managers at international agricultural research institutes: ICRISAT in India and CIAT in Colombia, respectively.

This book reviews our current knowledge of the following topics: the chemistry and mineralogy of Mo, the extraction of available Mo from various soils, analytical methods of determining Mo content in soils and plants, the biochemical role of Mo in crop production, the technology and application of Mo fertilizers for crops, the responses to Mo of various temperate and tropical crops, Mo deficiency and toxicity in various plant species, the interactions of Mo with other plant nutrients, and the distribution of Mo among plant parts. Factors affecting the availability of soil Mo to plants and Mo status in the semiarid and subhumid tropics are also discussed.

The editor is indebted to all the authors of the chapters of this book for their complete cooperation in the compilation of this comprehensive volume. It should prove to be a valuable reference tool to assist agricultural researchers, professors, and extension personnel.

My sincere thanks to my wife, Sharda Gupta, for her patience and understanding while I spent many, many hours at home on weekends and during weekday evenings working on this book. Thanks are due also to my three sons, Sharad, Kamal, and Subhas, for their unflagging support and encouragement during the preparation of this book.

Umesh C. Gupta
Charlottetown, P.E.I., Canada

1

Introduction

UMESH C. GUPTA

Molybdenum (Mo) is one of seven recently identified trace elements that are essential for plant growth. It is the only transition element in group VI in the periodic table that is essential for normal growth, metabolism, and reproduction of higher plants. The biological importance of Mo in plants is due to its highly beneficial action in the fixation of nitrogen, from the air, by the nitrogen-fixing bacterium (*Azotobacter chroococcum*). After the establishment of its essentiality by scientists in Australia more than half a century ago, its deficiency has been reported in several countries in a variety of crops. The agricultural researchers in Australia were able to overcome the symptoms of Mo deficiency in tomatoes (*Lycopersicon esculentum* Mill.) by addition of minute quantities of Mo in the nutrient solution. Some of the crops considered most sensitive to Mo deficiency are clovers (*Trifolium subterraneum* L.), cauliflower (*Brassica oleracea* var. *botrytis* L.), broccoli (*Brassica oleracea* L. Botrytis Group), rape (*Brassica napus* L.), beet (*Beta vulgaris* L.), spinach (*Spinacea oleracea* L.), lettuce (*Lactuca sativa* L.), and alfalfa (*Medicago sativa* L.).

Among the micronutrients, Mo is an exception in that it is readily translocated, and its deficiency symptoms generally appear on the whole plant. The deficiency symptoms for other micronutrients appear on the young leaves at the top of the plant because of their inability to translocate within the plant. Molybdenum deficiency emerges as general yellowing and stunting of the plant, interveinal mottling, and cupping of the older leaves, followed by necrotic spots at leaf tips and margins. The presence of large quantities of Mo in plants, on the order of 100–200 mg kg^{-1}, does not produce harmful effects on crop yields nor any abnormal symptoms on the plant foliage.

The absorption of Mo by plants is generally considered to take place by mass flow and root interception. Plant roots absorb Mo in the form of the anion MoO_4^{2-}. This anion form of Mo is mobile in the plant, and when applied to the primary leaves it can be transported to the stem and roots. Because $H_2MoO_4^{2-}$ is extensively dissociated in the pH range of 5–6, the anion MoO_4^{2-} will therefore be the predominant form of Mo in plant xylem, assuming that it does not associate with other plant constituents. It has been suggested that the form of Mo translocated in plants is still unknown, and the possibility of organic complexing cannot be excluded.

With regard to forms of Mo in soils, the matter has not been examined to any great extent by fractionation procedures. Generally, Mo has been found to occur in the following forms: (1) water-soluble Mo present in the soil solution, (2) Mo adsorbed by soil colloids, (3) Mo held in the crystal lattices of minerals, and (4) Mo present in organic matter. Studies on alluvial and desert soils have shown that 88–94% of the soil Mo is considered to be unavailable. Most Mo occurs in the amorphous Fe oxide fraction. The forms most available for plant use are the soluble forms present in the soil solution and Mo adsorbed by soil colloids. Highly weathered acid soils are apt to be more deficient in Mo. On the other hand, soils that are derived from granitic rocks, shells, slates, or argillaceous schists tend to be high in Mo. Alkaline and poorly drained soils with a high water table tend to produce plants high in Mo.

In the 1980s and 1990s, some progress has been made in the use of analytical techniques for determining Mo in soils and crops. It has not been researched as extensively as other micronutrients because its deficiency is not as widespread as those of the other micronutrients. In addition to the colorimetric methods used in the past, it can now be successfully analyzed by graphite-furnace atomic-absorption spectrometry and direct-current plasma-emission spectrometry.

Soil pH is one of the most important factors that affect the availability of Mo to plants. There are interactions between Mo and a number of nutrients, such as sulfur, nitrogen, phosphorus, and copper, that can affect its plant availability. Although large concentrations of Mo show no effects on crop yields of grains and forage crops, feeds containing Mo in excess of $10 \, mg \, kg^{-1}$, when fed to ruminants, can produce severe Mo toxicity (Mo-induced copper deficiency).

There have been some advances in our knowledge of Mo in several areas: analytical determination of soil and plant Mo, soil testing for Mo availability, establishing its deficiency levels and determining the responses to Mo in a variety of crops, interrelationships between Mo and other nutrients in plants and livestock, and factors affecting Mo uptake by crops. But there is a lack of such information assembled into a single publication. The objective of this book is to provide readers up-to-date knowledge of the various aspects of Mo in soil and crops and its relationship to livestock nutrition, as described by researchers from around the globe.

2

Chemistry and Mineralogy of Molybdenum in Soils

KATTA J. REDDY, LARRY C. MUNN,
and LIYUAN WANG

Introduction

Molybdenum (Mo) is important in ecosystems as a micronutrient for both plants and animals. It can also accumulate in the environment in toxic concentrations. Molybdenum is used widely in industrial societies and is an important fertilizer element in some agricultural systems. Soil Mo averages approximately $1.0–2.3\,mg\,kg^{-1}$ as a crustal constituent, making it 53rd in abundance (Krauskopf, 1979), but it can accumulate as a result of biogeochemical cycling to $300\,mg\,kg^{-1}$ or more in shales rich in organic matter. However, the common range of Mo concentrations in U.S. soils is $0.8–3.3\,mg\,kg^{-1}$ (dry weight) (Kubota, 1977). In soils, Mo can be found in four major fractions: (1) dissolved Mo in soil solution (water-soluble), (2) Mo occluded with oxides (e.g., Al, Fe, and Mn oxides), (3) Mo solid phases [e.g., molybdenite (MoS_2), powellite ($CaMoO_4$), ferrimolybdite ($Fe_2(MoO_4)_3$), wulfenite ($PbMoO_4$)], and (4) Mo associated with organic compounds.

Numerous processes take place in soil solution, including plant uptake, ion complexation, adsorption and desorption, and precipitation and dissolution (Figure 2.1). As shown in Figure 2.1, Mo solid phases dissolve upon contact with water and provide dissolved Mo in soil solution. The free molybdate ion (MoO_4^{2-}) reacts with metals to form complexes and ion pairs in soil solution. Plants absorb dissolved Mo, mainly as MoO_4^{2-}, from soil solution. Removal of MoO_4^{2-} by plants disrupts the electroneutrality of a soil solution. This causes desorption and adsorption of Mo by oxides, as well as dissolution and precipitation of Mo solid phases in soil solution, until charge balance is reached. The speciation of dissolved Mo in soil solutions must be understood in order to quantitatively describe the availability, toxicity, adsorption, and pre-

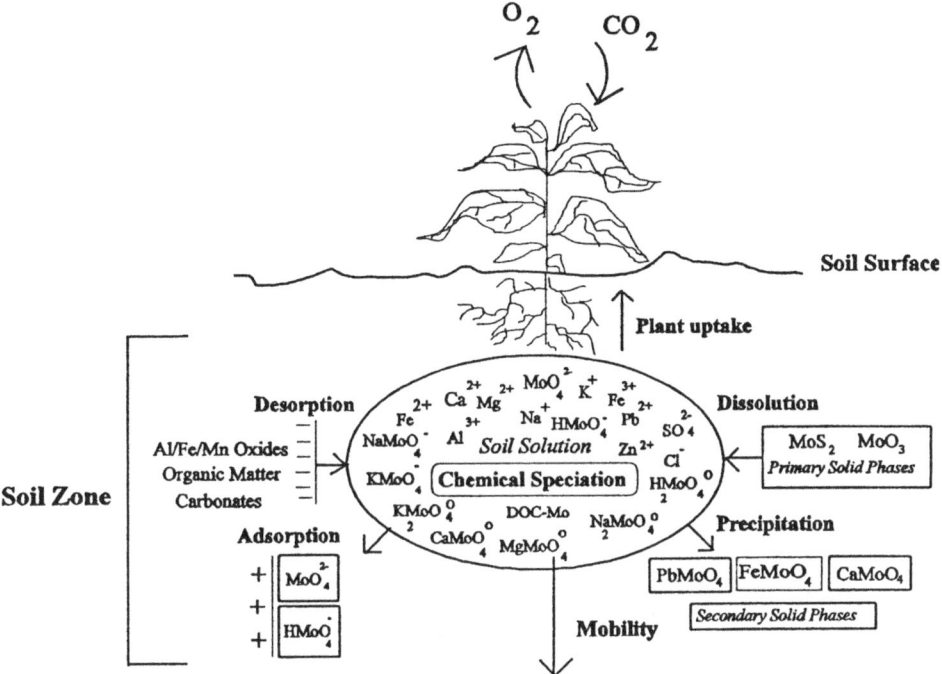

Figure 2.1. Schematic representation of a hypothetical soil zone and the various chemical processes that occur in a soil solution.

cipitation processes of Mo in soils. Oxidation of organic compounds and sulfides containing Mo can also contribute dissolved Mo to soil solution. For example, during surface coal mining, which is an important land use in the western United States, the soil material above the coal, exposed to the atmosphere, is removed. This process oxidizes organic compounds and sulfides containing Mo and contributes dissolved Mo to soil solution (Wang, Reddy, and Munn, 1994). Overall, the solubility, availability, and mobility of Mo in soil solution are functions of chemical form, pH, mineralogy, Mo saturation, adsorption, competing ions (primarily Fe and S), and precipitation (Vlek and Lindsay, 1977).

The purpose of this chapter is to review our present knowledge of the mineralogy and chemical processes controlling dissolved Mo in soils. This chapter emphasizes speciation, adsorption and desorption, and the precipitation and dissolution processes of dissolved Mo. In addition, the importance of dissolved organic carbon in these processes is discussed.

Mineralogy of Molybdenum in Soils

Molybdenum is found as a primary element in granites and granitic geologic terrains. It is also found in minerals such as molybdenite and ferrimolybdite, which are mined commercially (Adriano, 1986). Molybdenum also occurs as powellite, wulfenite, and ilsemannite (Mo_3O_8). In industry, Mo is used for the production of steels and alloys (84%) and as a chemical catalyst and in the manufacture of plastics, lubricants, and pigments (16%). The uses of Mo in the manufacture of plastics, pigments, and lubricants and as a catalyst have been increasing because of the relatively nontoxic properties of Mo as compared with potential substitutes: chromium (Cr), manganese (Mn), boron (B), and nickel (Ni) (Blossom, 1991).

Only a small portion of the Mo produced is used in agriculture. Molybdenum is used for fertilizer in the forms of molybdenite, molybdic oxide (MoO_3), and sodium and ammonium molybdates [($Na_2MoO_4 \cdot 2H_2O$ and ($NH_4)_2Mo_2O_7$)]. Molybdenum has a high affinity for iron (Fe) at high temperature and for lead (Pb) and calcium (Ca) at low temperature (Enzmann, 1972). In animals, Mo interacts with copper (Cu), which can result in Cu deficiency (molybdenosis) under some circumstances, as described in detail in Chapter 15.

Soil Mo bioaccumulates in the A horizons of well-drained soils and accumulates in the subsoil under poorly drained mineral soils and in Histosols (organic soils). Molybdenosis is referred to as "peat scour" by farmers in the United Kingdom (Thornton and Webb, 1980). Within soil profiles, Mo concentrations and forms vary with the chemical environment of the soil solution and with the nature of the soil adsorptive complex. For example, Mo in the A horizon might be associated with soil humic materials, whereas in the Bt horizon it might be adsorbed to iron oxide coatings on the soil mineral grains. Typically, high concentrations of Mo in soils have been reported to occur in soil horizons within reducing environments or within horizons of alkaline pH. In soils, the various forms of extractable Mo correlate poorly with total Mo if evaluated across a broad range of soils.

The soil content of Mo is dependent on the soil's parent material, the degree of weathering, the landscape position and internal drainage, and the soil's organic matter (Flemming, 1980; Adriano, 1986). Low Mo contents are reported for highly weathered and leached acid soils (e.g., Ultisols in Georgia in the United States). High Mo contents have been reported for alluvial soils with high water tables on granite parent materials in the western United States, as well as for alkaline soils (Kubota, 1977). In the United States, high soil contents of Mo are commonly reported from the western states ($6\,mg\,kg^{-1}$) and from areas in the southeast. In the eastern United States, a soil Mo content of $0.5\,mg\,kg^{-1}$ is typical (Kubota, 1977).

Soil Mo content varies spatially across the landscape and with depth in the profile. In a comprehensive Colorado study of Mo in soils and in waters associated with Mo ore deposits, Mo contents in natural waters from rivers were reported to range from 5 to $3,800\,\mu g\,L^{-1}$. Soil Mo values reported during that study ranged from 2.3 to $36.3\,mg\,kg^{-1}$ (Vlek, 1977). The high Mo content of dark shales indicates that the terrestrial Mo cycle has been unchanged for millions of years. The form of Mo in a soil is related to soil organic matter and its iron oxide content. A high content of phosphorus (P) in a soil increases Mo uptake, whereas a high content of sulfur (S) decreases Mo uptake (Barber, 1984).

Worldwide, Mo deficiency is widespread, whereas Mo toxicity tends to be rare in distribution (Adriano, 1986). Molybdenum deficiency can be related to both a low total Mo content in soil and, more commonly, to low availability of Mo. Total Mo concentrations of $0.4\text{–}3.5\,mg\,kg^{-1}$ have been reported to result in the production of high-Mo forage from alkaline soils, but not from acid soils. Molybdenum availability is often increased by adding lime to raise the pH of the plow layer; however, soils with low total Mo content may require addition of fertilizer Mo even after the soil pH has been adjusted for optimum plant growth (Flemming, 1980). As a fertilizer, Mo is notable for the small quantities that often can produce significant yield increases (quantities measured in grams of Mo per hectare).

Chemical Processes Controlling Dissolved Molybdenum in Soils

Speciation of Dissolved Molybdenum

Knowledge of the chemical speciation of a soil solution is fundamental for predicting the various biogeochemical processes that take place in such environments, as well as the consequences of those reactions for natural and anthropogenic constituents. If we consider a pure $Mo\text{-}H_2O$ system at high pH, the predominant dissolved species is aqueous MoO_4^{2-}, but at low pH, $HMoO_4^-$ and $H_2MoO_4^0$ are predominant. However, in soil solution, calcium (Ca), magnesium (Mg), sodium (Na), potassium (K), and other trace-element complexes and ion pairs may all be important (Figure 2.1). The formation of strong complexes and ion pairs in soil solution results in a lower degree of adsorption of MoO_4^{2-} ionic species to an oxide surface, which in turn increases the mobility of dissolved Mo in soil. Similarly, formation of complexes and ion pairs also affects the precipitation of Mo as metallic solid phases in soil. Moreover, it is well established that plant uptake is related to the activity of an individual ionic species, rather than to the total elemental concentration. For these reasons, knowledge of the speciation of dissolved Mo (free ionic species, complexes, and ion pairs) in soil solution is essential.

The chemistry of Mo in soils is complex, because Mo can exist in different oxidation states (II, III, IV, V, and VI). The complexation of Mo with metals (e.g., $CaMoO_4^0$, $MgMoO_4^0$) further complicates the chemistry of Mo in soils. Thermodynamic calculations indicate that solution species with an oxidation state VI should be predominant in soils. Lindsay (1979) reported that $HMoO_4^-$ and $H_2MoO_4^0$ are significant Mo solution species in acidic soil solutions. He also reported that MoO_4^{2-} is the major solution species in alkaline soil solutions. Reddy and Gloss (1993) showed that dissolved Mo species in alkaline soil solutions are dominated by $MgMoO_4^0$, followed by $CaMoO_4^0$ and MoO_4^{2-} (Table 2.1).

The data in Table 2.1 demonstrate that dissolved Mo in soil solutions contains not only MoO_4^{2-} but also other Mo solution species. Soil solutions also contain dissolved organic carbon (DOC) compounds. Therefore, one can expect that DOC-Mo species may exist in soil solutions as well. For example, Fio and Fujii (1990) and Abrams, Berau, and Zasoki (1990) reported that dissolved selenium (Se) in soils comprises both organic and inorganic species. Complexing with DOC is much more important in soil solutions than in surface water or groundwater, because soil solutions contain much higher concentrations of DOC because of biological acitrity.

Routinely, dissolved Mo in soil solutions is measured using atomic-absorption spectroscopy (AAS) coupled with a graphite furnace (GF). The GF-AAS method is capable of measuring low concentrations of dissolved Mo in soil solutions (micrograms per liter). However, this method determines the concentration (C) of all possible dissolved Mo species together:

$$
\begin{aligned}
\text{total dissolved Mo} = \sum & C MoO_4{}^{2-} + C HMoO_4{}^- + C H_2MoO_4{}^0 \\
& + C CaMoO_4{}^0 + C MgMoO_4{}^0 + C Na_2MoO_4{}^0 \\
& + C K_2MoO_4{}^0 + C NaMoO_4{}^- + C KMoO_4{}^- \\
& + C \text{DOC-Mo}
\end{aligned}
\tag{1}
$$

Table 2.1. *Speciation of dissolved Mo in soil solutions*

	Depth (cm)					
Species	10	20	30	50	60	70
Total dissolved Mo (mg L^{-1})	0.06	0.08	0.18	0.11	0.14	0.15
pH	7.95	8.10	8.10	8.06	8.07	8.12
MgMoO$_4^0$	40%	57%	66%	63%	61%	75%
CaMoO$_4^0$	37%	25%	17%	19%	21%	15%
MoO$_4^{2-}$	20%	16%	13%	15%	14%	13%
Other[a]	3%	2%	4%	3%	4%	NS[b]

[a] Na and K complexes.
[b] Not significant.
Source: Adapted from Reddy and Gloss (1993).

The concentration of an individual Mo solution species in a soil solution is calculated indirectly using chemical-speciation models, with inputs of total dissolved Mo concentration, pH, and the concentrations of the dissolved major cations and anions. If we did not consider both inorganic and organic dissolved Mo species in the chemical speciation of a soil solution, then we could overestimate the concentration of an individual Mo species. That, in turn, would lead to misinterpretation of the mechanisms controlling dissolved Mo in soil solutions.

Different approaches are available for determining the chemical speciation of a soil solution. Excellent discussions of this subject have been presented by Sparks (1984), Amacher (1984), Baham (1984), and Sposito (1984). The speciation of dissolved Mo in soil solutions is not well understood, partly because of the difficulty of measuring low concentrations of individual Mo species (at the level of micrograms per liter). For example, specific ion electrodes are often used to determine the concentrations of the ionic species in soil solutions. Specific ion electrodes work well when the concentrations of dissolved ionic species of interest are above 10 μM, but below that concentration, measurements with specific ion electrodes are not very reliable.

Adsorption and Desorption Processes of Dissolved Molybdenum

Introduction

We have discussed the importance of understanding the speciation of dissolved Mo in the plant uptake, adsorption, and precipitation processes in soil solutions. In this section we review current knowledge of the adsorption and desorption mechanisms of Mo in soils. Information regarding plant uptake is presented elsewhere in this book. Soil minerals such as Al, Fe, and Mn oxides, clay minerals, and carbonates can exhibit both positive and negative charges. The pH of the zero point of charge (ZPC) is a convenient reference point for predicting how charges will develop on mineral surfaces.

The ZPC is the pH at which the surface of a mineral is electrically neutral (Parks, 1965). At pH values above the ZPC the mineral surface is negatively charged, and at pH values below the ZPC the mineral surface is positively charged. A positive charge results in the mineral possessing anion exchange capacity, which is very important in MoO_4^{2-} retention. Positive charges are thought to arise from the protonation or addition of hydrogen ions (H^+) to hydroxyl groups. This mechanism depends on pH and the valence of the metal ions. It is usually important in Al and Fe oxides, but it is of less importance in Si oxides. Tropical and strongly weathered soils contain Al(III) and Fe(III) hydroxyoxides, whose negative charges are low and whose positive charges can be relatively high, especially at low pH. Under acidic conditions, these "variable-charge soils" can retain more anions than cations.

Theory

The adsorption of Mo in soils can be explained by the theory of specific adsorption, in which covalent bonds are formed to some degree between soil constituents and Mo ions. Another theory used to explain the strong adsorption of Mo to oxides is ligand exchange or anion penetration (Bohn, McNeal, and O'Connor, 1985). The hydroxyl ions on a hydrous oxide surface can be replaced by anions, which can enter into sixfold coordination with Al^{3+} or Fe^{3+} ions. This process is known as ligand exchange. Ligand exchange can occur on surfaces initially carrying a net negative, positive, or neutral charge. This contrasts with nonspecific anion adsorption, which occurs only when the surface carries a net positive charge. Ligand exchange may explain why weak-acid anions show maximum adsorption at pH values about equal to their pK values. At pH = pK, both the amount of anions (dissociated acid) available for ligand exchange and the amount of proton donor (undissociated acid) are greatest (Bohn et al., 1985).

Acidic soils contain high amounts of Fe and Al oxides and hydroxides. The MoO_4^{2-} ions reacting with these metal oxides and hydroxides form a series of soluble hydroxymolybdates. Another type of MoO_4^{2-} adsorption is the reaction between MoO_4^{2-} and aluminum silicate minerals. The MoO_4^{2-} ions react with octahedral Al by replacing the OH groups located on the surface plane of the mineral. This type of reaction is also prevalent under acidic conditions.

The Mo adsorption process can be studied by using adsorption isotherms. Langmuir and Freundlich equations are the two major types of isotherms used to describe the Mo adsorption process. The Langmuir equation is based on the kinetic theory of gaseous adsorption onto solids, but is often used to model the adsorption of ions from solution (Ellis and Knezek, 1972). A common form of the Langmuir equation is

$$\frac{x}{m} = \frac{KCb}{1 + KC} \tag{2}$$

where x is the amount adsorbed, m is the amount of adsorbent, K is a constant related to the binding strength, C is the equilibrium concentration of adsorbate, and b is the maximum amount of adsorbate that can be adsorbed. Equation (2) can be rearranged to the linear form

$$\frac{C}{x/m} = \frac{1}{Kb} + \frac{C}{b} \tag{3}$$

If the adsorption conforms to the Langmuir equation, plotting $C/(x/m)$ versus C should yield a straight line with slope $1/b$ and intercept $1/Kb$.

The Langmuir equation is limited to the range for which experimental data are available. An advantage of using the Langmuir equation for describing adsorption is that is defines a limit of adsorption on a given array of sites that meet the Langmuir model criteria (Bohn et al., 1985). However, the Langmuir equation implies that the energy of adsorption on a uniform surface is independent of surface coverage. Freundlich found that adsorption data from many dilute solutions could be fitted to an empirical equation of the form

$$\frac{x}{m} = KC^{1/n} \tag{4}$$

where K and n are constants and the other terms are as defined previously. The Freundlich equation implies that the energy of adsorption decreases logarithmically as the fraction of covered surface increases. The linear form of the Freundlich equation is

$$\log \frac{x}{m} = \frac{1}{n} \log C + \log K \tag{5}$$

The frequent good fit of adsorption data to this equation is undoubtedly influenced by the insensitivity of log-log plots and by the greater flexibility in curve fitting afforded by the two empirical constants (K and n) in the Freundlich equation. The Freundlich equation has no sound theoretical basis, but is an empirical relationship used to describe the adsorption of ions or molecules from liquid onto a solid phase. The major limitation of the Freundlich equation is that it does not predict a maximum adsorption capacity (Bohn et al., 1985).

Molybdenum Adsorption and Desorption Studies in Soils

Molybdenum adsorption by various soils and soil minerals has been studied by a number of soil scientists. The role of Fe(III) oxides and hydroxides as Mo adsorbents in soil has been emphasized by several investigators (Reisenauer, Tabikh, and Stout, 1962; Reyes and Jurinak, 1967; Taylor and Giles, 1970; Jarrell and Dawson, 1978). Other Mo-adsorbing minerals in soils that have been given some attention are metahalloysite, nontronite, kaolinite, illite, allophanes (Theng 1971) and the oxides of titanium and aluminum (Reisenauer et al., 1962). Soil organic matter has been considered an important adsorbent for Mo in humus-rich soils (Szilagyi, 1967).

Molybdenum adsorbents other than soil have been reviewed by Mikkonen and Tummavuori (1993a). Retention of MoO_4^{2-} has been calculated using various mathematical models (Bowden et al., 1980a,b; Sheindorf, Rebhun, and Sheintuch, 1981, 1982; Roy, Hassett, and Griffin, 1989). Bowden et al. (1980a) reported an extended version of a model of the adsorption process they had described earlier (Bowden et al., 1973, 1977). The adsorption equation for the extended version of the model can be written as

$$S = \frac{N_T \sum\limits^{i} K_i C_i \exp(-Z_i F\Psi_a/RT)}{1 + \sum\limits^{i} K_i C_i \exp(-Z_i F\Psi_a/RT)} \tag{6}$$

where S is the amount of anion adsorbed, N_T is the maximum adsorption, the subscript i refers to the individual species of ion present (e.g., $HMoO_4^-$ or MoO_4^{2-}), K is an affinity term, C is the concentration of ion i in solution, Z is the valence, F is the Faraday constant, Ψ_a is the electrical potential in the plane of adsorption, R is the universal gas constant, and T is the temperature in degrees Kelvin. The electrostatic potential is estimated by solving simultaneous equations that describe the changing behavior of the adsorbing material (Bowden et al., 1977), using the method described by Barrow et al. (1980). McKenzie (1983) applied this extended model to predict the adsorption of MoO_4^{2-} by an oxide surface and found that adsorption of MoO_4^{2-} on oxides was more or less proportionate to their surface areas; no special affinity was found between the MoO_4^{2-} ion and an iron oxide surface. Adsorption on goethite showed a maximum at pH 3.5 (McKenzie, 1983).

Sheindorf et al. (1981, 1982) developed a multicomponent Freundlich-type equation to describe the adsorption of binary solute mixtures containing arsenate and phosphate or arsenate and molybdate. The derivation of the Sheindorf-Rebhun-Sheintuch (SRS) equation was based on the assumption that there is an exponential distribution of adsorption energies available for each solute. The SRS equation can be written for the solute i from a binary solute mixture as

$$(x/m)_i{}^j = K_i C_i \left(C_i + a_{ij} C_j \right)^{(1/n_i)-1} \tag{7}$$

and for the adsorption of solute j from a binary mixture as

$$(x/m)_j{}^i = K_j C_j \left(C_j + a_{ji} C_j \right)^{(1/n_j)-1} \tag{8}$$

where $(x/m)_i^j$ is the amount of solute i adsorbed per unit mass of adsorbent in the presence of a competitive j, K_i and K_j are the single-solute Freundlich constants for solutes i and j, C_i and C_j are the equilibrium concentrations or activities of the solutes, n_i and n_j are the single-solute Freundlich exponents, and a_{ij} and a_{ji} are competitive coefficients.

Roy et al. (1986) applied the SRS model to study the competitive coefficients for adsorption of arsenate, molybdate, and phosphate mixtures in three soils. They found that adsorption of arsenate and MoO_4^{2-} by all three soils was significantly reduced by the presence of phosphate, which was attributed to competitive interactions. In the solute–soil system, two of the three soils studied were found to have reduced arsenate adsorption in the presence of MoO_4^{2-}, whereas arsenate did not compete strongly with MoO_4^{2-} adsorption. In contrast, the adsorption of arsenate by one soil was independent of MoO_4^{2-}, whereas the presence of arsenate lowered MoO_4^{2-} adsorption. Those authors concluded that the reliability of the model may depend on the relative proportions of the competing ions. The SRS model required the collection of competitive data to derive a competitive coefficient. The ability of the expression to describe the data was limited to situations where the ratios of the equilibrium concentrations of arsenate:phosphate and arsenate:molybdate were greater than 20:1. That limitation was partly attributed to the regression procedure used to calculate competitive coefficients (Roy et al., 1986).

The adsorption of Mo in soils is strongly pH-dependent. Reisenauer et al. (1962) reported that Mo adsorption increased with decreasing pH from 7.75 to 4.45. Possible explanations for that effect on adsorption in that pH range are that hydroxide and MoO_4^{2-} ions compete for adsorption sites or that Fe and A1 oxides become more active as pH decreases (Adriano, 1986). Recently, Mikkonen and Tummavuori (1993b) also found that maximum retention of Mo occurred below pH 4.5 for three Finnish mineral soils. Vlek (1977) studied Mo adsorption on several Colorado soils and developed the following equation to describe the adsorption of Mo as affected by pH:

$$soil + MoO_4^{2-} \rightleftharpoons soil - MoO_4^- + OH^- \tag{9}$$

This equation shows that MoO_4^{2-} activity decreases 10-fold as pH decreases one unit.

Karimian and Cox (1978) found that adsorption of Mo was positively correlated with Fe oxide content and organic-matter content. In acid soils, Fe oxides carry positive charges and can react with molybdate, but it is difficult to explain the adsorption of Mo by organic matter. However, on the basis of the content of organic matter and Fe in soil, it was suggested that Fe oxide bound to organic matter was actually responsible for the Mo adsorption (Karimian and Cox, 1978). Reisenauer et al. (1962) found that adsorption of MoO_4^{2-} onto soils at a fixed pH followed the Freundlich equation. Reyes and Jurinak (1967) and Theng (1971), in studying Mo adsorption onto hematite and soil at pH 4.0, found two adsorption reactions, each conforming to a Langmuir isotherm. The isotherms were interpreted as showing two energetically distinct binding sites for Mo.

Precipitation and Dissolution Processes of Dissolved Molybdenum

Introduction

In earlier sections we have discussed the speciation, adsorption, and desorption processes of dissolved Mo in soil solutions. In this section we review the principles of precipitation and dissolution processes and discuss the potential Mo solid phases that may control dissolved Mo in alkaline soil solutions.

During the weathering of soils, primary solid phases containing high amounts of free energy (G) dissolve as they seek lower energies. When doing so, some constituents may precipitate as secondary solid phases, some of which are crystalline, and others amorphous. Chemical reactions in soils seldom reach equilibrium, because weathering reactions are part of a continuous process. Additionally, some precipitation and dissolution reactions are very slow, particularly when biological processes are involved. Nevertheless, knowledge of precipitation and dissolution reactions will help in understanding the chemical status of soil solutions and the directions in which chemical reactions move.

Theory

The secondary Mo solid phases that precipitate during the weathering process of soils dissolve upon contact with water and supply dissolved Mo as it is removed from soil solution (Figure 2.1). In order to predict the potential solid phases controlling dissolved Mo in soil solutions, we must calculate the activities of MoO_4^{2-} and corresponding metal ions (e.g., Ca^{2+} or Pb^{2+}). From these calculations, ion activity products (IAPs) are determined and compared with the solubility products (K_{sp} values) of a range of Mo solid phases. The K_{sp} values are often derived from the Gibbs free energy of formation (G_f) or are determined from the solubility data. These calculations are used to calculate a saturation index (SI). For example:

$$\text{SI for } PbMoO_4 = \text{IAP}\left(PbMoO_4\right)\big/K_{sp}\left(PbMoO_4\right) \tag{10}$$

where $\text{IAP}(PbMoO_4) = aPb^{2+} \cdot aMoO_4^{2-}$ and a is the activity of the ion. If $SI = 1$ ($\log SI = 0$), then the dissolved Mo in the soil solution is predicted to be controlled by that Mo solid phase. If $SI > 1$, then the soil solution is supersaturated with respect to that solid phase, which will be predicted to precipitate. If $SI < 1$, then the soil solution is undersaturated with respect to that solid phase, which will be predicted to dissolve. However, because of uncertainties in the measurements of IAP and K_{sp}, often an assumption is made that an IAP within ± 0.50 log unit of the K_{sp} of a solid phase represents near saturation and is predicted to control the concentrations of ions involved.

Mathematical models are widely used to perform the foregoing calculations. The most commonly used models are GEOCHEM (Sposito and Mattigod, 1980) and MINTEQA2 (Brown and Allison, 1992). Soil-solution data are entered into these models in order to calculate (1) the chemical speciation of the soil solution, such as the free concentration and the activity of MoO_4^{2-}, by solving a series of mass-balance equations through an iterative procedure and (2) the state of saturation of the soil solution with respect to various solid phases.

However, one should be aware of the following three important issues when geochemical models are applied to determine the solid-phase control of soil solutions: (1) Solid-phase chemistry is based on the assumption of equilibrium. Therefore, soil solutions to be tested should be close to a steady-state condition (i.e., the condition in which little change has occurred in the major ions involved). (2) The mass-balance equation for each dissolved species should contain all possible solution species to ensure accurate calculation of the free concentration of the dissolved species. Omission of any significant solution species from the mass-balance equation will cause overestimation of the free concentration of the dissolved species. (3) Variations occur in the equilibrium constants for solution species and solid phases. All these factors could lead to misinterpretation of solid-phase equilibria in soil solutions.

Molybdenum Precipitation and Dissolution Studies in Soils

Our knowledge of Mo precipitation and dissolution reactions in soil solutions is very limited, partly because of lack of thermodynamic information on Mo solution species and solid phases. Lindsay (1979) reported the following sequence for the solubility of Mo solid phases in soils: $CuMoO_4 > ZnMoO_4 > MoO_3 > H_2MoO_4 > CaMoO_4 > PbMoO_4$. Among these solid phases, $PbMoO_4$ was predicted to be most stable in alkaline soils and was expected to control the dissolved Mo concentrations in alkaline soil solutions. Reddy et al. (1990) compiled thermodynamic data for different Mo species and described Mo reactions that are important in the soil chemistry of Mo.

Vlek and Lindsay (1977) examined the solubility of Mo solid phases in soils with pH values ranging from 5.5 to 7.7. Their data for alkaline soils suggest a state of near saturation with $PbMoO_4$. Reddy and Gloss (1993) examined the geochemical speciation of dissolved Mo in soil solutions as a function of depth to determine the potential solid phases controlling dissolved Mo in soil solutions (Figure 2.2). Using the reported $\log K_{sp}$ values for $PbMoO_4$, which ranged between -15.9 and -16.0, a comparison (Figure 2.2) between $\log K_{sp}$ values and IAPs suggested that IAPs for $PbMoO_4$ were very close to the $PbMoO_4$ solubility (a mean value of -15.77 ± 0.29), and $PbMoO_4$ was predicted to be the solid phase controlling the MoO_4^{2-} concentration in soil solutions.

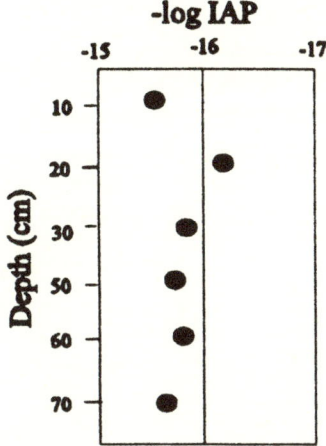

Figure 2.2. Comparison of ion activity products with the solubility product of wulfenite (lead molybdate) in soil solutions from Laramie Basin, Wyoming. [Reprinted from *Applied Geochemistry* (*Supplement*), vol. 2, K. J. Reddy and S. P. Gloss: Geochemical speciation as related to the mobility of F, Mo and Se in soil leachates, pp. 159–63. Copyright 1993, with kind permission from Elsevier Science Ltd., The Boulevard, Langford Lane, Kidlington OX5 1GB, United Kingdom.]

Wang et al. (1994) examined Mo solubility in a number of soils, including (1) soils in a surface coal mine, (2) soils near a coal mine, and (3) native soils. Initially, the IAPs for $PbMoO_4$ in those soil solutions suggested varying degrees of supersaturation, indicating that $PbMoO_4$ was not controlling the dissolved Mo concentrations in those samples. Wang et al. (1994) found that the DOC values for those soils ranged between 17.3 and 57.5 mgL^{-1}. DOC, in the form of fulvic acid, is known to complex with Pb^{2+} in natural waters (Gamble, 1970; Reuter and Perdue, 1977; Stevenson and Welch, 1979; Sarr and Webber, 1980; Sposito, Holtzclaw, and LeVesque-Madore 1981). Measurements of Pb in soil solutions using inductively coupled plasma (ICP) or atomic absorption (AA) include both inorganic and organic species of Pb. The Pb^{2+} activity calculated using the total concentration of Pb in a soil solution, without correcting for organic complexes, would result in overestimation of Pb^{2+} activity. When Wang et al. (1994) corrected total dissolved Pb for DOC-Pb^{2+} complexes, the resulting IAPs suggested a state close to saturation for $PbMoO_4$ (Table 2.2). On the other hand, the soil solutions in that study were highly undersaturated with respect to $CaMoO_4$ and $FeMoO_4$, suggesting that those solid phases were not controlling the dissolved Mo in soil solution.

Katta J. Reddy, Larry C. Munn, and Liyuan Wang

Table 2.2. *IAPs of PbMoO$_4$ in soil solutions from Wyoming*

Sample	DOC (mg L^{-1})	pH	(−log IAP) (−log K_{sp} = 16.0)	
			Without considering DOC-Pb^{2+} complexes	Considering DOC-Pb^{2+} complexes
Coal-mine soil 1	51.7	7.9	14.93	15.55
Coal-mine soil 2	46.4	7.9	14.00	15.65
Coal-mine soil 3	50.6	8.2	14.72	15.55
Soil near coal mine 1	54.7	7.1	13.86	15.70
Soil near coal mine 2	24.0	7.8	15.46	15.90
Soil near coal mine 3	17.3	8.3	15.61	15.80
Native soil 1	57.7	8.0	14.76	15.96
Native soil 2	35.3	8.2	14.56	16.11
Mean (±SD)			14.73 (±0.61)	15.77 (±0.20)

Source: Adapted from Wang et al. (1994).

In another study, Wang (1995) determined plant uptake and the potential solid phases controlling dissolved Mo in native (undisturbed) soil samples and disturbed surface coal-mine soils treated with Mo. The findings in that study suggested that dissolved Mo concentrations in native soil samples and soil samples treated with Mo at 1 mg kg^{-1} were close to saturation with PbMoO$_4$. However, when samples were treated with Mo at 3 and 5 mg kg^{-1}, the dissolved Mo concentrations approached saturation with CaMoO$_4$, which is slightly more soluble than PbMoO$_4$. Wang attributed the saturation with CaMoO$_4$ to the Mo treatment. The treatment with Mo increased the dissolved Mo concentrations above the saturation with CaMoO$_4$, which resulted in precipitation of CaMoO$_4$ in Mo-treated soils. The soils discussed earlier contained much higher amounts of Pb than of Mo, suggesting that an adequate source of Pb was available for the precipitation of PbMoO$_4$. Other studies have shown that PbMoO$_4$ can occur in soils of semiarid environments (Rosemeyer, 1990; Bideaux, 1990). Additionally, studies by Enzmann (1972) indicated that Mo has a high affinity for Ca and Pb at low temperatures. However, identification of minor solid phases like PbMoO$_4$ and CaMoO$_4$ in such samples with x-ray diffraction (XRD) or scanning electron microscopy

(SEM) would be difficult, because such techniques generally are successful only when the solid-phase content is greater than approximately 5%. If $PbMoO_4$ and $CaMoO_4$ control Mo solubility, the concentration and availability of Mo in these environments will increase as pH increases (Vlek and Lindsay, 1977). Such conditions can cause Mo toxicity in soils.

Summary

Dissolved Mo in soil solutions is the result of ion complexation, adsorption, desorption, precipitation, and dissolution processes. These chemical processes ultimately govern the solubility and availability, as well as the mobility, of Mo in soils. Currently, speciation of dissolved Mo in soils is not well understood. This is partially because of the difficulty of measuring low concentrations of dissolved individual Mo species. Molybdenum adsorption can be described by the adsorption isotherms (e.g., Langmuir and Freundlich), as well as in terms of solubility. Adsorption of Mo is highly pH-dependent, and it is also affected by the contents of soil Fe and Al oxides and organic matter. As pH decreases, the adsorption of Mo increases, and the maximum adsorption of Mo is found at pH 4.0. Adsorption of Mo is positively correlated with the content of Fe and Al oxides and negatively correlated with organic-matter content. Information derived from precipitation and dissolution studies suggests that dissolved Mo in alkaline soils may be controlled by $PbMoO_4$, when dissolved Pb^{2+} concentrations are corrected for DOC-Pb^{2+} complexes. When Mo fertilizers are applied to soils, the dissolved Mo may approach saturation with respect to $CaMoO_4$. Eventually, dissolved Mo in Mo-fertilized and unfertilized soils can approach saturation with respect to $PbMoO_4$, because this solid phase is the most insoluble Mo solid phase.

References

Abrams, M. M., Berau, R. G., and Zasoki, R. J. (1990). Organic selenium distribution in selected California soils. *Soil Sci. Soc. Am. J.* 54:979–82.

Adriano, D. C. (1986). *Trace Elements in the Terrestrial Environment*, pp. 329–61. Berlin: Springer-Verlag.

Amacher, M. C. (1984). Determination of ion activities in soil solutions and suspensions. Principal limitations. *Soil Sci. Soc. Am. J.* 48:519–24.

Baham, J. (1984). Prediction of ion activities in soil solutions. Computer equilibrium modeling. *Soil Sci. Soc. Am. J.* 48:525–31.

Barber, S. A. (1984). Molybdenum. In *Soil Nutrient Bioavailability*, pp. 338–45. New York: Wiley.

Barrow, N. J., Bowden, J. W., Posner, A. M., and Quirk, J. P. (1980). An objective method for fitting models of ion adsorption on variable charge surfaces. *Aust. J. Soil Res.* 18:37–47.

Bideaux, R. A. (1990). The desert mineral: wulfenite. *Rocks and Minerals* 65:11–30.

Blossom, J. W. (1991). Molybdenum. In *Minerals Yearbook. Vol. 1: Metals and Minerals*, pp. 1011–28. Washington, DC: U.S. Government Printing Office.

Bohn, H. L., McNeal, B. L., and O'Connor, G. A. (1985). Anion and molecular retention. In *Soil Chemistry*, pp. 184–207. New York: Wiley.

Bowden, J. W., Bolland, M. D. A., Posner, A. M., and Quirk, J. P. (1973). Generalized model for anion and cation adsorption at oxide surfaces. *Nature (Phys. Sci.)* 245:81–3.

Bowden, J. W., Nagarajah, S., Barrow, N. J., Posner, A. M., and Quirk, J. P. (1980a). Describing the adsorption of phosphate, citrate and selenite on a variable-charge mineral surface. *Aust. J. Soil Res.* 18:49–60.

Bowden, J. W., Posner, A. M., and Quirk, J. P. (1977). Ionic adsorption on variable charge mineral surfaces. Theoretical charge development and titration curves. *Aust. J. Res.* 15:121–36.

Bowden, J. W., Posner, A. M., and Quirk, J. P. (1980b). Adsorption and charging phenomena in variable charge soils. In *Soils with Variable Charges*, ed. B. K. G. Theng, pp. 147–66. Lower, NZ: New Zealand Society for Soil Science.

Brown, D. S., and Allison, J. D. (1992). *MINTEQA2: An Equilibrium Metal Speciation Model*. EPA/600/3-87/D12. Athens, GA: U.S. Environmental Protection Agency.

Ellis, B. G., and Knezek, B. D. (1972). Adsorption reactions of micronutrients in soils. In *Micronutrients in Agriculture*, ed. J. J. Mortvedt, P. M. Giordano, and W. L. Lindsay, pp. 59–78. Madison, WI: Soil Science Society of America.

Enzmann, R. D. (1972). Molybdenum: element and geochemistry. In *The Encyclopedia of Geochemistry and Environmental Sciences*, ed. R. W. Fairbridge, pp. 753–9. Stroudsburg, PA: Dowden, Hutchinson and Ross, Inc.

Fio, J. L., and Fujii, R. (1990). Selenium speciation methods and application to soil saturation extracts from San Joaquin Valley, California. *Soil Sci. Soc. Am. J.* 54:363–9.

Flemming, G. A. (1980). Essential micronutrients. I: Boron and molybdenum. In *Applied Soil Trace Elements*, ed. B. E. Davies, pp. 155–97. New York: Wiley.

Gamble, D. S. (1970). Titration curves of fulvic acid: the analytical chemistry of a weak acid polyelectrolyte. *Can. J. Chem.* 48:2662–9.

Jarrell, W. M., and Dawson, M. D. (1978). Sorption and availability of molybdenum in soils of western Oregon. *Soil Sci. Soc. Am. J.* 42:412–15.

Karimian, N., and Cox, F. R. (1978). Adsorption and extraction of molybdenum in relation to some chemical properties of soils. *Soil Sci. Soc. Am. J.* 42:757–61.

Krauskopf, K. B. (1979). *Introduction to Geochemistry*, 2nd ed. New York: McGraw-Hill.

Kubota, J. (1977). Molybdenum status of United States soils and plants. In *Molybdenum in the Environment*, vol. 2, ed. W. R. Chappell and K. K. Peterson, pp. 555–81. New York: Marcel Dekker.

Lindsay, W. L. (1979). *Chemical Equilibria in Soils*. New York: Wiley-Interscience.

McKenzie, R. M. (1983). The adsorption of molybdenum on an oxide surface. *Aust. J. Soil Res.* 21:503–13.

Mikkonen, A., and Tummavuori, J. (1993a). Retention of vanadium (V), molybdenum (VI) and tungsten (VI) by kaolin. *Acta Agric. Scand., Sect. B, Soil Plant Sci.* 43:11–15.

Mikkonen, A., and Tummavuori, J. (1993b). Retention of molybdate (VI) by three Finnish mineral soils. *Acta Agric. Scand., Sect. B, Soil Plant Sci.* 43:206–12.

Parks, G. A. (1965). Isoelectric points of solid oxides, solid hydroxides, and aqueous hydroxide complex systems. *Chem. Rev.* 65:177–98.

Reddy, K. J., and Gloss, S. P. (1993). Geochemical speciation as related to the mobility of F, Mo and Se in soil leachates. *Appl. Geochem.* (*Suppl.*) 2:159–63.

Reddy, K. J., Wang, L., and Lindsay, W. L. (1990). Molybdenum supplement to technical bulletin 134: *Selection of Standard Free Energies of Formation for Use in Soil Chemistry*. Fort Collins, CO: Agricultural Experiment Station, Colorado State University.

Reisenauer, H. M., Tabikh, A. A., and Stout, P. R. (1962). Molybdenum reactions with soils and the hydrous oxides of iron, aluminum and titanium. *Soil Sci. Soc. Am. Proc.* 26:23–7.

Reuter, J. H., and Perdue, E. M. (1977). Importance of heavy metal–organic matter interactions in natural waters. *Geochim. Cosmochim. Acta* 41:325–34.

Reyes, E. D., and Jurinak, J. J. (1967). A mechanism of molybdate adsorption in alpha Fe_2O_3. *Soil Sci. Soc. Am. Proc.* 37:637–41.

Rosemeyer, T. (1990). Wulfenite occurrences in Colorado. *Rocks and Minerals* 65:58–61.

Roy, W. R., Hassett, J. J., and Griffin, R. A. (1986). Competitive coefficients for the adsorption of arsenate, molybdate and phosphate mixture by soils. *Soil Sci. Soc. Am. J.* 50:1176–82.

Roy, W. R., Hassett, J. J., and Griffin, R. A. (1989). Quasi-thermodynamic basis of competitive-adsorption coefficients for anionic mixture in soils. *J. Soil Sci.* 40:9–15.

Sarr, R. A., and Webber, J. H. (1980). Conditional stability constants, solubility and implication for lead (II) mobility. *Environ. Sci. Technol.* 14:877–80.

Sheindorf, C., Rebhum, M., and Sheintuch, M. (1981). A Freundlich-type multicomponent isotherm. *J. Colloid Interface Sci.* 79:136–42.

Sheindorf, C., Rebhum, M., and Sheintuch, M. (1982). Organic pollutants adsorption from multicomponent systems modeled by Freundlich-type isotherm. *Water Res.* 16:357–62.

Sparks, D. L. (1984). Ion activities: an historical and theoretical overview. *Soil Sci. Soc. Am. J.* 48:514–18.

Sposito, G. (1984). The future of an illusion: ion activities in soil solutions. *Soil Sci. Soc. Am. J.* 48:531–6.

Sposito, G., Holtzclaw, K. M., and LeVesque-Madore, C. S. (1981). Trace metal complexation by fulvic acid extracted from sewage sludge. I. Determination of stability constants and linear correlation analysis. *Soil Sci. Soc. Am. J.* 45:464–8.

Sposito, G., and Mattigod, S. V. (1980). *GEOCHEM: A Computer Program for the Calculation of Chemical Equilibria in Soil Solutions and Other Natural Water Systems.* Riverside, CA: The Kearney Foundation of Soil Science, University of California, Riverside.

Stevenson, F. J., and Welch, L. F. (1979). Migration of applied lead in a field soil. *Environ. Sci. Technol.* 13:1255–9.

Szilagyi, M. (1967). Sorption of molybdenum by humus materials. *Geochem. Internat.* 4:1165–7.

Taylor, R. M., and Giles, J. B. (1970). The association of vanadium and molybdenum with iron oxides in soils. *J. Soil Sci.* 21:203–15.

Theng, B. K. G. (1971). Adsorption of molybdate by some crystalline and amorphous soil clays. *N.Z. J. Sci.* 14:1040–56.

Thornton, I., and Webb, J. S. (1980). Regional distribution of trace element problems in Great Britain. In *Applied Soil Trace Elements*, ed. B. E. Davies, pp. 399–406. New York: Wiley.

Vlek, P. L. G. (1977). The chemistry, availability and mobility of molybdenum in Colorado soils. Ph.D. dissertation, Department of Plant and Soil Science, Colorado State University, Fort Collins.

Vlek, P. L. G., and Lindsay, W. L. (1977). Thermodynamic stability and solubility of molybdenum minerals in soils. *Soil Sci. Soc. Am. J.* 41:42–6.

Wang, L. (1995). Solubility and bioavailability of molybdenum in mine spoils and soils of Wyoming. Ph.D. dissertation, Department of Plant, Soil, and Insect Sciences. University of Wyoming, Laramie.

Wang, L., Reddy, K. J., and Munn, L. C. (1994). Geochemcial modeling for predicting potential solid phases controlling the dissolved molybdenum in coal overburden, Powder River Basin, WY, U.S.A. *Appl. Geochem.* 9:37–43.

3

Distribution and Mobility of Molybdenum in the Terrestrial Environment

KATHLEEN S. SMITH, LAURIE S. BALISTRIERI,
STEVEN M. SMITH, and R. C. SEVERSON

Introduction

Molybdenum (Mo) is an essential element for many plants and animals (Newton and Otsuka, 1980). Because of its chemical properties, Mo readily provides sites for reactions and catalysis in biochemical systems (Haight and Boston, 1973). It is therefore important to understand the processes that control the distribution, speciation, and behavior of Mo in the surficial environment. These processes will affect the bioavailability of Mo and ultimately its passage into the food chain.

In this chapter we discuss the distribution of Mo in the terrestrial environment and examine the factors that control its mobility.

General Chemical Properties of Molybdenum

Molybdenum is a transition element and a member of the $4d$ series of metals in period 5 of the periodic table. In elemental form, these metals generally are very hard and have high melting temperatures. They exhibit a wide range of oxidation states in their compounds, and they form bonds of high covalent character (Parish, 1977). Other elements that exhibit typical $4d$ chemistry include zirconium (Zr), niobium (Nb), technetium (Tc), ruthenium (Ru), rhodium (Rh), and palladium (Pd). Molybdenum is also a member of group VIB, along with chromium (Cr) and tungsten (W). There are many chemical similarities between Mo and W, but few similarities between Mo and Cr. The electronic configuration of the free atom of Mo is $[Kr]4d^5 5s^1$. Cotton and Wilkinson (1988) provided an in-depth discussion of the inorganic chemistry of Mo and stated that Mo reactions are among the most complex reactions involving any of the chemical elements.

Molybdenum can occur in all oxidation states from II to VI. The most common oxidation state is VI. In its higher (III–VI) oxidation states, Mo generally behaves as a class A metal, having an affinity for oxides and oxygen-containing groups and for the lighter halide elements. Molybdenum also has an affinity for sulfur-containing groups.

The affinity of Mo for oxygen-containing groups is the reason for its predominant presence as dissolved anionic species in aqueous systems and accounts for much of the behavior and mobility of Mo in terrestrial systems. The details of this behavior will be discussed in later sections.

There are some similarities in the chemical behaviors of the molybdate anion (MoO_4^{2-}) and the sulfate anion (SO_4^{2-}). Both are binegative and tetrahedral in structure. They can compete for sorption sites, and in biological systems they are taken up, transported, and excreted along many of the same routes (Haight and Boston, 1973).

Distribution of Molybdenum in the Terrestrial Environment

Geologic Sources of Molybdenum

Molybdenum is a fairly rare element, with a crustal abundance of about $1.2 \, mg \, kg^{-1}$ (Fortescue, 1992). Molybdenum does not occur in nature in its native state, but instead occurs combined with other chemical elements (Blossom, 1994). The most common Mo mineral, molybdenite (MoS_2), is usually found in granites (Palache, Berman, and Frondel, 1944). Wulfenite ($PbMoO_4$), the next most common Mo mineral, is often found in the oxidized zones of Pb- and Mo-containing mineral deposits (Palache, Berman, and Frondel, 1951). Molybdenum is known to accumulate in many geologic environments, but 95% of the world's Mo supply has been mined from porphyry deposits related to intrusive igneous rocks (King et al., 1973). Coal, phosphorite, bedded sandstone uranium (U) deposits, and lignite and lignitic sandstone (King et al., 1973), as well as marine black shales, are also known to contain large amounts of Mo. The United States (e.g., Colorado, New Mexico, Utah), China, Chile, Canada, and Russia accounted for 87% of the world's Mo production in 1993 (Blossom, 1994).

Molybdenum can be found associated with various chemical elements in several types of mineral deposits (Levinson, 1980):

Mo, W, Re, Cu, Sn, Be, B, P, F, Zn, pegmatites
  Bi, and Fe
Mo, Bi, W, F, and Be greisens
Mo, Cu, Re, Ag, Au, and Zn porphyry copper deposits
Mo, U, Se, V, and Cu sandstone-type U deposits

Because of its almost universal association with porphyry copper deposits and its mobility characteristics, Mo has been used extensively as a pathfinder element for geochemical prospecting for porphyry copper deposits (Rose, Hawkes, and Webb, 1979). Dispersion of Mo in mineralized areas has been shown to occur by both hydromorphic (Bradshaw, 1974) and mechanical (Hansuld, 1966) processes.

Molybdenum can substitute for ferric iron, titanium (Ti), aluminum (Al), and possibly silicon (Si) in the lattices of several minerals. It is found in feldspars, biotite, amphiboles, pyroxenes, and magnetite-ilmenite. Molybdenite may weather to secondary minerals such as ferrimolybdite $[Fe_2(MoO_4)_3]$, powellite $(CaMoO_4)$, ilsemannite (Mo_3O_8), wulfenite, lindgrenite $[Cu_3(MoO_4)_2(OH)_2]$, Mo-rich jarosite $[KFe_3(SO_4)_2(OH)_6]$, and Mo-rich limonite $[FeO \cdot OH \cdot nH_2O]$ (Kaback, 1977).

Black shales often contain high levels of trace elements, including Mo (Plant and Raiswell, 1983). Kim and Thornton (1993) reported high concentrations of Mo in soils developed on uraniferous black shales in Korea. They found that plant uptake of Mo from those soils was dependent on soil pH, with more uptake at higher pH.

Distribution and Concentration of Molybdenum in the Surficial Environment

Books and review articles commonly paraphrase or partially quote Mitchell (1964) when describing the source of Mo in soils. His introductory paragraph on the sources of trace elements states that

the trace element content of a soil is dependent almost entirely on that of the rocks from which the soil parent material was derived and on the processes of weathering, both geochemical and pedochemical, to which the soil-forming materials have been subjected. The more mature and older the soil, the less may be the influence of the parent rock. The effects of human interference are generally of secondary importance. [Mitchell, 1964, p. 321]

The initial amount of Mo in a soil is dependent on the primary minerals in the source rocks, or parent material, but this amount will be altered by many processes, both physical and chemical. The classic soil-forming factors (climate, organisms, topography) that act over time, the transport processes (colluvial, alluvial, eolian, glacial, etc.) that redistribute mineral grains, and the agrochemical and industrial additions are all superimposed to alter a soil's Mo content from that of the initial source rock.

The Mo concentration measured in a soil also is affected by the influence of geochemical processes on its mobility, transport, and deposition. The extent of its mobilization and transport from source rocks is determined by mineral stability, which in turn is affected by the weathering environment. Once it is mobilized from the mineral source, the transport, deposition, and availability of Mo to organisms are dependent on its interactions with other soil components (such as clays, organic matter, microbes, and Fe and Mn oxyhydroxides) and the chemistry (pH, Eh, and other ion concentrations) of the soil solution. Molybdenum associated with clay minerals, oxyhydroxides, and organic matter represents the "available" fraction.

Although only very small amounts of Mo are required by plants, deficiencies of Mo have been reported from around the world for more than 40 higher plant species (Adriano, 1986). Those deficiencies commonly are corrected by raising the soil pH by addition of lime (Gupta and Lipsett, 1981). Soils commonly deficient in Mo include the following: highly podzolized soils, because of either low total Mo or sequestering of Mo by oxyhydroxides; extensively weathered soils, in which secondary minerals may fix Mo; soils with pH values below 6, in which the Mo is unavailable; sandy, well-drained soils, in which the total Mo content is low (Severson and Shacklette, 1988).

Molybdenum toxicity to plants has been induced in the laboratory but has not been observed under field conditions (Adriano, 1986; Gupta and Lipsett, 1981). In addition to geologic sources with high contents of Mo, elevated amounts of Mo in soils are generally associated with wet conditions, alkaline reactions, and high concentrations of organic matter (Fleming, 1980; Gupta and Lipsett, 1981). In the United States, there is greater concern with excess Mo and trace-element imbalances in animals than with Mo deficiencies (Gupta and Lipsett, 1981). Neuman, Shrack, and Gough (1987) cited examples of molybdenosis (resulting from high concentrations of Mo and S and low amounts of Cu in forage) being induced near a lignite ashing plant, in an area of uranium mining, near clay mining, and at a uranium-bearing lignite area, where Mo problems were not known to exist before mining and mineral processing began.

Irrigation of alkaline soils developed from Cretaceous-age shales in the western United States may also increase the mobility and transport of Mo, resulting in higher amounts of Mo in agricultural soils, crops, and wetland sediments and biota. Data on stream, lake, and wetland sediments from more than 20 irrigation projects in the western United States show a range of total Mo from less than $2\,mg\,kg^{-1}$ to $120\,mg\,kg^{-1}$ (Severson, Wilson, and McNeal, 1987; Harms et al., 1990; Stewart et al., 1992). Shacklette and Boerngen (1984) reported a geometric mean of $0.85\,mg\,kg^{-1}$ and a range of less than $3\,mg\,kg^{-1}$ to $7\,mg\,kg^{-1}$ total Mo for soils from the western United States. Although Mitchell's (1964) generalization that the trace-element content of a soil is largely determined by its parent material remains valid, industrial, mining, and agricultural activities have been shown to affect local geochemical environments, resulting in greater mobility, transport, and availability of trace elements, especially Mo. However, exceptions have been noted in which identical plant species growing on native soils or adjacent reclaimed coal-mine spoil have shown little differences in Mo concentrations (Erdman and Ebens, 1979; Gough and Severson, 1981, 1983), even though Gough and Severson (1995) have generalized that exposure by surface mining of previously reduced organic-rich and biologically inactive strata results in dramatic shifts in the biogeochemistry, lithogeochemistry, and hydrogeochemistry of the disturbed zone.

The distributions and concentrations of Mo in the surficial environment have been addressed in many papers and books. The average Mo concentration in surface waters in the United States is about $1\,\mu g\ L^{-1}$ (Hem, 1989). Mannheim (1978) reported that river and lake waters from areas not affected by pollution generally have Mo concentrations less than $1\,\mu g\ L^{-1}$. Some studies have noted higher Mo concentrations in water (e.g., Durfor and Becker, 1964; Barnett, Skougstad, and Miller, 1969), and Voegeli and King (1969) and Kaback (1976) reported that Mo concentrations above $5\,\mu g\ L^{-1}$ in surface waters in Colorado appeared to be anomalous.

Data on the average Mo content of various rock types were published by Turekian and Wedepohl (1961). Aubert and Pinta (1977) presented tabular data based on geographic location, parent rock, and soil type for many areas of the world. There are compilations of data from several studies (Connor and Shacklette, 1975; Ebens and Shacklette, 1982) to bring together data on a single geographic region or to address a societal concern such as environmental pollution. Kabata-Pendias and Pendias (1984) presented extensive tabular information on Mo concentrations in soils and plants worldwide. Although these data tabulations provide valuable information on Mo in different geographic regions, different rock types, different plants, and different soil types, they do not address the spatial distribution of Mo on the landscape.

The first generation of maps showing the distribution classes for Mo and other trace elements in legumes across the United States was prepared in 1976 by Kubota (1980). Those maps showed general patterns and described the principal factors responsible for trace-element distributions. Synoptic maps with somewhat better resolution, showing actual trace-element contents of soils, were prepared by Shacklette and Boerngen (1984) for the conterminous United States and by Gough, Severson, and Shacklette (1988) for Alaska. Those studies were conducted to establish estimates of the averages and ranges of element concentrations in soil, as well as to prepare geochemical maps displaying broad patterns. Advances in computer technology allowed publication of synoptic geochemical maps of increasing resolution for whole countries (Webb et al., 1978; Fauth et al., 1985). Geochemical mapping studies, costly in both time and resources, require long-term sampling and analytical commitments. Geochemical maps provide data useful in a number of different disciplines. Such data provide a baseline description of the element composition of the surficial environment at a point in time and can be useful in dealing with environmental problems, mineral-resource exploration, health-related investigations, nutrition of domestic and wild animals, and epidemiological studies.

Derivative geochemical maps have been prepared for northern Europe to show the current soil weathering rates and degrees of soil acidification and to predict future patterns (Johansson and Savolainen, 1991). Derivative maps detailing selenium (Se) transport and accumulation in the San Joaquin Valley in California have been prepared by Tidball et al. (1986). We are not aware of any such derivative geochemical maps for Mo. However, similar derivative maps predicting areas of Mo (and other trace-element) accumulation and potential toxicity, based on known physical and chemical conditions and processes that reflect Mo sources, transport, and accumulation, can be prepared using geographic information systems (GIS) based on compilations of data (e.g., Ebens and Shacklette, 1982) for a certain geographic area. Data bases for geochemical data are beginning to appear on compact disks (Hoffman and Marsh, 1994). These kinds of data bases, when coordinated with GIS manipulation, should make the preparation of derivative or special-purpose geochemical maps much more feasible.

Distribution of Molybdenum in Surficial Materials in the United States

A geochemical data base for sediments and waters in the United States was created as part of the National Uranium Resource Evaluation (NURE) program of the U.S. Atomic Energy Commission, now the U.S. Department of Energy (Hoffman and Buttleman, 1994). The NURE program, designed to identify and assess uranium resources, systematically sampled about 60% of the conterminous United States on the basis of $1° \times 2°$ quadrangles in the late 1970s. Although the focus of the study was uranium, many of the samples were also analyzed for several other elements, including Mo.

Figure 3.1 shows the distribution of Mo in sediment samples from the NURE study. No consistent sample type was collected throughout the country. Sediment samples that were analyzed for Mo came from a variety of sources: 78% stream sediments, 19% soils, 2% lake sediments, and 1% spring sediments. At the scale of Figure 3.1, no systematic patterns can be detected based on these various sampled sources.

Two NURE laboratories analyzed 132,667 samples for Mo by inductively coupled plasma/atomic-emission spectrometry (ICP-AES), by spectrochemical analysis (dc-arc source), or by atomic-absorption analysis. Because the lower limits for detection of Mo with those methods varied between 2 and $5\,mg\,kg^{-1}$, and the average crustal abundance of Mo is usually given as $1–1.5\,mg\,kg^{-1}$, 84% of the Mo samples were below the limit of analytical detection, and 97% of the Mo contents were below $5\,mg\,kg^{-1}$. Some of the data showed strong biases that may have been due to instrument drift or the use of different laboratories, analytical methods, or sampling techniques. The Idaho Falls $1° \times 2°$ quadrangle in southeastern Idaho and the Holbrook $1° \times 2°$ quadrangle in central Arizona are two areas that showed erroneously high Mo values, due primarily to analytical bias. At least six other quadrangles have similar problems.

Given that we can use only the upper 3% of the data and the scale of Figure 3.1, only a few observations can be made regarding the distribution of Mo. Most of the occurrences represent a small number of samples, or occasionally just one sample, reported to have elevated amounts of Mo. The most prominent example is the "bull's-eye" in west central Texas that represents one sample in a sparsely sampled area. In contrast, the smaller bull's-eye in northern New Mexico contains 13 closely spaced samples in the vicinity of the Questa Mo porphyry deposit. Other similar

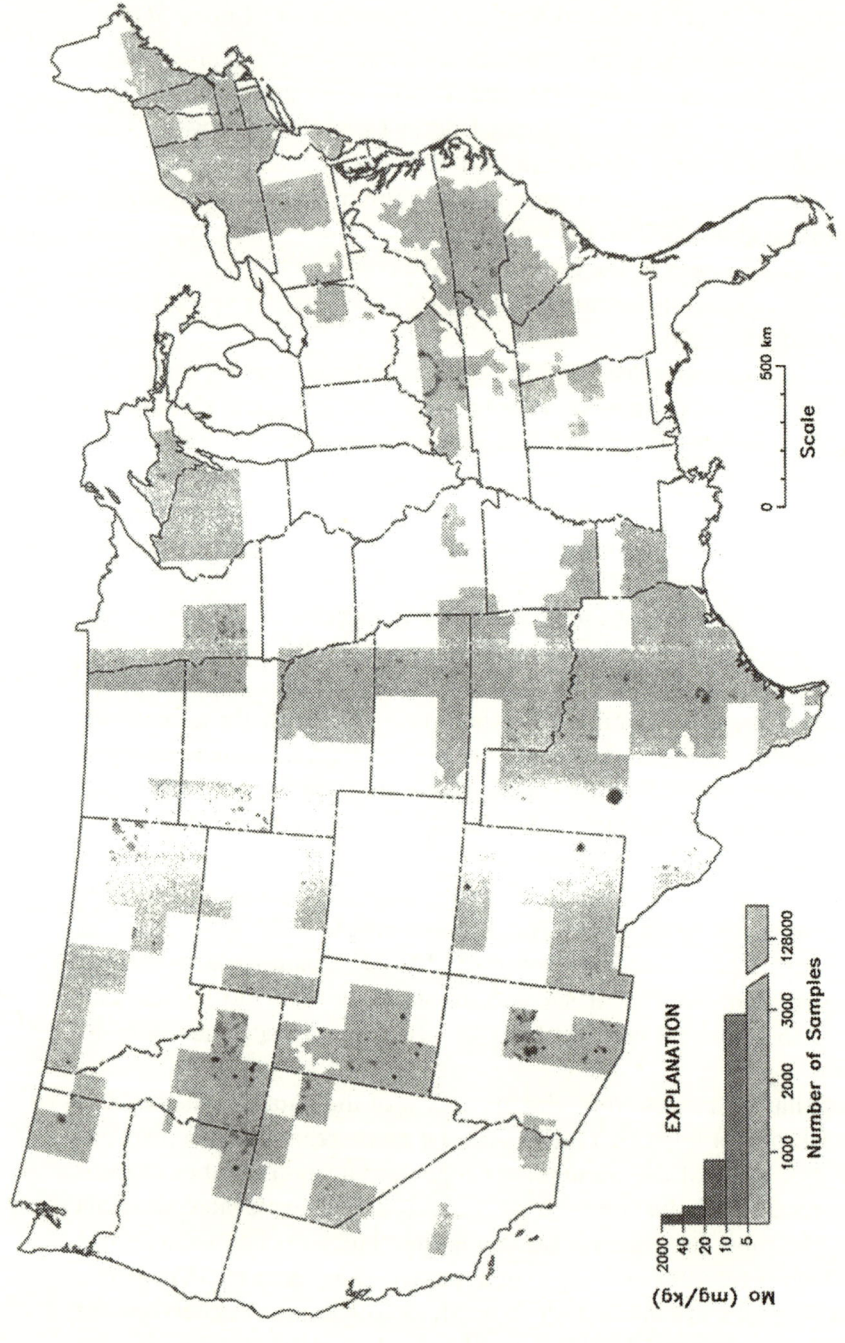

Figure 3.1. Distribution of Mo in surficial sediment samples collected for the NURE program from the conterminous United States. NURE data on Mo obtained from Hoffman and Buttleman (1994).

bull's-eye features in Arizona, Idaho, Utah, and Washington identify samples collected near known Mo-bearing deposits or mining districts.

The reported high concentration of Mo in southeast Texas, shaped like an inverted comma, is enigmatic. Inspection of the original data suggests that those samples may have been contaminated during collection, sample preparation, or the analytical stage and may not actually represent an area of Mo enrichment. The arcuate area of high Mo concentration reported in central Kentucky has been verified. This pattern closely corresponds to outcrops of organic-rich Upper Devonian/Lower Mississippian marine black shales. Connor (1981) reported 88 shale samples from that unit with a geometric mean Mo content of $76 \, mg \, kg^{-1}$.

Factors that Influence the Mobility of Molybdenum in the Terrestrial Environment

Aqueous Speciation

Molybdenum generally forms dissolved anionic species in aqueous solution. The predominant aqueous species of Mo in most natural systems is the molybdate anion (MoO_4^{2-}). This species is thermodynamically stable under most natural conditions and has a low pK_a value of about 4. Molybdate does not form strong aqueous complexes with major ions such as Na, K, Mg, or Ca (Turner, Whitfield, and Dickson, 1981).

The solution species that are generally found in soils are MoO_4^{2-}, $HMoO_4^-$, and $H_2MoO_4^0$ (Chojnacki and Oleksyn, 1965). Cruywagen and De Wet (1988) stated that MoO_4^{2-} is the dominant aqueous species at pH > 4, whereas at lower pH (3.5–4), $Mo(OH)_6^0$, $HMoO_4^-$, and $HMo_2O_7^-$ are the prevailing forms.

Polymerization occurs at Mo concentrations greater than 10^{-4} M at lower pH values (Baes and Mesmer, 1976; Mannheim, 1978). In most natural systems, however, concentrations of Mo will not be high enough for polymeric species to be of importance (Jenkins and Wain, 1963).

Complex ions of Mo in all oxidation states from II to VI are known to exist in aqueous solution (Haight and Boston, 1973), but probably are not important in most natural systems. Dolukhanova (1960) proposed that Mo occurs as a complex with sulfate ($MoO_2SO_4^0$) in systems that contain high sulfate concentrations, such as the waters draining many ore deposits.

There have been several conflicting reports in the literature on Mo speciation and behavior in aqueous systems. Haight and Boston (1973) stated that because of the versatile chemistry of Mo, clear-cut definitions of its species and their behaviors in aqueous solution present a difficult experimental problem. Haight and Boston (1973) also gave a summary of the aqueous-solution chemistry of Mo.

Molybdenum has been found to be complexed with and fixed by natural organic matter, especially humic and fulvic acids (Jenne, 1968). Molybdenum has been found to be associated with organic matter in several different environments, including (1) sediments in the Black Sea and Mediterranean Sea (Baturin, Kochenov, and Shimkus 1967), (2) reduced sediments in a fjord in British Columbia (Presley et al., 1972), and (3) many black shales (e.g., Manskaya and Drozdova, 1968). Yamazaki and Gohda (1990) reported that a significant amount of Mo is scavenged as an organically associated species in coastal water and oceanic surface water. Szilagyi (1967) found that Mo(VI) is strongly sorbed on particulate humic substances and is released as soluble Mo(V) humic complexes. Contreras et al. (1978), Brumsack and Gieskes (1983), and Malcolm (1985) reported that Mo is strongly associated with organic matter in some marine interstitial pore waters and proposed that dissolved organic matter plays an important role in the mobility of Mo in early diagenesis, where Mo(VI) is reduced to Mo(V) and is complexed by organic matter. Bibak and Borggaard (1994) found that the Mo adsorption capacity of humic acid is high at pH 3.5–4, but decreases with increasing pH. Cruywagen and De Wet (1988) stated that Mo reacts directly with organics containing carboxyl and phenol groups.

pH

Concentrations of Mo in water and soil solutions are generally lower under acid conditions than under near-neutral or alkaline conditions (Moore and Patrick, 1991). This behavior of Mo in the surficial environment is related mainly to its tendency to form dissolved anionic species. The availability of Mo to plants is largely dependent on soil pH, in that the availability of Mo in soils is greatest under alkaline conditions and least under acidic conditions.

Above pH 4.23, MoO_4^{2-} is the major solution species. The solution species generally decrease in the order MoO_4^{2-} > $HMoO_4^-$ > $H_2MoO_4^0$ > $MoO_2(OH)^+$ > MoO_2^{2+} (Lindsay, 1979). Reddy, Wang, and Lindsay (1990) gave the following equilibrium reactions at 25°C and 1 atm:

$$MoO_4{}^{2-} + H^+ \leftrightarrow HMoO_4{}^- \qquad \left(\log K = 4.23\right)$$

$$MoO_4{}^{2-} + 2H^+ \leftrightarrow H_2MoO_4{}^0 \qquad \left(\log K = 8.23\right)$$

$$MoO_4{}^{2-} + 3H^+ \leftrightarrow MoO_2(OH)^+ + H_2O \qquad \left(\log K = 8.17\right)$$

$$MoO_4{}^{2-} + 4H^+ \leftrightarrow MoO_2{}^{2+} + 2H_2O \qquad \left(\log K = 8.64\right)$$

Sorption of anions is a function of pH, and a decrease in pH will favor anion sorption. Sorption reactions are often implicated as the controlling factors in trace-element concentrations in natural aqueous systems (Reyes and Jurinak, 1967; Jenne, 1968, 1977). Sorption reactions are discussed in detail in a following section.

Redox Conditions

The mobility of Mo is strongly influenced by redox conditions. Increased mobility of Mo occurs in oxic systems relative to anoxic systems, as predicted by thermodynamic calculations and as observed in environments with different redox conditions.

Although Mo can exist in the oxidation states II, III, IV, V, and VI, its dominant oxidation states in nature are IV and VI (Adriano, 1986). Thermodynamic calculations indicate that for oxic and neutral pH conditions, Mo exists in the VI oxidation state as a mobile, dissolved oxyanion ($MoO_4{}^{2-}$). In anoxic systems, Mo is predicted to be in the IV oxidation state as the insoluble sulfide mineral molybdenite (MoS_2) (Brookings, 1987).

Profiles of dissolved Mo in oxic freshwater and seawater indicate that Mo is conservative (Collier, 1985; van der Weijden et al., 1990; Magyar, Moor, and Sigg, 1993). In oxic seawater, the ratio of dissolved Mo concentration (micromoles per kilogram) to chlorinity (grams per gram) is constant at 5.47×10^{-7} (Quinby-Hunt and Turekian, 1983). Therefore, removal processes (e.g., sorption or precipitation) for Mo in oxic systems appear to be much slower than mixing processes. Concentrations of dissolved Mo in anoxic, sulfidic water columns and pore water tend to decrease with depth (Lahann, 1977; Pedersen, 1985; Shaw, Gieskes, and Jahnke, 1990; van der Weijden et al., 1990; Domagalski, Eugster, and Jones, 1990; Emerson and Huested, 1991; Magyar et al., 1993; Balistrieri, Murray, and Paul, 1994). In addition, Mo is strongly enriched in organic-rich sediments in anoxic basins (Pilipchuk and Volkov, 1974; François,

1988) and in sediments that have overlying waters containing low oxygen concentrations (Brumsack, 1986; Shaw et al., 1990). These observations suggest an efficient transfer mechanism for Mo from dissolved phase to particulate phase in anoxic environments. Decreases in dissolved Mo concentrations in the presence of hydrogen sulfide generally have been attributed to precipitation of Mo sulfide or coprecipitation of Mo with Fe sulfide (Bertine, 1972; Bertine and Turekian, 1973; Lahann, 1977; Pedersen, 1985; Domagalski et al., 1990; Shaw et al., 1990; van der Weijden et al., 1990; van der Sloot et al., 1990; Magyar et al., 1993; Amrhein, Mosher, and Brown, 1993; Balistrieri et al., 1994). The extraction studies of anoxic marine sediments by Huerta-Diaz and Morse (1992) showed that pyrite (FeS_2) is an important sedimentary sink for Mo. Differences in the chemical behavior of Mo between oxic and anoxic systems have been proposed or used to identify the redox state of past depositional environments as found in the sedimentary record (Emerson and Huested, 1991; Hatch and Leventhal, 1992).

Sorption Reactions

The sorption characteristics of anions, particularly plant-nutrient anions (such as phosphate, sulfate, and molybdate) and toxic elements (such as As and Se), have been extensively studied. Several excellent reviews summarizing anion sorption have been published (Parfitt, 1978; Hingston, 1981; Mott, 1981; Barrow, 1985). Emphasis in both laboratory and field studies has been on the affinity of anions for metal (e.g., Fe, Mn, or Al) oxide phases.

Strongly binding anions, such as molybdate, sorb on oxides by a ligand-exchange mechanism that involves the exchange of an oxide-surface hydroxyl group (SOH) for an aqueous anion (A^{2-}) (Stumm, Kummert, and Sigg, 1980; Sposito, 1984; Stumm, 1992). This reaction results in the formation of an inner sphere complex and is illustrated as follows:

$$SOH + A^{2-} + H^+ \leftrightarrow SA^- + H_2O$$

This reaction indicates that sorption of an anion is a function of pH and the concentrations of surface sites and anion. A decrease in pH and an increase in surface site or anion concentration should favor the removal of anions from solution. Laboratory studies have indicated that sorption of Mo by pure metal oxide phases and soils is highly pH-dependent (e.g., Theng, 1971; Katz and Runnells, 1974; Vlek and Lindsay, 1977;

McKenzie, 1983; Balistrieri and Chao, 1990; Bibak and Borggaard, 1994). The maxima in Mo sorption as a function of pH on oxide phases tend to occur between pH 4 and 5, which is consistent with the tendency for maximum sorption of anions to occur at pH values near their acidity constants (pK_a) (Hingston et al., 1967, 1972; Theng, 1971; Stumm et al., 1980; Bibak and Borggaard, 1994).

Several models have been developed to describe reactions between aqueous ions and solid surfaces. These models tend to fall into two categories: (1) empirical partitioning models, such as distribution coefficients and isotherms (e.g., Langmuir and Freundlich isotherms), and (2) surface-complexation models (e.g., constant-capacitance, diffuse-layer, or triple-layer model) that are analogous to solution complexation with corrections for the electrostatic effects at the solid–solution interface (Davis and Kent, 1990). These models have been described in numerous articles (Westall and Hohl, 1980; Morel, Yeasted, and Westall, 1981; James and Parks, 1982; Barrow, 1983; Westall, 1986; Davis and Kent, 1990; Dzombak and Morel, 1990). Travis and Etnier (1981) provided a comprehensive review of the partitioning and kinetic models typically used to define sorption of ions by soils. The reader is referred to the cited articles for details of the models.

The sorption of dissolved Mo onto soils has been well studied because of the probable role of sorption in controlling the availability of Mo to plants. Research has focused on defining the characteristics of Mo sorption on soils and soil components and on the competition of anionic plant nutrients, such as phosphate, sulfate, and molybdate, for surface sites. Various empirical and surface-complexation models have been used to interpret such data.

Laboratory Studies of Molybdenum Sorption

Katz and Runnells (1974) found that the maximum sorption capacity of acidic alpine soils for Mo, as predicted by Langmuir isotherms, was 20 times greater than that of alkaline desert soils. They attributed this difference to pH and, possibly, the higher organic-carbon content of the alpine soils. LeGendre and Runnells (1975) found that dissolved Mo in wastewater could be removed by Fe oxyhydroxides at low pH. Their field studies showed that Fe precipitates also were responsible for the removal of dissolved Mo in a stream impacted by mining wastes. The work of Phelan and Mattigod (1984) indicated that kaolinite has a strong affinity for Mo at pH 7. Stollenwerk (1991) successfully modeled the sorption of

Mo on natural aquifer sediments using the diffuse-layer model and the characteristics of well-defined ferrihydrite.

Chemical extractions of soils have been used to identify specific soil components that influence the sorption of Mo. The work of Theng (1971) and Jarrell and Dawson (1978) suggested important roles for Fe and Al oxides in soil as sequestering agents for Mo. For 32 soils in the southeastern United States, Karimian and Cox (1978) reported the amount of Mo sorbed to be closely correlated to soil pH, dithionite-citrate-extractable Fe, organic matter, and acid-extractable phosphorus. All of their data were modeled with Freundlich isotherms. Bibak and Borggaard (1994) measured the sorption of Mo on pure metal (Fe and Al) oxides, humic acid, and one soil. They found good agreement between the measured Mo sorption on the soil and the predicted Mo sorption using an additivity model that incorporated the sorption capacities of the pure phases, the oxalate- and dithionite-extractable Al and Fe, and the total C contents of the soil.

Competition between Mo and Other Anions

Ryden, Syers, and Tillman (1987) examined anion sorption on hydrous ferric oxide and found the following affinity sequence at pH 6.5 in single sorbate systems:

$$P > As > Se(IV) > Si > Mo > SO_4 > Se(VI) > Cl > NO_3$$

As would be expected from this affinity sequence, phosphate outcompeted Mo for surface sites in multisorbate systems. The work of Balistrieri and Chao (1990) indicated that at pH 7, phosphate has greater affinity for amorphous Fe oxide than does molybdate, whereas the reverse is true for the affinity sequence on manganese dioxide. This difference also was reflected in the abilities of phosphate and molybdate to compete with selenite at pH 7 on the two oxides.

Most studies of anion competition for soil-surface sites have involved phosphate. The studies of Theng (1971), Barrow (1974), Karimian and Cox (1978), Roy, Hassett, and Griffin (1986a,b), Xie and Mackenzie (1991), and Xie, MacKenzie, and Lou (1993) revealed Mo desorption or decreased Mo sorption in the presence of phosphate. The data on anion competition of Roy et al. (1986a,b) have been modeled using various empirical sorption models (Roy et al., 1986b, 1989; Barrow, 1989).

Field Observations of Molybdenum Sorption

Berrang and Grill (1974) found strong correlations between particulate Mo and particulate manganese (Mn) concentrations in Saanich Inlet, British Columbia, and suggested that dissolved Mo is sorbed or scavenged by particulate Mn oxide phases. Takematsu et al. (1985) and Koide et al. (1986) reported that scavenging of Mo by Mn oxides accounts for enrichment of this element in ferromanganese minerals. Strong correlations between solid-phase Mo and Mn concentrations in marine sediments off Baja California also suggest uptake of Mo by Mn oxide phases (Shimmield and Price, 1986). Van der Weijden et al. (1990) suggested that large decreases in dissolved Mo concentrations across the oxic–anoxic interfaces in the Tyro and Bannock basins (eastern Mediterranean Sea) may be due in part to sorption of Mo by Mn oxides just above the interface. Model calculations defining transport mechanisms for Mo in seasonally anoxic Lake Greifen (Switzerland) included adsorption of dissolved Mo onto Mn-enriched particles (Magyar et al., 1993).

Precipitation Reactions

The most important minerals containing Mo include molybdenite (MoS_2), powellite ($CaMoO_4$), ferrimolybdite [$Fe_2(MoO_4)_3$], wulfenite ($PbMoO_4$), and ilsemannite (Mo_3O_8) (Vlek and Lindsay, 1977; Adriano, 1986).

The thermodynamic calculations and associated solubility diagrams of Lindsay (1979) indicate that the solubilities of Mo minerals in soils not affected by redox generally should decrease in the following order:

$$CuMoO_4(c) > ZnMoO_4(c) > MoO_3 > CaMoO_4(c) > PbMoO_4(c)$$

For soils affected by redox, Mo solubility is influenced by other ions that are redox-sensitive, particularly sulfur. Calculations indicate that molybdenite (MoS_2) should control Mo solubility in strongly reducing soils. Other minerals whose ions are also affected by redox [e.g., $MnMoO_4(c)$ or $FeMoO_4(c)$] are too soluble to precipitate in soils. Reddy et al. (1990) have provided an up-to-date comprehensive compilation of thermodynamic data for Mo mineral phases.

Several studies have used thermodynamic calculations and laboratory leach experiments to assess the presence of Mo minerals in soils. The work of Follett and Barber (1967) indicated that $CaMoO_4(c)$ and $FeMoO_4(c)$ are too soluble to control Mo concentrations in a sandy loam soil. Solubility studies of Colorado soils showed that in only 1 of 13 soils could Mo concentrations be explained by equilibrium with $PbMoO_4(c)$ (Vlek and Lindsay, 1977); sorption appeared to control Mo concentrations in soils where no Mo minerals were present.

The solubility of Mo minerals in soils and stream sediments in mining-impacted areas also has been examined. Kaback and Runnells (1980) used thermodynamic calculations and the dissolved composition of stream water near a Mo deposit in Colorado to determine that ferrimolybdite, ilsemannite, molybdenite, and powellite were undersaturated for normal flow conditions. They argued that dissolved Mo in the stream was controlled by sorption on amorphous Fe oxyhydroxides. On the other hand, wulfenite may be present in mine tailings in semiarid environments (Bideaux, 1990; Rosemeyer, 1990) and may control dissolved Mo concentrations in some alkaline soils (Reddy and Gloss, 1993). Active formation of wulfenite on fracture surfaces and in stream-bed sediments has been observed during field studies in arid mineralized areas (S. M. Smith, personal communication, U.S. Geological Survey, 1995). Wang, Reddy, and Munn (1994) found that dissolved Mo concentrations in coal-mining spoils and nearby soils in the Powder River Basin, Wyoming, could be explained by equilibrium with $PbMoO_4(c)$ if complexes between Pb^{2+} and dissolved organic carbon were included in the thermodynamic calculations. Under acidic reducing conditions, ilsemannite is expected to be the stable solid phase of Mo (Vlek and Lindsay, 1977). Moore and Patrick (1991) found that ferrous molybdate, calcium molybdate, manganous molybdate, and zinc molybdate were all undersaturated in the low-pH soils they studied. They also found that ilsemannite was supersaturated in many of their soils and that wulfenite is likely to be the stable Mo mineral under acidic conditions.

Summary

Molybdenum generally forms dissolved anionic species in aqueous solution. The predominant aqueous species of Mo in most natural systems is the molybdate anion. The behavior of Mo in terrestrial environments is strongly influenced by its tendency to form aqueous anionic species. Molybdenum generally is least soluble in acid soils and is readily mobilized in alkaline soils.

Under oxidizing alkaline conditions, Mo is a relatively mobile chemical element. The mobility of Mo in the weathering zone can be limited by (1) the dissolution rate of molybdenite and other Mo-containing minerals, (2) acidic conditions, (3) fixation by natural organic matter, (4) sorption onto hydrous oxide minerals, (5) reducing conditions, (6) precipitation of ferrimolybdite in an iron-containing environment, (7) precipitation of powellite in a carbonate-rich environment, (8) precipitation of ilsemannite in an iron-poor environment, and (9) precipitation of wulfenite or lindgrenite if sufficient lead or copper is present.

References

Adriano, D. C. (1986). *Trace Elements in the Terrestrial Environment.* New York: Springer-Verlag.

Amrhein, C., Mosher, P. A., and Brown, A. D. (1993). The effects of redox on Mo, U, B, V, and As solubility in evaporation pond soils. *Soil Sci.* 155:249–55.

Aubert, H., and Pinta, M. (1977). *Developments in Soil Science. vol. 7: Trace Elements in Soils.* New York: Elsevier.

Baes, C. F., Jr., and Mesmer, R. E. (1976). *The Hydrolysis of Cations.* New York: Wiley.

Balistrieri, L. S., and Chao, T. T. (1990). Adsorption of selenium by amorphous iron oxyhydroxide and manganese dioxide. *Geochim. Cosmochim. Acta* 54:739–51.

Balistrieri, L. S., Murray, J. W., and Paul, B. (1994). The geochemical cycling of trace elements in a biogenic meromictic lake. *Geochim. Cosmochim. Acta* 58:3993–4008.

Barnett, P. R., Skougstad, M. W., and Miller, K. J. (1969). Chemical characterization of a public water supply. *Am. Water Works Assoc. J.* 61:61–7.

Barrow, N. J. (1974). On the displacement of adsorbed anions from soil. 1: Displacement of molybdate by phosphate and by hydroxide. *Soil Sci.* 116:423–31.

Barrow, N. J. (1983). A mechanistic model for describing the sorption and desorption of phosphate. *J. Soil Sci.* 34:733–50.

Barrow, N. J. (1985). Reaction of anions and cations with variable-charge soils. *Adv. Agron.* 38:183–230.

Barrow, N. J. (1989). Testing a mechanistic model. IX. Competition between anions for sorption by soils. *J. Soil Sci.* 40:415–25.

Baturin, G. N., Kochenov, A. V., and Shimkus, K. M. (1967). Uranium and rare metals in the sediments of the Black and Mediterranean Seas. *Geochem. Internat.* 4:29.

Berrang, P. G., and Grill, E. V. (1974). The effect of manganese oxide scavenging on molybdenum in Saanich Inlet, British Columbia. *Marine Chemistry* 2:125–48.

Bertine, K. K. (1972). The deposition of molybdenum in anoxic waters. *Marine Chemistry* 1:43–53.

Bertine, K. K., and Turekian, K. K. (1973). Molybdenum in marine deposits. *Geochim. Cosmochim. Acta* 37:1415–34.

Bibak, A., and Borggaard, O. K. (1994). Molybdenum adsorption by aluminum and iron oxides and humic acid. *Soil Sci.* 158:323–8.

Bideaux, R. A. (1990). The desert mineral: wulfenite. *Rocks and Minerals* 65:11–30.

Blossom, J. W. (1994). *Molybdenum.* Bureau of Mines annual report 1993. Washington, DC: U.S. Department of the Interior.

Bradshaw, P. M. D. (1974). Conceptual models in exploration geochemistry. *J. Geochem. Explor.* 4:213.

Brookins, D. G. (1987). *Eh–pH Diagrams for Geochemistry.* New York: Springer-Verlag.

Brumsack, H. J. (1986). The inorganic geochemistry of Cretaceous black shales (DSDP Leg 41) in comparison to modern upwelling sediments from the Gulf of California. In *North Atlantic Paleoceanography*, ed. C. P. Summerhayes and N. J. Shackleton, pp. 447–62. Geological Society special publication no. 21. London: Geological Society.

Brumsack, H. J., and Gieskes, J. M. (1983). Interstitial water trace-metal chemistry of laminated sediments from the Gulf of California, Mexico. *Marine Chemistry* 14:89–106.

Chojnacki, J., and Oleksyn, B. (1965). Polymerization of molybdates in dilute aqueous solutions. *Rocz. Chem. Ann. Soc. Chem. Polonarum.* 39:1141–4.

Collier, R. W. (1985). Molybdenum in the northeast Pacific Ocean. *Limnol. Oceanogr.* 30:1351–3.

Connor, J. J. (1981). *Geochemical Analyses and Summaries of Shale from Kentucky.* U.S. Geological Survey open-file report 81–1097.

Connor, J. J., and Shacklette, H. T. (1975). *Background Geochemistry of Some Rocks, Soils, Plants, and Vegetables in the Conterminous United States.* U.S. Geological Survey professional paper 574-F.

Contreras, R., Fogg, T. R., Chasteen, N. D., Gaudette, H. E., and Lyons, W. B. (1978). Molybdenum in pore waters of anoxic marine sediments by electron paramagnetic resonance spectroscopy. *Marine Chemistry* 6:365–73.

Cotton, F. A., and Wilkinson, G. (1988). *Advanced Inorganic Chemistry*, 5th ed. New York: Wiley.

Cruywagen, J. J., and De Wet, H. F. (1988). Equilibrium studies of the adsorption of molybdenum (VI) on activated carbon. *Polyhedron* 7:547–56.

Davis, J. A., and Kent, D. B. (1990). Surface complexation modeling in aqueous geochemistry. In *Mineral–Water Interface Geochemistry*, ed. M. F. Hochella and A. F. White, pp. 177–260. Washington, DC: Mineralogical Society of America.

Dolukhanova, N. I. (1960). An experiment in the application of hydrochemical survey to copper and molybdenum deposits in the Armenian S.S.R. *Internat. Geol. Rev.* 2:20–42.

Domagalski, J. L., Eugster, H. P., and Jones, B. F. (1990). Trace metal geochemistry of Walker, Mono, and Great Salt Lakes. In *Fluid–Mineral Interactions: A Tribute to H. P. Eugster*, ed. R. J. Spencer and I. M. Chou, pp. 315–53. Geochemical Society special publication no. 2. London: Geochemical Society.

Durfor, C. N., and Becker, E. (1964). *Public Water Supplies of the 100 Largest Cities in the United States, 1962.* U.S. Geological Survey water-supply paper 1812.

Dzombak, D. A., and Morel, F. M. M. (1990). *Surface Complexation Modeling*. New York: Wiley.

Ebens, R. J., and Shacklette, H. T. (1982). *Geochemistry of Some Rocks, Mine Spoils, Stream Sediments, Soils, Plants, and Waters in the Western Energy Region of the Conterminous United States*. U.S. Geological Survey professional paper 1237.

Emerson, S. R., and Huested, S. S. (1991). Ocean anoxia and the concentrations of molybdenum and vanadium in seawater. *Marine Chemistry* 34:177–96.

Erdman J. A., and Ebens, R. J. (1979). Element content of crested wheatgrass grown on reclaimed coal spoils and on soils nearby. *J. Range Management* 32:159–61.

Fauth, H., Hindel, R., Siewers, U., and Zinner, J. (1985). *Geochemischer Atlas Bundesrepublik Deutschland*. Hannover: Bundesanstalt für Geowissenschaften und Rohstoffe.

Fleming, G. A. (1980). Essential micronutrients. I: Boron and molybdenum. In *Applied Soil Trace Elements*, ed. Brian E. Davis, pp. 155–97. New York: Wiley.

Follett, R. F., and Barber, S. A. (1967). Molybdate phase equilibria in soil. *Soil Sci. Soc. Am. Proc.* 31:26–9.

Fortescue, J. A. C. (1992). Landscape geochemistry: retrospect and prospect – 1990. *Appl. Geochem.* 7:1–53.

François, R. (1988). A study on the regulation of the concentrations of some trace metals (Rb, Sr, Zn, Pb, Cu, V, Cr, Ni, Mn and Mo) in Saanich Inlet sediments, British Columbia, Canada. *Marine Geology* 83:285–308.

Gough, L. P., and Severson, R. C. (1981). *Biogeochemical Variability of Plants at Native and Altered Sites, San Juan Basin, New Mexico*. U.S. Geological Survey professional paper 1134-D.

Gough, L. P., and Severson, R. C. (1983). Rehabilitation materials from surface-coal mines in the western U.S.A. II. Biogeochemistry of wheatgrass, alfalfa, and fourwing saltbush. *Reclam. Reveg. Res.* 2:103–22.

Gough, L. P., and Severson, R. C. (1995). Mine-land reclamation: the fate of trace elements in arid and semi-arid areas. In *Environmental Aspects of Trace Elements in Coal*, ed. D. J. Swaine and F. Goodarzi, pp. 275–307. Dordrecht: Kluwer.

Gough, L. P., Severson, R. C., and Shacklette, H. T. (1988). *Element Concentrations in Soils and Other Surficial Materials of Alaska*. U.S. Geological Survey professional paper 1458.

Gupta, U. C., and Lipsett, J. (1981). Molybdenum in soils, plants, and animals. *Adv. Agron.* 34:73–115.

Haight, G. P., and Boston, D. R. (1973). Molybdenum species in aqueous solution – a brief summary. In *First International Conference on the Chemistry and Uses of Molybdenum, Proceedings of a Conference Held at the University of Reading, England, September 17–21, 1973*, ed. P. C. H. Mitchell, pp. 48–51. London: Climax Molybdenum Company, Ltd. (AMAX).

Hansuld, J. A. (1966). Eh and pH in geochemical exploration. *Can. Mining Metall. Bull.* 59:315–22.

Harms, T. F., Stewart, K. C., Briggs, P. H., Hageman, P. L., and Papp, C. S. E. (1990). *Chemical Results for Bottom Material for Department of the Interior Irrigation Drainage Task Group Studies 1988–1989*. U.S. Geological Survey open-file report 90-50.

Hatch, J. R., and Leventhal, J. S. (1992). Relationship between inferred redox potential of the depositional environment and geochemistry of the Upper Pennsylvanian (Missourian) Stark Shale Member of the Dennis Limestone, Wabaunsee County, Kansas, U.S.A. *Chemical Geology* 99:65–82.

Hem, J. D. (1989). *Study and Interpretation of the Chemical Characteristics of Natural Water*, 3rd ed. U.S. Geological Survey water-supply paper 2254.

Hingston, F. J. (1981). A review of anion adsorption. In *Adsorption of Inorganics at Solid–Liquid Interfaces*, ed. M. A. Anderson and A. J. Rubin, pp. 51–90. Ann Arbor: Ann Arbor Science.

Hingston, F. J., Atkinson, R. J., Posner, A. M., and Quirk, J. P. (1967). Specific adsorption of anions. *Nature* 215:1459–61.

Hingston, F. J., Posner, A. M., and Quirk, J. P. (1972). Anion adsorption by goethite and gibbsite. I: The role of the proton in determining adsorption envelopes. *J. Soil Sci.* 23:177–92.

Hoffman, J. D., and Buttleman, K. (1994). National geochemical data base: national uranium resource evaluation data for the conterminous United States, with MAPPER display software by R. A. Ambroziak and MAPPER documentation by C. A. Cook. U.S. Geological Survey digital data series DDS-0018-A CD-ROM.

Hoffman, J. D., and Marsh, S. P. (1994). National geochemical data base. In *USGS Research on mineral resources – 1994. Part A: Program and Abstracts*, ed. L. M. H. Carter, M. I. Toth, and W. C. Day, pp. 47–48. U.S. Geological Survey circular 1103-A.

Huerta-Diaz, M. A., and Morse, J. W. (1992). Pyritization of trace metals in anoxic marine sediments. *Geochim. Cosmochim. Acta* 56:2681–702.

James, R. O., and Parks, G. A. (1982). Characterization of aqueous colloids by their electrical double layer and intrinsic surface chemical properties. *Surface Colloid Science* 12:119–217.

Jarrell, W. M., and Dawson, M. D. (1978). Sorption and availability of molybdenum in soils of western Oregon. *Soil Sci. Soc. Am. J.* 42:412–15.

Jenkins, I. L., and Wain, A. G. (1963). Molybdenum species in acid solution. *J. Appl. Chem. (London)* 13:561–4.

Jenne, E. A. (1968). Controls on Mn, Fe, Co, Ni, Cu, and Zn concentrations in soils and water: the significant role of hydrous Mn and Fe oxides. In *Trace Inorganics in Water*, ed. R. F. Gould, pp. 337–87. American Chemical Society advances in chemistry series 73.

Jenne, E. A. (1977). Trace element sorption by sediments and soils – sites and processes. In *Molybdenum in the Environment*, vol. 2, ed. W. R. Chappell and K. K. Petersen, pp. 425–553. New York: Marcel Dekker.

Johansson, M., and Savolainen, I. (1991). Regional estimation of future forest soil acidification. In *Environmental Geochemistry in Northern Europe*, ed. E. Pulkkinen, pp. 303–16. Special paper 9. Espoo, Finland: Geological Survey of Finland.

Kaback, D. S. (1976). Transport of molybdenum in mountainous streams, Colorado. *Geochim. Cosmochim. Acta* 40:581–2.

Kaback, D. S. (1977). The geochemistry of molybdenum in stream waters and sediments, central Colorado. Ph.D. thesis, University of Colorado at Boulder.

Kaback, D. S., and Runnells, D. D. (1980). Geochemistry of molybdenum in some stream sediments and waters. *Geochim. Cosmochim. Acta* 44:447–56.

Kabata-Pendias, A., and Pendias, H. (1984). *Trace Elements in Soils and Plants*. Boca Raton, FL: Chemical Rubber Company Press.

Karimian, N., and Cox, F. R. (1978). Adsorption and extractability of molybdenum in relation to some chemical properties of soils. *Soil Sci. Soc. Am. J.* 42:757–61.

Katz, B. G., and Runnells, D. D. (1974). Experimental study of sorption of molybdenum by desert, agricultural and alpine soils. In *Trace Substances in Environmental Health*, vol. 8, ed. D. D. Hemphill, pp. 107–14. Proceedings of the University of Missouri's 8th Annual Conference on Trace Substances in Environmental Health.

Kim, K.-W., and Thornton, I. (1993). Influence of uraniferous black shales on cadmium, molybdenum and selenium in soils and crop plants in the Deog-Pyoung area of Korea. *Environ. Geochem. Health* 15:119–33.

King, R. U., Shawe, D. R., and MacKevett, E. M., Jr. (1973). Molybdenum. In *United States Mineral Resources*, ed. D. A. Brobst and W. P. Pratt, pp. 425–35. U.S. Geological Survey professional paper 820.

Koide, M., Hodge, V. F., Yang, J. S., Stallard, M., Goldberg, E. G., Calhoun, J., and Bertine, K. K. (1986). Some comparative marine chemistries of rhenium, gold, silver and molybdenum. *Appl. Geochem.* 1:705–14.

Kubota, J. (1980). Regional distribution of trace element problems in North America. In *Applied Soil Trace Elements*, ed. Brian E. Davis, pp. 441–66. New York: Wiley.

Lahann, R. W. (1977). Molybdenum and iron behavior in oxic and anoxic lake water. *Chemical Geology* 20:315–323.

LeGendre, G. R., and Runnells, D. D. (1975). Removal of dissolved molybdenum from wastewaters by precipitates of ferric iron. *Environ. Sci. Technol.* 9:744–9.

Levinson, A. A. (1980). *Introduction to Exploration Geochemistry*, 2nd ed. Wilmette, IL: Applied Publishing.

Lindsay, W. L. (1979). *Chemical Equilibria in Soils*. New York: Wiley.

McKenzie, R. M. (1983). The adsorption of molybdenum on oxide surfaces. *Aust. J. Soil Res.* 21:505–13.

Magyar, B., Moor, H. C., and Sigg, L. (1993). Vertical distribution and transport of molybdenum in a lake with a seasonally anoxic hypolimnion. *Limnol. Oceanogr.* 38:521–31.

Malcolm, S. J. (1985). Early diagenesis of molybdenum in estuarine sediments. *Marine Chemistry* 16:213–25.

Mannheim, F. T. (1978). Molybdenum, natural waters and atmospheric precipitation. In *Handbook of Geochemistry*, vol. 2, part 4, ed. K. H. Wedepohl, pp. 42-H-1–8. Berlin: Springer-Verlag.

Manskaya, S. M., and Drozdova, T. V. (1968). *Geochemistry of Organic Substances*. New York: Pergamon Press.

Mitchell, R. L. (1964). Trace elements in soils. In *Chemistry of the Soil*, 2nd ed., ed. F. E. Bear, pp. 320–68. American Chemical Society monograph series no. 160. New York: Van Nostrand Reinhold.

Moore, P. A., Jr., and Patrick, W. H., Jr. (1991). Aluminum, boron and molybdenum availability and uptake by rice in acid sulfate soils. *Plant Soil* 136:171–81.

Morel, F. M. M., Yeasted, J. G., and Westall, J. C. (1981). Adsorption models: a mathematical analysis in the framework of general equilibrium calculations. In *Adsorption of Inorganics at Solid–Liquid*

Interfaces, ed. M. A. Anderson and A. J. Rubin, pp. 263–94. Ann Arbor: Ann Arbor Science.

Mott, C. J. B. (1981). Anion and ligand exchange. In *The Chemistry of Soil Processes*, ed. D. J. Greenland and M. H. B. Hayes, pp. 179–219. New York: Wiley.

Neuman, D. R., Shrack, J. L., and Gough, L. P. (1987). Copper and molybdenum. In *Reclaiming Mine Soils and Overburden in the Western United States, Analytical Parameters and Procedures*, ed. R. D. Williams and G. E. Shuman, pp. 215–32. Ankeny, IA: Soil Conservation Society of America.

Newton, W. E., and Otsuka, S. (1980). *Molybdenum Chemistry of Biological Significance*. New York: Plenum Press.

Palache, C., Berman, H., and Frondel, C. (1944). *Dana's System of Mineralogy. Vol. I: Elements, Sulfides, Sulfosalts, Oxides*, 7th ed. New York: Wiley.

Palache, C., Berman, H., and Frondel, C. (1951). *Dana's System of Mineralogy. Vol. II: Halides, Nitrates, Borates, Carbonates, Sulfates, Phosphates, Arsenates, Tungstates, Molybdates*, 7th ed. New York: Wiley.

Parfitt, R. L. (1978). Anion adsorption by soils and soil materials. *Adv. Agron.* 30:1–50.

Parish, R. V. (1977). *The Metallic Elements*. New York: Longman.

Pedersen, T. F. (1985). Early diagenesis of copper and molybdenum in mine tailings and natural sediments in Rupert and Holberg inlets, British Columbia. *Can. J. Earth Sci.* 22:1474–84.

Phelan, P. J., and Mattigod, S. V. (1984). Adsorption of molybdate anion (MoO_4^{2-}) by sodium-saturated kaolinite. *Clays Clay Miner.* 32:45–8.

Pilipchuk, M. F., and Volkov, I. I. (1974). Behavior of molybdenum in processes of sediment formation and diagenesis. In *The Black Sea: Geology, Chemistry and Biology*, ed. E. T. Degens and D. A. Ross, pp. 542–52. American Association of Petroleum Geologists, memoir, no. 20.

Plant, J. A., and Raiswell, R. (1983). Principles of environmental geochemistry. In *Applied Environmental Geochemistry*, ed. I. Thornton, pp. 1–39. London: Academic Press.

Presley, B. J., Kolodny, Y., Nissenbaum, A., and Kaplan, I. R. (1972). Early diagensis in a reducing fjord, Saanich Inlet, British Columbia. II: Trace element distribution in interstitial water and sediment. *Geochim. Cosmochim. Acta* 36:1073–90.

Quinby-Hunt, M. S., and Turekian, K. K. (1983). Distribution of elements in sea water. *EOS, Trans. Am. Geophys. Union* 64:130–1.

Reddy, K. J., and Gloss, S. P. (1993). Geochemical speciation as related to the mobility of F, Mo and Se in soil leachates. *Appl. Geochem.* 8:159–63.

Reddy, K. J., Wang, L., and Lindsay, W. L. (1990). Molybdenum supplement to technical bulletin 134: *Selection of Standard Free Energies of Formation for Use in Soil Chemistry*. Technical bulletin LTB90-4. Fort Collins: Department of Agronomy, Colorado State University.

Reyes, E. D., and Jurinak, J. J. (1967). A mechanism of molybdate adsorption on alpha Fe_2O_3. *Soil Sci. Soc. Am. Proc.* 31:637–41.

Rose, A. W., Hawkes, H. E., and Webb, J. S. (1979). *Geochemistry in Mineral Exploration*, 2nd ed. New York: Academic Press.

Rosemeyer, T. (1990). Wulfenite occurrences in Colorado. *Rocks and Minerals* 65:58–61.

Roy, W. R., Hassett, J. J., and Griffin, R. A. (1986a). Competitive interactions of phosphate and molybdate on arsenate adsorption. *Soil Sci.* 142:203–10.

Roy, W. R., Hassett, J. J., and Griffin, R. A. (1986b). Competitive coefficients for the adsorption of arsenate, molybdate, and phosphate mixtures by soils. *Soil Sci. Soc. Am. J.* 50:1176–82.

Roy, W. R., Hassett, J. J., and Griffin, R. A. (1989). Quasi-thermodynamic basis of competitive-adsorption coefficients for anionic mixtures in soils. *J. Soil Sci.* 40:9–15.

Ryden, J. C., Syers, J. K., and Tillman, R. W. (1987). Inorganic anion sorption and interactions with phosphate sorption by hydrous ferric oxide gel. *J. Soil Sci.* 38:211–17.

Severson, R. C., and Shacklette, H. T. (1988). *Essential Elements and Soil Amendments for Plants: Sources and Use for Agriculture.* U.S. Geological Survey circular 1017.

Severson, R. C., Wilson, S. A., and McNeal, J. M. (1987). *Analysis of Bottom Material Collected at Nine Sites in the Western United States for the Department of the Interior Irrigation Drainage Task Group.* U.S. Geological Survey open-file report 87-490.

Shacklette, H. T., and Boerngen, J. G. (1984). *Element Concentrations in Soils and Other Surficial Materials of the Conterminous United States.* U.S. Geological Survey professional paper 1270.

Shaw, T. J., Gieskes, J. M., and Jahnke, R. A. (1990). Early diagenesis in differing depositional environments: the response of transition metals in porewater. *Geochim. Cosmochim. Acta* 54:1233–46.

Shimmield, G. B., and Price, N. B. (1986). The behavior of molybdenum and manganese during early sediment diagenesis – offshore Baja California, Mexico. *Marine Chemistry* 19:261–80.

Sposito, G. (1984). *The Surface Chemistry of Soils.* Oxford University Press.

Stewart, K. C., Fey, D. L., Briggs, P. H., Hageman, P. L., Kennedy, K. R., Love, A. H., McGregor, R. E., Papp, C. S. E., Peacock, T. R., Sharkey, J. D., Vaughn, R. B., and Welsch, E. P. (1992). *Results of Chemical Analysis for Sediments from Department of the Interior National Irrigation Water Quality Program Studies, 1988–1990.* U.S. Geological Survey open-file report 92–443.

Stollenwerk, K. G. (1991). *Simulation of Molybdate Sorption with the Diffuse Layer Surface-Complexation Model.* U.S. Geological Survey water-resources investigations report 91-4034:47–52.

Stumm, W. (1992). *Chemistry of the Solid–Water Interface.* New York: Wiley.

Stumm, W., Kummert, R., and Sigg, L. (1980). A ligand exchange model for the adsorption of inorganic and organic ligands at hydrous oxide interfaces. *Croat. Chem. Acta* 53:291–312.

Szilagyi, M. (1967). Sorption of molybdenum by humus preparations. *Geochem. Internat.* 4:1165–7.

Takematsu, N., Sato, Y., Okabe, S., and Nakayama, E. (1985). The partition of vanadium and molybdenum between manganese oxides and sea water. *Geochim. Cosmochim. Acta* 49:2395–9.

Theng, B. K. G. (1971). Adsorption of molybdate by some crystalline and amorphous soil clays. *N.Z. J. Sci.* 14:1040–56.

Tidball, R. R., Severson, R. C., Gent, C. A., and Riddle, G. A. (1986). Element associations in soils of the San Joaquin Valley of California. In *Selenium in Agriculture and the Environment*, ed. L. W. Jacobs, pp. 179–

93. Soil Science Society of America special publication no. 23. Madison, WI: American Society of Agronomy.

Travis, C. C., and Etnier, E. L. (1981). A survey of sorption relationships for reactive solutes in soil. *J. Environ. Qual.* 10:8–17.

Turekian, K. K., and Wedepohl, K. H. (1961). Distribution of the elements in some major units of the Earth's crust. *Geol. Soc. Am. Bull.* 72:175–92.

Turner, D. R., Whitfield, M., and Dickson, A. G. (1981). The equilibrium speciation of dissolved components in freshwater and seawater at 25°C and 1 atm pressure. *Geochim. Cosmochim. Acta* 45:855–81.

van der Sloot, H. A., Hoede, D., Hamburg, G., Woittiez, J. R. W., and van der Weijden, C. H. (1990). Trace elements in suspended matter from the anoxic hypersaline Tyro and Bannock Basins (eastern Mediterranean). *Marine Chemistry* 31:187–203.

van der Weijden, C. H., Middleburg, J. J., De Lange, G. J., van der Sloot, H. A., Hoede, D., and Woittiez, J. R. W. (1990). Profiles of the redox-sensitive trace elements As, Sb, V, Mo and U in the Tyro and Bannock Basins (eastern Mediterranean). *Marine Chemistry* 31:171–86.

Vlek, P. L. G., and Lindsay, W. L. (1977). Thermodynamic stability and solubility of molybdenum minerals in soils. *Soil Sci. Soc. Am. J.* 41:42–6.

Voegeli, P. T., Sr., and King, R. U. (1969). *Occurrence and Distribution of Molybdenum in the Surface Water of Colorado.* U.S. Geological Survey water-supply paper 1535-N.

Wang, L., Reddy, K. J., and Munn, L. C. (1994). Geochemical modeling for predicting potential solid phases controlling the dissolved molybdenum in coal overburden, Powder River Basin, WY, U.S.A. *Appl. Geochem.* 9:37–43.

Webb, J. S., Thornton, I., Howarth, R. J., Thompson, M., and Lowenstein, P. L. (1978). *The Wolfson Geochemical Atlas of England and Wales.* Oxford University Press.

Westall, J. C. (1986). Reactions at the oxide–solution interface: chemical and electrostatic models. In *Geochemical Processes at Mineral Surfaces*, ed. J. A. Davis and K. F. Hayes, pp. 54–78. American Chemical Society symposium series, vol. 323. Washington, DC: ACS.

Westall, J., and Hohl, H. (1980). A comparison of electrostatic models for the oxide/solution interface. *Adv. Colloid Interface Sci.* 12:265–294.

Xie, R. J., and MacKenzie, A. F. (1991). Molybdate sorption–desorption in soils treated with phosphate. *Geoderma* 48:321–33.

Xie, R. J., MacKenzie, A. F., and Lou, Z. J. (1993). Causal modeling pH and phosphate effects on molybdate sorption in three temperate soils. *Soil Sci.* 155:385–97.

Yamazaki, H., and Gohda, S. (1990), Distribution of dissolved molybdenum in the Seto Inland Sea, the Japan Sea, the Bering Sea and the Northwest Pacific Ocean. *Geochem. J.* 24:273–81.

4

Biochemical Significance of Molybdenum in Crop Plants

P. C. SRIVASTAVA

Introduction

Many early studies reported the molybdenum (Mo) requirements of bacteria (Bortels, 1930; Horner et al., 1942), fungi (Steinberg, 1936, 1937), green algae (Arnon et al., 1955), and higher plants (Arnon and Stout, 1939) long before recognition of its various biochemical and physiological functions. Because of subsequent advances in biochemical and physiological research techniques and instrumentation, Mo has been found to be an essential constituent of several important enzymes, besides having a range of nonspecific roles in plant metabolism. This chapter reviews the information available on the biochemical and physiological roles of Mo in crop plants.

Chemical Basis for the Roles of Molybdenum in Biochemical Reactions

It is worthwhile to examine the chemistry of Mo to emphasize what makes it suitable for catalyzing many unique biochemical reactions. Molybdenum is a transition element with an atomic number of 42 and electronic configurations of $1s^2$, $2s^22p^6$, $3s^23p^63d^{10}$, $4s^24p^64d^5$, and $5s^1$. The outermost $4d$ electrons can be easily removed to yield several oxidation states of Mo, ranging from zero to VI. However, the oxidation states IV, V, and VI are more common and involve only low potentials. Higher oxidation states of Mo have a tendency to form oxocomplexes. As Mo(VI) has no electron in its d orbital, it accepts two e⁻ from an O^{2-} anion to form a sigma (σ) bond. As the Mo—O bond length is short, the d orbital of Mo suffers sidewise weak overlap with the p orbitals of an O atom, to yield a slightly double-bond (Mo=O°) character. Because of a lower electron density on the O atom, it works as an electrophilic group

47

Table 4.1. *Molybdoenzymes in organisms*

Enzymes	Reaction catalyzed	Occurrence
Nitrate reductase	$NO_3^- + NADH + H^+ \longrightarrow NO_2^- + NAD + H_2O$	Bacteria, fungi, higher plants
Nitrogenase	$N_2 \xrightarrow[6H^+]{6e^-} 2NH_3$	Symbiotic and asymbiotic N-fixing bacteria
Xanthine oxidase	$XH \xrightarrow[H_2O]{O_2} XOH + H_2O_2$	Bacteria, insects, birds, mammals
Xanthine dehydrogenase	$XH \xrightarrow{H_2O} XOH$	Bacteria, insects, birds, mammals, higher plants(?),[a] dehydrogenase
Aldehyde oxidase	$RCHO \xrightarrow[H_2O]{O_2} RCOOH + H_2O_2$	Mammals
Sulfite oxidase	$SO_3^{2-} + H_2O \longrightarrow SO_4^{2-} + 2H^+ + 2e^-$	Birds, mammals, highers plants(?)[a]
Formate dehydrogenase	$HCOO^- \longrightarrow CO_2$	Bacteria

[a] Occurrence yet to be proved in most higher plants.

to accept a pair of electrons from nucleophilic agents. On the other hand, the Mo(IV) state has two electrons left in the d orbital, and this reduces its tendency to accept electrons from an O atom. The oxocomplexes of Mo(IV) are very limited and less stable.

Thus, by virtue of its relatively low potential for multiredox states, Mo is capable of mediating many oxidation–reduction reactions in biological systems.

Several Mo-containing enzymes are known to occur in various organisms (Table 4.1). All these molybdoenzymes are multicenter electron-transfer proteins, with the Mo moiety serving as both substrate binding site and redox site. Stiefel, Coucouvanis, and Newton (1993) have reviewed Mo enzymes in detail.

Biochemical and Physiological Roles of Molybdenum in Crop Plants

Nitrogen Metabolism

Being required for both nitrate reduction and biological nitrogen fixation, Mo has crucial roles to play in the nitrogen metabolism of crop plants.

Nitrate Reduction

Nitrate is a common source of nitrogen (N) for most higher plants. After absorption, it is first reduced to nitrite and then to ammonia. The first reduction ($NO_3^- \rightarrow NO_2^-$) is catalyzed by a molybdoenzyme, nitrate reductase. The essentiality of Mo for most crop plants is accounted for by this step. It has been observed that cauliflower (*Brassica oleracea* var. *Botrytis* L.) supplied with N as NH_4 under sterile conditions shows no apparent Mo requirement (Hewitt and Gundry, 1970), but under nonsterile conditions Mo is required if the source of N is NO_3^-, or else the nitrification process remains uninhibited (Agarwala and Hewitt, 1955b). High concentrations of NO_3^- have been found to increase the Mo requirements of cauliflower plants (Agarwala and Hewitt, 1945a). A suboptimal supply of Mo in the nutrient medium drastically decreases the activity of nitrate reductase in plant leaves (Figure 4.1). Candela, Fisher, and Hewitt (1957) demonstrated that infiltration of an Mo solution at 5×10^{-5} M through deficient cauliflower leaves increased the nitrate reductase activity. A good proportional relationship exists between enzyme activity and Mo concentration in plants, especially at low Mo concentrations (Hew and Chai, 1984). That provides the basis for diagnosing Mo deficiency in crop plants by monitoring the activity of nitrate reductase in the plant foliage.

Nitrate reductase activity
(μmol NO_2^-/g fresh weight/h)

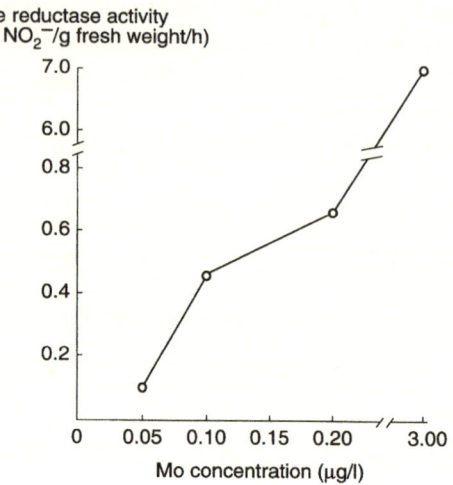

Mo concentration (μg/l)

Figure 4.1. Effect of Mo supply on nitrate reductase activity in spinach leaves. [Plotted from the data of Kaplan and Lips, 1984, by permission of Laser Pages Publishing (1992) Ltd.]

Nitrate reductase from higher plants is a tetramer with molecular weight of 160,000–500,000, with 0.04–0.06% of the Mo present as a prosthetic group (Notton and Hewitt, 1979; Hewitt and Notton, 1980). Each enzyme molecule has two flavine-adenine dinucleotides (FADs), two hemes, and one Mo atom. The nitrate reductases present in lower plants differ slightly in molecular weight, protein subunits, and number of flavine or heme units (Hewitt and Notton, 1980).

The enzyme activity is induced by both NO_3^- and Mo when in the presence of each other. The induction of enzyme activity by NO_3^- is a slow process and requires mRNA-dependent synthesis of apoprotein, whereas the induction of enzyme activity by Mo is much faster, as it involves only rapid activation of the apoprotein by Mo (Jones et al., 1976, 1978). Notton and Hewitt (1979) showed that the Mo-free apoenzyme could be activated by addition of Mo complex obtained from acid washings of the native enzyme. Tungsten (W) can substitute for Mo in nitrate reductase, but the enzyme activity is decreased (Heimer, Wray, and Filner, 1969), as the formation of an active Mo cofactor is prevented. In an experiment with W-treated tobacco (*Nicotiana tabacum*) plants supplied with N as NO_3^-, Deng, Moureaux, and Caboche (1989) reported a decrease in nitrate reductase activity, but several-fold increases in the accumulation of nitrate reductase apoprotein and corresponding mRNA because of excessive expression of a nitrate reductase structural gene.

That deregulation in nitrate reductase gene expression could possibly have been due to a modified transcription rate for the nitrate reductase gene or a modified translation rate for nitrate reductase mRNA, or it could have been due to interference with nitrate reduction, which could have limited reduced N metabolites to low levels. Further studies are needed to examine these possibilities.

A scheme for Mo-mediated reduction of nitrate by nitrate reductase is as follows:

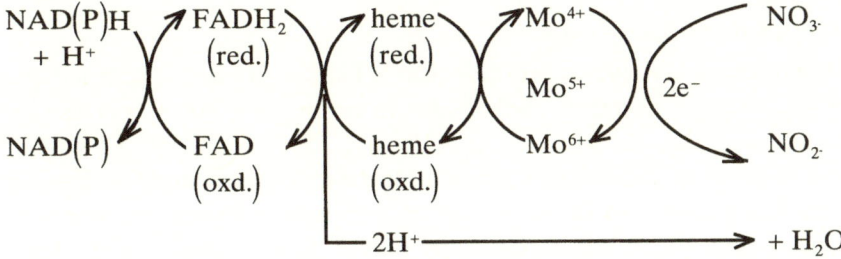

Hewitt and Notton (1980) reviewed some possible mechanisms of nitrate reduction. According to the model proposed by Hewitt, Notton, and Garner (1979), the reducing power for the reaction comes from NADH generated by the oxidation of glyceraldehyde 3-phosphate released from chloroplasts as a result of photosynthetic activity or produced in glycolysis. The NAD(P)H reduces one FAD center to $FADH_2$, and with the second FAD it produces two FADH semi-quinones, which in turn reduce two heme centers and release $2H^+$. One heme center reduces $O{=}Mo^{6+}{=}O$, and with the substrate and a proton that results in an oxo-Mo^{5+}-OH group liganded to NO_3^-. That further reacts with the second reduced heme and a proton to produce oxo-Mo^{4+} with a coordinated water molecule and liganded NO_3^-. Water is displaced by ligand exchange to form a bidentate nitrato complex. This nitrate is reduced by the transfer of two e^- from the oxo-Mo^{4+} couple to produce a dioxo-Mo^{6+}-nitrito complex. Nitrite is released or displaced by NO_3^- for cyclic reaction.

In the presence of Mo deficiency, low nitrate reductase activity results in high accumulations of NO_3^- in plants, with many physiological implications. Disorders such as "scald" disease of beans (*Phaseolus vulgaris*) (Wilson, 1949) and leaf burn/scorch of clementine (*Citrus sinensis*) trees (Triboi, 1978; Cassin et al., 1982) have been related to nitrate accumulation in the presence of a short supply of Mo. Premature sprouting of maize (*Zea mays* L.) seeds having low Mo content ($<0.03\,mg\,kg^{-1}$) has also been reported and attributed to accumulation of nitrates (Tanner, 1978). A heavy late-side application of ammonium nitrate will only increase the severity of the problem, but Mo application will reduce it. Das Gupta and Basuchaudhuri (1977) observed that Mo fertilization promoted nitrate reductase activity in a rice (*Oryza sativa* L.) crop receiving high applications of N fertilizers and favored a higher concentration of reduced N, which could build up a concentration gradient for the uptake and greater assimilation of N in the tissues.

Biological Nitrogen Fixation

Molybdenum is required for biological N fixation, an activity carried out by root-nodule bacteria (*Rhizobium*) in leguminous crops and also by asymbiotic bacteria such as *Azotobacter, Rhodospirulum, Klebsiella*, and blue-green algae. Biochemical reduction of atmospheric dinitrogen becomes possible in the presence of a molybdoenzyme, nitrogenase. Recent reviews by Smith and Eady (1992), Stacey, Burris, and Evans (1992), Stiefel et al. (1993), Kim and Rees (1994), and Rees, Chan, and Kim (1994) on the structure and functioning of nitrogenase are worth consulting.

Basically, the nitrogenase enzyme system consists of two oxygen-sensitive metalloproteins, a larger Mo-Fe protein (molybdoferredoxin) and a smaller Fe protein (azoferredoxin). Alternative nitrogenases containing vanadium (V) have been reported in *Azotobacter chroococcum* (Arber et al., 1987) and *Anabaena variabilis* (Yakunin, Chanvan, and Gogotov, 1991).

The Mo-Fe protein is an $\alpha_2\beta_2$ tetramer with a molecular weight of 220,000–240,000. It has two types of centers, the Fe-Mo cofactor (M center) and an Fe-S cluster (P cluster). The Fe-Mo factor consists of approximately one Mo atom, six to eight Fe atoms, eight to nine sulfur (S) atoms, and one homocitrate group (Nelson, Levy, and Orme-Johnson, 1983; Hoover et al., 1989; Burgess, 1990), and the P cluster has two 4Fe:4S clusters. The overall composition of Mo-Fe protein (2Mo:30Fe:30S) suggests that two copies each of the Fe-Mo cofactor and the P cluster are present. The detailed structures of Fe-Mo cofactor and P clusters have not been firmly established. However, on the basis of crystallographic analysis of nitrogenase from *Azotobacter vinelandii*, a structural model for Mo-Fe protein was recently proposed by Kim and Rees (1992). According to this model, the Fe-Mo cofactor is a dimer of $M\mathrm{Fe_3S_3}$ (M = Fe or Mo) partial cubanes, linked by two bridging sulfides and a third ligand Y possibly resulting from a well-ordered O or N

species or a less well ordered S species, or due to compositional hetero-
geneity (Figure 4.2a). The Mo atom in the cofactor shows approximately octahedral geometry. It is linked to three S in the cofactor, two O
from homocitrate, and the imidazole side chain of histidine. With the Fe-
Mo cofactor, three bridging nonprotein ligands are also present. An
alternative model for the Fe-Mo cofactor proposed by Orme-Johnson
(1992) envisaged two MFe_3S_3 clusters linked by a common hexa-
coordinate S atom. Consistent with the Kim and Rees model, Chan, Kim,
and Rees (1993) envisioned that a cavity (~2.5Å Fe–Fe distance between
bridged iron sites) exists in the interior of the Fe-Mo cofactor for
substrate (N_2) binding. Though the size of this cavity seems quite small
to accommodate a nitrogen molecule, it may do so under the highly

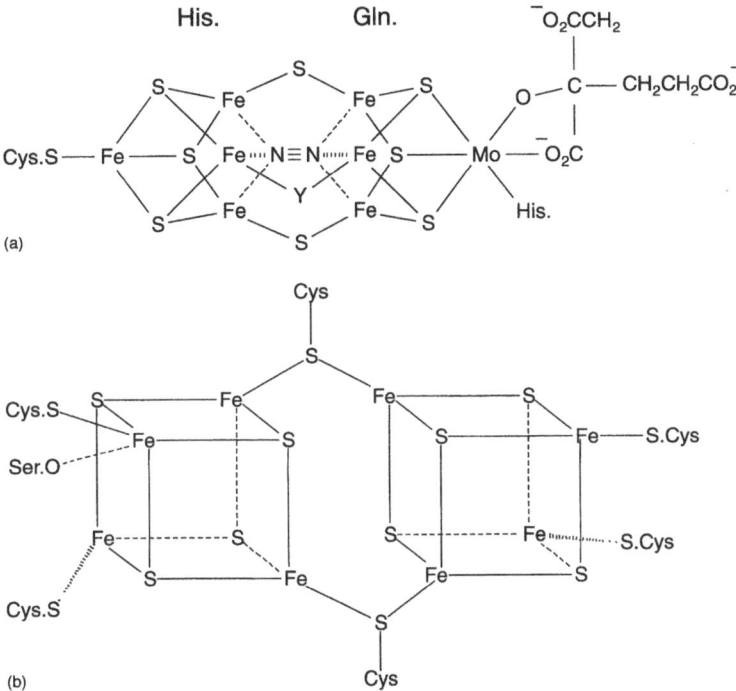

Figure 4.2. Schematic models of Fe-Mo cofactor with possible binding of
dinitrogen substrate (a) and P cluster (b). Y indicates a bridging ligand. (Part a
adapted, with permission, from Kim, J., and Rees, D. C.: Structural models for
metal centres in the nitrogenase molybdenum-iron protein. *Science* 257:1677–82.
Copyright 1992, American Association for the Advancement of Science. Part b
adapted, with permission, from Chan, M. K., Kim, J., and Rees, D. C.: The
nitrogenase Fe-Mo-cofactor and P-cluster pair: 2.2Å resolution structures. *Science* 260:792–4. Copyright 1993, American Association for the Advancement of
Science.)

reduced state of the cofactor, when the separation distance between bridged Fe–Fe sites is likely to increase.

The P cluster has two Fe_4S_4 cubane clusters bridged by two cysteine thiolate groups and a disulfide bridge (Figure 4.2b). The presence of the disulfide bridge also explains the susceptibility of the Mo-Fe protein to oxidation. The remaining Fe sites are linked to amino acids.

The Fe protein of nitrogenase (molecular weight 64,000) has two subunits and a single Fe_4S_4 unit. Each subunit folds as a single α/β-type domain. It binds with magnesium adenosine triphosphate (MgATP).

The reduction of dinitrogen by nitrogenase can be depicted as

$$N_2 + 8H^+ + 8e^- + 16MgATP \xrightarrow{\text{nitrogenase}} 2NH_3 + H_2 + 16MgADP + 16ip$$

In this reaction, Fe protein is first reduced by accepting electrons from ferredoxin or flavodoxin. The reduced Fe protein combines with MgATP, lowering its redox potential. The complex of Fe protein with MgATP transfers an electron to the Mo-Fe protein through a P-cluster pair, with concomitant hydrolysis of MgATP to MgADP. The reduced Mo-Fe protein transfers electrons to dinitrogen ($N\equiv N$) bound to the Fe-Mo cofactor. Ochiai (1977, 1987) and Deng and Hoffman (1993) have reviewed different mechanisms and possible geometries for the binding and reduction of N_2 bound to the Fe-Mo cofactor. According to the scheme of Chatt (1979), electrons are accepted stepwise by N_2 bound to nitrogenase, and the strong bonds between N atoms are broken by protonation, as follows:

$$\begin{array}{c}
\dfrac{\text{flavodoxin}}{\text{ferredoxin}} \xrightarrow{\quad e^- \quad} \quad
\begin{array}{c}
\text{Fe protein} \\
\text{(reduced)} \\
\downarrow e- \\
\text{Mo} \\
\text{(Mo-Fe protein)} \\
\big| \; (N\equiv N)
\end{array}
\end{array}$$

$$
\begin{array}{ccc}
\text{Mo} \vdots N\equiv N & \qquad & \text{Mo} + NH_3 \\
+H^+ \Big\downarrow e^- & & +2H^+ \Big\uparrow 2e^- \\
\text{Mo} \vdots N \cong NH & & \text{Mo} \cong NH + NH_3 \\
+H^+ \Big\downarrow e^- & & +H^+ \Big\uparrow e^- \\
\text{Mo} \equiv M \cdots NH_2 & \xrightarrow[\quad e^- \quad]{+H^+} & \text{Mo} = NH\!-\!NH_2 \\
\end{array}
$$

The exact oxidation states of the reduction sites are not fully known, but it has been suggested that electrons to reduce N probably are supplied by both Fe and Mo.

Being essential for N-fixing activity, the absence of Mo results in poor growth and activity of root nodules in leguminous plants. In the absence of Mo, pea plants (*Pisum sativum* L.) produce small pale-yellow-brown nodules of low N-fixing activity in large numbers (Mulder, 1948). In Mo-deficient red clover (*Trifolium pratense* L.), root-nodule bacteroids show some morphological signs of accelerated senescence, and canal-type bacteroids are formed (Fedorova and Potatueva, 1984). Such bacteroids have large numbers of electron-transparent canals formed by invaginations of the bacteroid cell membrane and cell wall. Application of Mo has been found to increase leghemoglobin content and nitrogenase activity in root nodules of red clover, in addition to decreasing the amount of energy lost due to hydrogen evolution (Table 4.2). In a field study of soybean [*Glycine max* (L.) Merr.], Hashimoto and Yamasaki (1976) demonstrated that Mo application did not increase the number of root nodules, but did ensure better growth of each nodule (Table 4.3). In the Mo-treated crop, root nodules, on average, were larger in size and had greater dry weight during the latter half of the

Table 4.2. *Effects of Mo application on dry matter, leghemoglobin content, and N fixation by red clover plants during flowering phase*

Parameter	Mo application to soil (mg kg^{-1})	
	0	1
Dry matter (g)	9.53	14.10
Leghemoglobin content[a]	7.99	17.7
N$_2$ fixation activity[b]	23.9	78.7
Relative efficiency of N$_2$ fixation $(1 - H_2/C_2H_4)$	0.60	0.75
Energy loss due to H$_2$ evolution (%)	40.0	24.4

[a] Milligrams per gram of nodule dry weight.
[b] Micromoles of C_2H_2 per gram of nodule dry weight per hour.
Source: Adapted from Fedorova and Potatueva (1984), with permission of Plenum Publishing Corporation.

Table 4.3. *Effects of Mo treatment on nodule number per plant, dry weight of nodules, average nodule dry weight, and N fixation in soybean (cv. Kitami Shiro) at various stages of sampling (year, 1973)*

	Number of nodules		Dry weight of nodules		Average nodule dry weight		N_2-fixing activity per plant[b]		N_2-fixing activity per gram dry nodule[b]	
Days	$-Mo^a$	$+Mo^a$	$-Mo$	$+Mo$	$-Mo$	$+Mo$	$-Mo$	$+Mo$	$-Mo$	$+No$
32	—	—	—	—	—	—	0.1	0.1	25.8	38.0
40	22.2	20.5	12.7	14.1	0.60	0.67	0.7	1.2	48.0	78.7
60	70.4	62.8	106	162	1.58	2.44	6.5	9.4	59.6	59.8
82	59.7	61.1	191	229	3.06	3.78	8.4	9.2	45.5	38.7
101	166.4	134.3	466	496	2.89	3.95	24.4	26.4	54.4	50.1

[a] $-Mo$, no Mo treatment of seeds; $+Mo$, 130 mg Mo as sodium molybdate per liter of seed.
[b] N_2-fixing activity by acetylene-reduction assay (micromoles of ethylene produced). Data averaged for seed treatment with cobalt (Co) at 15 mg per liter of seed and for control ($-Mo$) and treatment ($+Mo$) at 130 mg Mo + 15 mg Co per liter of seed.
Source: Adapted from Hashimoto and Yamasaki (1976), by permission of the Japanese Society of Soil Science and Plant Nutrition.

growing season. Molybdenum supply promoted the synthesis of nitrogenase in root nodules. Those workers attributed the Mo-enhanced N_2 fixation activity per plant to higher activity per unit weight of nodule during the early growth stages, but to greater weight of nodules per plant during the latter half of the growth. Sharma, Srivastava, and Johri (1991) reported that the endomycorrhizal (*Glomus macrocarpum*) association in *Sesbania aculeata* ensured higher nitrogenase activity by increasing Mo uptake by the plant.

Organic Nitrogen Content

In leguminous plants, the first detectable compound formed during N fixation is ammonia, and subsequently glutamic acid is synthesized from the ammonia. Application of Mo has been found to increase the contents of free glutamic acid, α-alanine, γ-aminobutyric acid, and asparagine in lucerne (*Medicago sativa* L.) nodules (Mulder, Bakema, and Van Veen, 1959) and the contents of arginine, threonine, and valine in lucerne and red clover (Gruhn, 1961) by improving N nutrition. Bozova and Kozarova (1973) reported that application of Mo increased the contents of free amino acids in lucerne, without influencing their qualitative compositions.

In nonlegumes, Mo deficiency hampers NO_3^- reduction and decreases the amounts of most amino acids. Addition of Mo to deficient plants has been found to increase the contents of glutamic acid, glutamine, α-alanine, serine, and aspartic acid in spinach (*Spinacea oleracea* L.), cauliflower, tomato (*Lycopersicon esculentum* Mill.) (Mulder et al., 1959), and maize (Berducou and Mache, 1963). However, decreases in the contents of some amino acids and amides during later stages of growth of Mo-fertilized crops can result from their incorporation into proteins or from subsequent metabolic reactions such as transamination reactions or conversion to amides (Possingham, 1957).

Protein Content

Because Mo promotes the utilization of absorbed nitrates in plants and promotes N fixation in leguminous plants, it is likely to influence the protein content of plants. Applications of Mo have been shown to increase the contents of soluble protein in maize (Figure 4.3) (Agarwala et al., 1978) and beans (Domska, Benedycka, and Krauze, 1989).

Besides that, a direct role for Mo in the synthesis of proteins is also suspected. Ratner and Akimochkina (1962) demonstrated that addition of Mo at moderate concentrations to legumin hydrolysate and its introduction into lettuce (*Lactuca sativa* L.) leaves by vacuum infiltration increased protein synthesis from the hydrolysate. Poor development of growing points and generative organs in Mo-deficient cauliflower has been related to improper regulation and maintenance of protein synthesis because of low degrees of methylation of the cytosine residue (5-methylcytosine) in DNA (Bozhenko and Belyaeva, 1977). A low degree of methylation of DNA possibly could be attributed to a low concentration of methionine (the main source of methyl groups) in Mo-deficient plants. It has been suggested that methylation of certain cytosine residues can prevent readily degradation of chains of DNA by endonucleases. A low content of 5-methylcytosine in the presence of Mo deficiency is therefore likely to disturb the replication and transcription of DNA, with consequent effects on protein synthesis. Agarwala et al. (1978) encountered increased activity of ribonuclease enzyme but decreased alanine aminotransferase activity in Mo-deficient corn plants (Figure 4.3). Supply of Mo to the deficient plants produced a significant

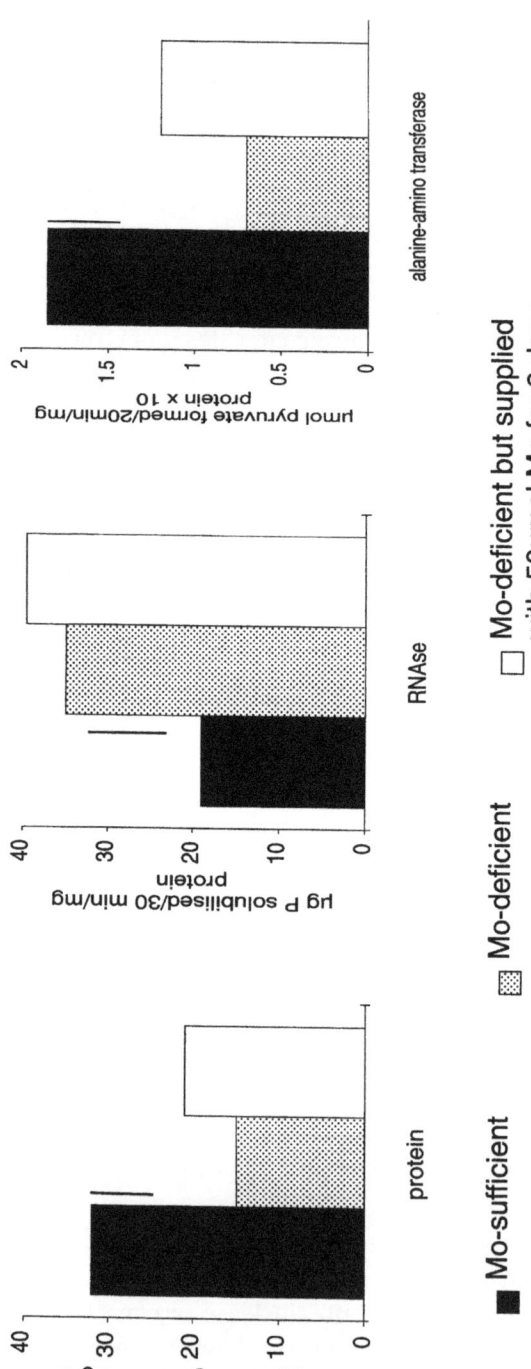

Figure 4.3. Effect of Mo deficiency on soluble-protein content and activities of ribonuclease and alanine aminotransferase in corn (cv. T41). The enzyme activities are at 30°C. The vertical lines depict LSD ($p = 0.05$). [Adapted from Agarwala et al. (1978), with permission from the *Canadian Journal of Botany*. Copyright 1978, Dr. C. Chatterjee.]

increase in the specific activity of alanine aminotransferase, without significantly increasing ribonuclease activity. This evidence suggests some role for Mo in protein synthesis at steps other than merely the reduction of nitrates.

Purine Catabolism

In many legumes, transportation of N from root to shoot occurs in the form of ureids, allantoin, and allantoic acid, which are synthesized from uric acid, an oxidation product of purine (xanthine). Poor growth of legumes in the presence of Mo deficiency can be ascribed in part to poor upward transport of N because of disturbed xanthine catabolism. In plants, oxidation of xanthine is mediated by another molybdoenzyme, xanthine dehydrogenase (Mendel and Muller, 1976; Nguyen and Feierabend, 1978). This enzyme has a constitution similar to that of the xanthine oxidase found in animals. It has two identical subunits, and each unit contains one Mo atom, one FAD, and four Fe-S groups.

Regarding the reaction mechanism, according to Ochiai (1987), $Mo=O$ picks up the hydrogen from xanthine in this reaction and Mo^{VI} is reduced to Mo^{IV}:

$$\left[Mo^{VI}\text{-}O^{2-} \leftrightarrow Mo^{IV}{=}O\right] + XH \rightarrow Mo^{IV}\text{-}OH + X^+$$

The X^+ temporarily combines with a nucleophilic site (—S— or —S—S—) before reacting with water. The Mo(IV) is oxidized back to Mo(V) and then to Mo(VI) by the Fe-S unit. The electrons are accepted by FAD.

Sulfur Metabolism

Biochemical oxidation of sulfite to sulfate is mediated by a molybdoenzyme, sulfite oxidase. The enzyme activity has been reported in animals, plants, and certain bacteria. The enzyme contains Mo and heme Fe in the ratio 1:1. Molybdenum plays roles in both substrate binding and transfer of the O atom (Ochiai, 1987).

$$\left[Mo^{VI}-O^{2-} \leftrightarrow Mo^{IV}{::}O\right] + {:}S{=}O \rightarrow Mo^{IV}{::}O{:}S{=}O \rightarrow$$

$$Mo^{IV}+O{=}S{=}O$$

Keeping in view the importance of this enzyme in S metabolism, Mo is likely to be important for oil-seed crops.

Regulation of Other Enzyme Activities and Biochemicals

Molybdenum is known to alter the activities of several enzymes for which it is not an essential constitutent. Molybdenum deficiency decreases the activities of indoleacetic acid oxidase (Fujiwara and Tsutsumi, 1955), aldolase, succinic dehydrogenase, alanine aminotransferase (Agarwala et al., 1978), glycolate dehydrogenase and glycolate oxidase (Kaplan and Lips, 1984), catalase, and cytochrome c oxidase (Agarwala, Nautiyal, and Chatterjee, 1986), but increases the activities of peroxidase, ribonuclease, and acid phosphatase (Chatterjee, Nautiyal, and Agarwala, 1985; Agarwala et al., 1986, 1988), in many field and horticultural crops. Resumption of an Mo supply to deficient plants increases the activities of catalase, alanine aminotransferase, peroxidase (Agarwala et al., 1978), and aldolase, but decreases the activity of acid phosphatase, for which it acts as an inhibitor (Table 4.4) (Agarwala et al., 1988).

Molybdenum application, especially at higher rates (sodium molybdate at $10.1\,kg\,ha^{-1}$), was shown to decrease the activity of polyphenol (catechol) oxidase and the contents of total phenolic com-

Table 4.4. *Effects of Mo deficiency and recovery therefrom on the activities of certain enzymes in sorghum (cv. Deshi)*

Enzyme	Enzyme activities following Mo supply $(mg\,L^{-1})$		
	1×10^{-5}	2×10^{-2}	1×10^{-5} + added Mo[a]
Catalase (μ moles H_2O_2 decomposed)	130 ± 8[b]	167 ± 3	160 ± 3
Peroxidase (change in optical density)	1.30 ± 0.04	3.81 ± 0.19	2.21 ± 0.06
Aldolase (μg triose phosphate formed)	216 ± 3	296 ± 7	247 ± 5
Acid phosphatase (μg phosphorus liberated)	53 ± 2.6	32 ± 0.8	37 ± 0.7

[a] Supplied with 5 mg Mo per liter for 48 hours.
[b] ±standard error.
Source: Adapted from Agarwala et al. (1988), by permission of *Tropical Agriculture.* Copyright 1988, Dr. C. Chatterjee.

pounds, chlorogenic acid, and tyrosine in potato (*Solanum tuberosum* L.) tubers (Munshi and Mondy, 1988). Mondy and Munshi (1993) also reported that Mo fertilization of a potato crop, especially foliar spray at 13 weeks after planting, significantly reduced the contents of some naturally occurring toxicants such as total glycoalkaloids (α-solanine and α-chaconine) and nitrates. Molybdenum deficiency has been found to decrease the contents of deoxyribonucleic acid (DNA) and ribonucleic acid (RNA) in mustard (*Brassica campestris* L. var. Sarson) leaves (Chatterjee et al., 1985).

Molybdenum thus appears to regulate the levels of several biochemicals and the activities of many non-molybdoenzymes in crop plants through its indirect effects on plant metabolism, possibly by altering the bioavailability of other nutrients synergistically or antagonistically. Molybdenum deficiency is known to reduce the availability of Fe and phosphorus (P) and the utilization of N, but to increase the contents of S, copper (Cu), and manganese (Mn) in plants (Chatterjee, 1992).

Carbon Assimilation and Carbohydrate Metabolism

Molybdenum influences both carbon assimilation and the utilization of carbohydrates in crop plants. An adequate supply of Mo will reduce the chlorophyll content of leaves in cauliflower (Agarwala and Hewitt, 1954b) and rice (Table 4.5) (Das Gupta and Basuchaudhuri, 1977). Agarwala et al. (1978) reported a marked increase in the leaf chlorophyll content of Mo-deficient corn plants within 3 days of resuming adequate Mo supply. In the presence of Mo deficiency, poor organization of chloroplasts, with reduced grana stacking and distorted expansion of intrathylakoid space, has been observed in cauliflower (Fido et al., 1977). The damage to chloroplasts has been attributed to superoxide production under Mo deficiency. The Mo supply also influences the contents of carotenoids in the leaves of sugar beets (*Beta vulgaris* L.) (Godnev and Lipskaya, 1965) and grapevines (*Vitis vinifera* L.) (Lakiza, 1980). Because Mo is not directly involved in the synthesis of plant pigments, poor concentration of pigments in Mo-deficient plants can be related to a secondary effect of disrupted N metabolism.

Das Gupta, Basuchaudhuri, and Gupta (1976) noted that Mo application to a rice crop enhanced CO_2 assimilation by the crop, especially at a high N dosage, by increasing both the efficiency and size of the assimilatory surface (Table 4.4). Similar effects have been noted in grapevines (Lakiza, 1980). Molybdenum supply leads to complex changes in the

Table 4.5. *Effects of Mo supply (foliar application) on chlorophyll content and mean carbon concentration in different organs of rice (cv. IR8) grown under varying N rates*

N rate (mg L^{-1})[a]	Mo per plant (mg)	Chlorophyll content (mg per gram fresh weight)[b]	Mean ^{14}C content (cpm per 50 mg dry weight)	
			Leaf	Stem
40	0	1.92	595 ± 8[c]	42 ± 4
	30	2.05	802 ± 9	73 ± 5
120	0	2.20	1,426 ± 13	104 ± 9
	30	3.04	1,804 ± 15	139 ± 8

[a] N source: ammonium nitrate.
[b] Data from Das Gupta and Basuchaudhuri (1977), reprinted by permission of Kluwer Academic Publishers (modified from Das Gupta et al., 1976).
[c] ±standard error.

concentrations of total sugar and reducing sugars in the leaves and stems of cauliflower plants (Agarwala and Hewitt, 1995a). In the presence of Mo deficiency, low sugar concentrations are related to failure of photosynthesis and nonutilization of carbohydrates for plant growth.

Reproductive Physiology

In many crop plants, the reproductive phase is more sensitive to Mo deficiency than is the vegetative phase, and such deficiency will result in poor crop yields in the absence of Mo application. In corn, Mo deficiency results in delayed emergence of cobs, which are severely condensed; the styles are poorly developed and fail to emerge out of the leaf sheath (Agarwala et al., 1978). Molybdenum deficiency also inhibits tasseling, anthesis, and development of sporogenous tissues (Agarwala et al., 1979). It has been observed that in Mo-deficient corn the anthers formed are fewer in number and have but few small, nonviable pollen grains (Table 4.6). Pollen sterility in Mo-deficient corn is attributed to poor capacity to utilize sucrose because of lowered activities of invertase and starch phosphorylase. In the presence of Mo deficiency, the activity of ribonuclease in the sporogenous tissue also increases, particularly at postmitotic stages (Agarwala et al., 1979).

Table 4.6. *Effects of Mo supply on Mo concentration in pollen grains, pollen-producing capacity, and pollen size and viability in maize plants*

Mo supply ($mg\,L^{-1}$)	Mo concentration in pollen ($\mu g\,g^{-1}$)	No. of pollen grains per anther	Pollen diameter (μm)	Pollen germination (%)
1×10^{-5}	0.017	$1,300 \pm 230$[a]	67.9 ± 2.4	27.3 ± 12.0
1×10^{-4}	0.061	$1,937 \pm 138$	85.0 ± 2.2	51.2 ± 9.5
2×10^{-2}	0.092	$2,437 \pm 145$	93.8 ± 1.3	85.7 ± 5.6

[a] $\pm$ standard error.
Source: Adapted from Agarwala et al. (1979), with permission from the *Canadian Journal of Botany*. Copyright 1979, Dr. C. Chatterjee.

Accumulation of Organic Anions

As nitrates accumulate in the cells of Mo-deficient plants, the synthesis of organic anions to maintain electrical neutrality inside the cells is also hampered. In Mo-deficient tomato plants, the magnitudes of the decreases in the contents of organic anions vary with the type of anion, and the decreases are especially high for malate and citrate (Merkel, Witt, and Jungk, 1975). Decreases in organic anion concentrations generally are greater in the presence of both NH_4^+ and NO_3^- as nitrogen sources, as compared with NO_3^- alone. Angelov and Peneva (1977) reported subnormal concentrations of organic salts in Mo-deficient sunflower (*Helianthus annuus* Mill.) plants that showed yellowing, with marked reductions in growth.

Molybdenum deficiency sharply decreases the contents of ascorbic acid (vitamin C) in several crops, such as cauliflower (Agarwala and Hewitt, 1954b), leguminous plants (Avdonin and Arens, 1966), and potato tubers (Munshi and Mondy, 1988). The decreases in the content of ascorbic acid are more pronounced in the presence of NO_3^- as the N source, but they also occur with NH_4^+ and other N sources (Hewitt and McCready, 1956).

Rhizosphere Effects

Molybdenum indirectly stimulates microbial activity in the plant rhizosphere by promoting root exudation. Foliar application of Mo was found to increase the concentrations of free amino acids, reducing sugars (arabinose), and organic acids in the rhizosphere soil of a rice crop

(Dey and Ghosh, 1986a) and to stimulate the proliferation of free-living N-fixing bacteria and blue-green algae in the rhizosphere soil (Dey and Ghosh, 1986b). Soil application of Mo increased the population of *Azotobacter* and *Rhizobium* in the rhizosphere of a soybean crop (Tripathi and Edward, 1978).

Plant Water Relations

Molybdenum fertilization increases the drought resistance of plants by maintaining a higher degree of hydration of cell colloids (Vasil'eva and Startseva, 1959). The effect is ascribed to the higher contents of total N and protein N and starch in Mo-sufficient plants. Molybdenum deficiency decreases the resistance of plants to stresses such as low temperature and waterlogging (Vunkova-Radeva et al., 1988).

An adequate Mo supply has been shown to decrease the transpiration rate in pea leaves (Zaslonkin, 1968). This effect may be related to the low content of free water in Mo-sufficient plant cells, which may lower the suction force and osmotic pressure of cellular fluid.

Plant Disease Incidence

Like other micronutrients, Mo has been found to influence the incidence of plant diseases. Molybdenum application has been reported to decrease leaf-spot infection caused by *Septoria sojina* in soybeans (Girenko, 1975) and to decrease *Verticillium* wilt in tomatoes (Dutta and Bremner, 1981) and cotton (*Gossypium* spp.) (Miller and Becker, 1983). It also reduces the production of roridin E toxin by *Myrothecium roridum* (Fernando, Jarvis, and Bean, 1986) and zoosporangia formation by *Phytophthora* spp. (Halsall, 1977). Haque and Mukhopadhyay (1983) observed that soil application of Mo caused some reduction in the population of parasitic nematodes. Whether these effects are due to some specific role of Mo in plant disease resistance or are indirect effects of Mo through plant metabolism is not certain.

On the other hand, Mo appears to facilitate the development of some plant diseases. Joham (1953) reported that the severity of *Ascochyta* leaf spot on cotton increased with increasing concentrations of Mo in the nutrient solution. An optimum supply of Mo has been found essential for multiplication of sun hemp (*Crotolaria juncea*) mosaic virus (Sastry, 1962). Pirone and Pound (1962) reported that the reduced amounts of tobacco mosaic virus in Mo-deficient plant cells could be reversed by supplying high amounts of N as NH_4^+ to the plants, indicating that multiplication of a virus depends on the protein-synthesis apparatus of the host plant, which is likely to be impaired when there is Mo deficiency.

Summary

As an essential constituent of several important molybdoenzymes, Mo has specific roles in the metabolism of N and S in crop plants. Further studies are needed to determine the detailed structures of these molybdoenzymes and the involvement of different valence states of Mo in their reaction mechanisms. Furthermore, Mo has a wide range of nonspecific effects involving the regulation of many enzymes other than molybdoenzymes, carbohydrate metabolism, reproductive physiology, anion balance, root exudation, plant water relations, and disease incidence in plants. In the future, attempts must be made to understand the precise roles of Mo in DNA stability, protein synthesis, carbohydrate metabolism, plant stress, and disease resistance. Research is also needed on the effects of Mo nutrition on the quality of food grains and vegetable and fruit crops.

References

Agarwala, S. C., Chatterjee, C., and Nautiyal, N. (1988). Effect of induced molybdenum deficiency on growth and enzyme activity in sorghum. *Trop. Agric.* 65:333–6.

Agarwala, S. C., Chatterjee, C., Sharma, P. N., Sharma, C. P., and Nautiyal, N. (1979). Pollen development in maize plants subjected to molybdenum deficiency. *Can. J. Bot.* 57:1946–50.

Agarwala, S. C., and Hewitt, E. J. (1954a). Molybdenum as a plant nutrient. III. The interrelationships of molybdenum and nitrate supply in the growth and molybdenum content of cauliflower plants grown in sand culture. *J. Hort. Sci.* 29:278–90.

Agarwala, S. C., and Hewitt, E. J. (1954b). Molybdenum as a plant nutrient. IV. The interrelationships of molybdenum and nitrate supply in chlorophyll and ascorbic acid fractions in cauliflower plants grown in sand culture. *J. Hort. Sci.* 29:291–300.

Agarwala, S. C., and Hewitt, E. J. (1955a). Molybdenum as a plant nutrient. V. The interrelationships of molybdenum and nitrate supply in the concentration of sugars, nitrate and organic nitrogen in cauliflower plants grown in sand culture. *J. Hort. Sci.* 30:151–62.

Agarwala, S. C., and Hewitt, E. J. (1955b). Molybdenum as a plant nutrient. VI. Effect of molybdenum supply on the growth and composition of cauliflower plants given different sources of nitrogen supply in sand culture. *J. Hort. Sci.* 30:163–80.

Agarwala, S. C., Nautiyal, B. D., and Chatterjee, C. (1986). Manganese, copper and molybdenum nutrition of papaya. *J. Hort. Sci.* 61:397–405.

Agarwala, S. C., Sharma, C. P., Farooq, S., and Chatterjee, C. (1978). Effect of molybdenum deficiency on the growth and metabolism of corn plants raised in sand culture. *Can. J. Bot.* 56:1905–8.

Angelov, A. P., and Peneva, D. B. (1977). The disrupted ion equilibrium in sunflower plants grown on grey forest soil (in Bulgarian). *Pochvoznanie i Agrokhimiya* 12:49–52.

Arber, J. M., Dobson, B. R., Eady, R. R., Stevens, P., Hasnain, S. S., Garner, C. D., and Smith, B. E. (1987). Vanadium K-edge X-ray absorption spectrum of the VFe protein of the vanadium nitrogenase of *Azotobacter chroococcum*. *Nature* 325:372–4.

Arnon, D. I., Ichioka, P. S., Wessel, G., Fujiwara, A., and Woolley, J. T. (1955). Molybdenum in relation to nitrogen metabolism. I. Assimilation of nitrate nitrogen by *Scendesmus*. *Physiol. Plant.* 8:538–51.

Arnon, D. I., and Stout, P. R. (1939). Molybdenum as an essential element for higher plants. *Plant Physiol.* 14:599–602.

Avdonin, N. S., and Arens, I. P. (1966). Effect of molybdenum on biochemical processes in plants and crop quality (in Russian). *Agrokhimiya* 3:70–9.

Berducou, J., and Mache, R. (1963). Effect of molybdenum on nitrogen content and amino-acid distribution in maize seedlings (in French). *C. R. Acad. Sci. Paris* 257:229–31.

Bortels, H. (1930). Molybdan als Katalysator bei der biologischen Stickstoffbindung. *Arch. Mikrobiol.* 1:333–42.

Bozhenko, V. P., and Belyaeva, V. N. (1977). Content of 5-methyl cytosine in total DNA of plants under conditions of molybdenum and zinc deficiency. *Sov. Plant Physiol.* 24:281–6.

Bozova, L., and Kozarova, M. (1973). The accumulation of free amino acids in two-year-old lucerne after molybdenum fertilizing (in Bulgarian). *Rasteniev" dni-Nauki* 10:57–65.

Burgess, B. K. (1990). The iron-molybdenum cofactor of nitrogenase. *Chem. Rev.* 90:1377–406.

Candela, M. I., Fisher, E. G., and Hewitt, E. J. (1957). Molybdenum as a plant nutrient. X. Some factors affecting the activity of nitrate reductase in cauliflower plants grown with different nitrogen sources and molybdenum levels in sand culture. *Plant Physiol.* 32:280–8.

Cassin, P. J., Blondel-Triboi, A. M., Marchal, J., Favreau, P., Perrier, X., Juste, C., Brun, P., and Lossois, P. (1982). A molybdenum deficiency in citrus in Corsica aggravated by sulphate applications (in French). *Fruits* 37:77–85.

Chan, M. K., Kim, J., and Rees, D. C. (1993). The nitrogenase Fe-Mo-cofactor and P-cluster pair: 2.2Å resolution structures. *Science* 260:792–4.

Chatt, J. (1979). Problems of dinitrogen reduction and its prospects. In *Nitrogen Assimilation of Plants*, ed. E. J. Hewitt and C. V. Cutting, pp. 17–26. London: Academic Press.

Chatterjee, C. (1992). Interaction of copper or molybdenum with other nutrients. In *Management of Nutrient Interactions in Agriculture*, ed. H. L. S. Tandon, pp. 78–96. New Delhi: Fertilizer Development and Consultation Organisation.

Chatterjee, C., Nautiyal, N., and Agarwala, S. C. (1985). Metabolic changes in mustard plants associated with molybdenum deficiency. *New Phytol.* 100:511–18.

Das Gupta, K. D., and Basuchaudhuri, P. (1977). Molybdenum nutrition of rice under low and high nitrogen level. *Plant Soil* 46:681–5.

Das Gupta, K. D., Basuchaudhuri, P., and Gupta, P. S. (1976). Effect of Mo on carbon assimilation by the rice plant. *Indian Agriculturist* 20:51–4.

Deng, H., and Hoffman, R. (1993). How nitrogen activation by the nitrogenase iron-molybdenum cofactor can take place. *Angew. Chem.* 105:1125–8.

Deng, M., Moureaux, T., and Caboche, M. (1989). Tungstate and molybdenum analog inactivating nitrate reductase, deregulates the expression of the nitrate reductase structural gene. *Plant Physiol.* 91:304–9.

Dey, B. K., and Ghosh, A. (1986a). Effect of molybdenum on the free amino acid, reducing sugars and organic acid composition of rhizosphere soils of rich. *Indian Agriculturist* 30:53–8.

Dey, B. K., and Ghosh, A. (1986b). Effect of molybdenum on nitrogen fixation in rhizosphere soils of rice. *J. Indian Soc. Soil Sci.* 34:493–7.

Domska, D., Benedycka, Z., and Krauze, A. (1989). Residual effect of molybdenum fertilization on the content of nitrogen compounds in beans (in Polish). *Acta Academiae Agriculturae ac Technicae olstenensis, Agricultura* 48:99–106.

Dutta, B. K., and Bremner, E. (1981). Trace elements as plant chemotherapeutants to control *Verticillium* wilt. *Z. Pflanzen Krankh. Pflanzenschutz* 88:405–12.

Fedorova, E. E., and Potatueva, Y. A. (1984). Ultrastructure and nitrogen-fixing activity of red clover nodules on application of molybdenum. *Sov. Plant Physiol.* 31:876–81.

Fernando, T., Jarvis, B. B., and Bean, G. (1986). Effects of micro-elements on production of Roridin E by *Myrothecium roridum*, a strain pathogenic to muskmelon (*Cucumis melo*). *Trans. Br. Mycol. Soc.* 86:273–7.

Fido, R. J., Gundry, C. S., Hewitt, E. J., and Notton, B. A. (1977). Ultrastructural features of molybdenum deficiency and whiptail of cauliflower leaves: effects of nitrogen source and tungsten substitution for molybdenum. *Aust. J. Plant Physiol.* 4:675–89.

Fujiwara, A., and Tsutsumi, M. (1955). Biochemical studies of microelements in green plants. I. Deficiency symptoms of microelements on barley plants and changes in indoleacetic acid oxidase activity (in Japanese). *J. Sci. Soil Tokyo* 26:259–62.

Girenko, L. T. (1975). Effect of trace elements on yield of soybeans (in Russian). *Nauchnye Trudy, Ukrainskaya Sel'skokhozyaistvennaya Akademiya* 163:149–52.

Godnev, T. N., and Lipskaya, G. A. (1965). Accumulation of chlorophyll in the chloroplasts of sugar beet under the influence of cobalt, molybdenum and zinc (in Russion). *Fiziol. Rast.* 12:1012–16.

Gruhn, K. (1961). The effect of molybdenum fertilizing on some nitrogen fractions in lucerne and red clover (in German). *Z. Pflanzenernaehr. Dueng. Bodenkde.* 95:110–18.

Halsall, D. M. (1977). Effect of certain cations on the formation and infectivity of *Phytophthora* zoospores. 2. Effect of copper, boron, cobalt, manganese, molybdenum, and zinc ions. *Can. J. Microbiol.* 23:1002–10.

Haque, M. S., and Mukhopadhyay, M. C. (1983). Influence of some micro-nutrients on *Rotylenchulus reniformis*. *Indian J. Nematol.* 13:115–16.

Hashimoto, K., and Yamasaki, S. (1976). Effects of molybdenum application on the yield, nitrogen nutrition and nodule development of soybeans. *Soil Sci. Plant Nutr.* 22:435–43.

Heimer, Y. M., Wray, J. L., and Filner, P. (1969). The effect of tungstate on nitrate assimilation in higher plant tissues. *Plant Physiol.* 44:1197–9.

Hew, C. S., and Chai, B. W. (1984). Effect of light intensity, nitrate and molybdenum levels on nitrate assimilation in choy sam (*Brassica chinensis*). In *Proceedings of the Sixth International Congress on Soilless Culture*, pp. 255–72. Wageningen, Netherlands: International Society for Soilless Culture.

Hewitt, E. J., and Gundry, C. S. (1970). Molybdenum requirement of plants in relation to nitrogen supply. *J. Hort. Sci.* 45:351–8.

Hewitt E. J., and McCready, C. C. (1956). Molybdenum as a plant nutrient. VII. The effects of different molybdenum and nitrogen supplies on yields and composition of tomato plants grown in sand culture. *J. Hort. Sci.* 31:284–90.

Hewitt, E. J., and Notton, B. A. (1980). Nitrate reductase systems in eukaryotic and prokaryotic organisms. In *Molybdenum and Molybdenum Containing Enzymes*, ed. M. P. Coughlan, pp. 273–325. Oxford: Pergamon Press.

Hewitt, E. J., Notton, B. A., and Garner, C. S. (1979). Nitrate reductases: properties and possible mechanism. *Proc. Biochem. Soc.* 7:629–33.

Hoover, T. R., Imperial, J., Ludden, P. W., and Shah, V. K. (1989). Homocitrate is a component of the iron-molybdenum cofactor of nitrogenase. *Biochemistry* 28:2768–71.

Horner, C. K., Burk, D., Allison, F. E., and Sherman, F. S. (1942). Nitrogen fixation by *Azotobacter* as influenced by molybdenum and vanadium. *J. Agric. Res.* 65:173–93.

Joham, H. E. (1953). Accumulation and distribution of molybdenum in the cotton plant. *Plant Physiol.* 28:275–80.

Jones, R. W., Abbott, A. J., Hewitt, E. J., Best, G. R., and Watson, E. F. (1978). Nitrate reductase activity in Paul's scarlet rose suspension cultures and the differential role of nitrate and molybdenum in induction. *Planta* 141:183–9.

Jones, R. W., Abbott, A. J., Hewitt, E. J., James, D. M., and Best, G. R. (1976). Nitrate reductase activity and growth in Paul's scarlet rose suspension cultures in relation to nitrogen source and molybdenum. *Planta* 135:27–34.

Kaplan, D., and Lips, S. H. (1984). A comparative study of nitrate reduction and oxidation of glycolate. *Israel J. Bot.* 33:1–11.

Kim, J., and Rees, D. C. (1992). Structural models for metal centres in the nitrogenase molybdenum-iron protein. *Science* 257:1677–82.

Kim, J., and Rees, D. C. (1994). Nitrogenase and biological nitrogen fixation. *Biochemistry* 33:389–97.

Lakiza, E. N. (1980). Effect of molybdenum on photosynthesis in grapevines (in Russian). *Mikroelement Vokruzh. Srede Kiev, Ukrainian SSR.* 93–6.

Mendel, R. R., and Muller, A. J. (1976). A common genetic determinant of xanthine dehydrogenase and nitrate reductase in *Nicotiana tabacum*. *Biochem. Physiol. Pfl.* 170:538–41.

Merkel, D., Witt, H. H., and Jungk, A. (1975). Effect of molybdenum on the cation–anion balance of tomato plants at different nitrogen nutrition. *Plant Soil* 42:131–43.

Miller, V. R., and Becker, Z. E. (1983). The role of microelements in cotton resistance to *Verticillium* wilt. *Sel'skokhoz. Biol.* 11:54–6.

Mondy, N. I., and Munshi, C. B. (1993). Effect of soil and foliar application of molybdenum on the glyco-alkaloid and nitrate concentration of potatoes. *J. Agric. Food Chem.* 41:256–8.

Mulder, E. G. (1948). Importance of molybdenum in the nitrogen metabolism of microorganisms and higher plants. *Plant Soil* 1:94–119.

Mulder, E. G., Bakema, K., and Van Veen, W. L. (1959). Molybdenum in symbiotic nitrogen fixation and in nitrate assimilation. *Plant Soil* 10:319–34.

Munshi, C. B., and Mondy, N. I. (1988). Effect of soil applications of sodium molybdate on the quality of potatoes: polyphenol oxidase activity, enzymatic discoloration, phenol and ascorbic acid. *J. Agric. Food Chem.* 36:919–22.

Nelson, M. J., Levy, M. A., and Orme-Johnson, W. H. (1983). Metal and sulphur composition of iron-molybdenum cofactor of nitrogenase. *Proc. Natl. Acad. Sci. USA* 80:147–50.

Nguyen, J., and Feierabend, J. (1978). Some properties and subcellular localization of xanthine dehydrogenase in pea leaves. *Plant Sci. Lett.* 13:125–32.

Notton B. A., and Hewitt, E. J. (1979). Structure and properties of higher plant nitrate reductase especially *Spinacea oleracia*. In *Nitrogen Assimilation of Plants*, ed. E. J. Hewitt and C. V. Cutting, pp. 227–44. London: Academic Press.

Ochiai, E.-I. (1977). *Bioinorganic Chemistry*. Boston: Allyn & Bacon.

Ochiai, E.-I. (1987). *General Principles of Biochemistry of the Elements*. New York: Plenum Press.

Orme-Johnson, W. H. (1992) Nitrogenase structure: where to now? *Science* 257:1639–40.

Pirone, T. P., and Pound, G. S. (1962). Molybdenum nutrition of *Nicotiana tabacum* in relation to multiplication of tobacco mosaic virus. *Phytopathology* 52:822–7.

Possingham, J. V. (1957). The effect of mineral nutrition on the contents of free amino acids and amides in tomato plants. II. A study of the effect of molybdenum nutrition. *Aust. J. Biol. Sci.* 10:40–9.

Ratner, E. I., and Akimochkina, T. A. (1962). Role of molybdenum and vitamins in utilization of nitrate nitrogen by plants (in Russian). *Fiziol. Rast.* 9:663–73.

Rees, D. C., Chan, M. K., and Kim, J. (1994). Structure and function of nitrogenase. *Adv. Inorg. Chem.* 40:89–119.

Sastry, K. S. M. (1962). Effect of molybdenum nutrition on multiplication and concentration of sunn hemp mosaic virus. *Curr. Sci.* 31:347–9.

Sharma, A. K., Srivastava, P. C., and Johri, B. N. (1991). Influence of *Glomus macrocarpum* on growth and physiology of *Sesbania aculeata*. Presented at the second national conference on *Mycorrhiza*, Bangalore, India, Nov. 21–3.

Smith, B. E., and Eady, R. R. (1992). Metalloclusters of the nitrogenases. *Eur. J. Biochem.* 205:1–5.

Stacey, G., Burris, R. H., and Evans, H. J. (1992). *Biological Nitrogen Fixation*. New York: Chapman & Hall.

Steinberg, R. A. (1936). Relation of accessory growth substances to heavy metals, including molybdenum, in the nutrition of *Aspergillus niger*. *J. Agric. Res.* 52:439–48.

Steinberg, R. A. (1937). Role of molybdenum in the utilization of ammonium and nitrate nitrogen by *Aspergillus niger*. *J. Agric. Res.* 55:891–902.

Stiefel, E. I., Coucouvanis, D., and Newton, W. E. (1993). *Molybdenum Enzymes, Cofactor and Model Systems*. Washington, DC: American Chemical Society.

Tanner, P. D. (1978). A relationship between premature sprouting on the cob and the molybdenum and nitrogen status of maize grain. *Plant Soil* 49:427–32.

Triboi, A. M. (1978). Nitrate reductase activity: an indicator of molybdenum deficiency (in French). *Fruits* 33:831.

Tripathi, S. K., and Edward, J. C. (1978). Response of *Rhizobium* culture inoculation, zinc and molybdenum application on rhizosphere and phyllosphere microbial population of soybean (*Glycine max* Merril). *Curr. Sci.* 47:503–4.

Vasil'eva, I. M., and Startseva, A. V. (1959). Effect of the trace elements boron, molybdenum, copper and zinc on the water conditions of red clover leaves (in Russian). *Izv. Kazan. Fil. Akad. Nauk Ser. Biol. Nauk* 7:39–47.

Vunkova-Radeva, R., Schiemann, J., Mendel, R. R., Salcheva, G., and Georgieva, D. (1988). Stress and activity of molybdenum containing complex (molybdenum cofactor) in winter wheat seeds. *Plant Physiol.* 87:533–5.

Wilson, R. D. (1949). Molybdenum in relation to the scald disease of beans. *Aust. J. Sci.* 11:209–11.

Yakunin, A. F., Chanvan, N., and Gogotov, I. N. (1991). Effect of molybdenum, vanadium, and tungsten on the growth of *Anabena variabilis* and its synthesis of nitrogenases. *Microbiology* 60:52–6.

Zaslonkin, V. P. (1968). Effect of molybdenum on rate of transpiration by pea leaves (in Russian). *Fiziol. Rast.* 15:921–3.

5

Soil and Plant Factors Affecting Molybdenum Uptake by Plants

UMESH C. GUPTA

Introduction

A number of factors can affect the availability of molybdenum (Mo) to crops. The most important ones include the nature of the parent rock, soil pH, the organic matter in the soil, drainage, interactions with other nutrients, and plant species, plant part, and stage of plant growth at sampling. This chapter attempts to review the information available on the soil and plant factors that affect Mo uptake by plants in different parts of the world.

Soil Factors

Molybdenum Occurrences in Parent Rocks

The concentrations and forms of Mo in rocks and soils tend to vary according to the particular origins and conditions of formation. Molybdenum is a versatile element insofar as valence is concerned, and it can precipitate under either oxidizing (Mo^{6+} predominant) or reducing (Mo^{4+}) conditions (Manheim and Landergren, 1978). Consequently, there may be local enrichments or depletions, and recent work has largely been concerned with elucidating the sequences of occurrence, mobilization, and deposition in particular situations.

Occurrences in Igneous and Metamorphic Rocks

Igneous rocks make up some 95% of the crust of the earth (Mitchell, 1964), and Mo occurs in both acid and basic igneous rocks. Manheim and Landergren (1978) suggested an overall Mo content of nearly 2.0 ppm for granitic rocks and somewhat lower values for basalts.

Although the occurrences of Mo in metamorphic rocks have not been widely studied, metamorphism would be expected to alter the form and site of occurrence rather than the amount of Mo present. New minerals may be formed that must undergo weathering (again, in the case of sedimentary parent material) before the Mo can become available to plants.

Occurrences in Sedimentary Rocks

The sedimentary rocks that are formed following weathering and transport usually retain some of the Mo of the parent material. Concentrations of Mo may be high if the rocks have been formed under conditions favoring accumulation and precipitation of Mo, as, for example, at great depths in the ocean or in the presence of carbon (coals, oil shales, some limestones). Manheim and Landergren (1978) suggested less than 1 ppm Mo overall, but Norrish (1975) suggested 2 ppm in the sediments. The lowest values are found in sandstones that contain stable minerals and have undergone high drainage losses.

Studies in Kentucky have shown that soils derived from the same parent rock, a black fissle shale of Devonian age, can differ greatly in their Mo concentrations (Massey and Lowe, 1961). For example, soils from the Colyer series classed as lithosols, which are shallow, with deep broken topography, contain much higher amounts of Mo than do the red-yellow and gray-brown podzolic intergrade soils that have gentle slopes and are moderately well drained. Some soils in the Colyer series, upon liming, produce Mo concentrations in alfalfa that are toxic to livestock (Massey and Lowe, 1961).

Weathering and Occurrences in Water and Soil

Molybdenum is released from rocks by the processes of weathering (Mitchell, 1964), which involve one or more cycles of solution, oxidation, and precipitation before the Mo from a given rock either appears in the soil formed from that rock or is transported to ocean sediments as part of the sedimentary cycle. Molybdenum is fairly readily released from primary minerals by weathering, and, as compared with other metals, it remains relatively mobile as potentially soluble molybdates (Mo^{6+}). Consequently, transport of Mo via leaching is likely unless iron, aluminum, or manganese oxides interfere under conditions appropriate for occlusion by those minerals (Davies, 1956).

On land, the Mo content of a soil will reflect the amount in its parent rock. That and the effect of mobility are clearly seen near ore bodies, where the influence of mineralization can be detected some distance away from the ore body. Measurements of the Mo contents of soils have been used to locate deposits of Mo and associated metals such as Cu (Manheim and Landergren, 1978). Thornton and Webb (1973) used the Mo contents of stream sediments to map likely patterns of high-Mo soils in England and Wales. The soils of highest content are underlain by marine black shales. In general, soils that overlie sediments will vary more widely in Mo content than will those that overlie igneous rocks, because the latter are more uniform.

Soils formed on sandstones, which lack suitable minerals and may already have experienced heavy losses, are most likely to be deficient in total amount of Mo. There can also be deficiencies that probably are due to secondary reactions that reduce the availability of Mo.

Chemistry of Soil Molybdenum

Occurrence as the Anion

It is now well accepted that Mo, having been released from crystal lattices by weathering, is found in well-aerated agricultural soils (pH in water >5.0) mainly as molybdate ions, MoO_4^{2-} (Lindsay, 1972). This predominance of the anionic form makes Mo unique among metallic plant nutrients. It leads to known interactions with the other major anions SO_4^{2-} and PO_4^{3-}, notably the effect of SO_4^{2-} to reduce uptake of Mo.

Effect of pH on the Solubility of Soil Molybdenum

It is now well known that the availability of soil Mo to plants is, in general, increased by raising the pH, and reduced by lowering it. However, a high content of organic matter in soil may in some way protect Mo at low pH (Mitchell, 1964), which may explain the relative infrequency (Anderson, 1956) of Mo dressings applied to many of the Australian soils on which Mo deficiency was first observed. On those soils, clover (*Trifolium* spp.) growth was made possible by applying Mo with superphosphate, and soil organic matter increased substantially. Evidently, the Ca added in the superphosphate was not enough to saturate the new exchange capacity, and thus soil acidity increased (Donald and Williams, 1954). Nevertheless, the availability of the applied Mo was apparently maintained, possibly because of protection by the organic matter. In that way, the use of lime was postponed (Anderson, 1956).

Forms of Molybdenum in Soils

The forms in which Mo occurs in soil have not yet been completely described, because of the difficulties in dealing with small amounts of relatively labile material. Five potential fractions are recognized:

1. primary crystalline material
2. water-soluble molybdates in the soil solution
3. organically complexed Mo
4. molybdate adsorbed on positively charged surfaces
5. discrete, secondary compounds, either crystalline or amorphous

A full description of soil Mo would include the extents of occurrence of the materials in these five categories, the transformations between them, and the plant availability of the Mo in each category. It appears that high amounts of available Mo probably result from the high Mo contents in fractions 2 and 3; thus, wet soils high in organic matter commonly yield plants with high contents of Mo (Davies, 1956; Kubota, Lemon, and Allaway, 1963). Low Mo contents may reflect a low content in the soil overall, but can also occur because of the presence or formation of the fractions 4 and 5. The relative roles of these fractions are the main concerns in the chemistry of Mo-deficient soils and in the correction of such deficiency.

Effects of Iron and Aluminum Oxides

Iron oxides found in acid soils carry positive charges and can react with MoO_4^{2-}. Jones (1957) showed that soils high in free iron oxide sorbed the largest quantities of MoO_4^{2-} from aqueous solution. Likewise, Karimian and Cox (1978) showed that Mo adsorption on mineral soils increased with increasing content of free iron oxide. Aluminum oxides are also capable of removing Mo from aqueous solution, but their effectiveness is less than that of iron oxide under the same conditions (Jones, 1957). Adsorption of Mo by Fe_2O_3 is pH-dependent, as reported by Jones (1956), who showed that adsorption of Mo by Fe_2O_3 in a solution containing 100 µg of Mo decreased from 98 µg at pH 7 to 22 µg at pH 9 after shaking for 15 hours with 100 mg of amorphous Fe_2O_3.

Jarrell and Dawson (1978) reported that the Mo content of plants correlated poorly with the amounts of Mo sorbed by the soil and NaF-released hydroxyls. Their work showed that Mo adsorption was related to the $(NH_4)_2C_2O_4$-extractable Fe in soil, but not to other fractions of extractable Fe or to extractable Al forms. The amount of $(NH_4)_2C_2O_4$-extractable Mo has also been found to be correlated with the amorphous and free iron oxide contents, and that probably is a result of the extensive dissolution of iron oxide by oxalate (Karimian and Cox, 1979).

Soil pH

Soil pH is one of the most important factors affecting the uptake of Mo by plants. The beneficial effects of liming acid soils to increase Mo solubility are obvious. The concentration of MoO_4^{2-}, which is the form most readily available to plants, increases 100-fold for each unit increase in pH (Lindsay, 1972). With increasing pH, the amount of the soluble MoO_4^{2-} species in equilibrium with soil Mo is much greater than those for $HMoO_4^-$ and H_2MoO_4. At a pH of 5 or 6, the ion $HMoO_4^-$ becomes dominant, and at very low pH values the un-ionized acid H_2MoO_4 and the cation MoO_2^{2+} are the principal species present (Krauskopf, 1972). The MoO_4^{2-} anion exists in an exchangeable form in the soil. Thus, the fact that Mo availability to plants increases with increasing pH may possibly be explained by an anion exchange of the type $2OH^- \rightleftharpoons MoO_4^{2-}$ (Berger and Pratt, 1965). Even wulfenite ($PbMoO_4$), the least soluble of the possible soil compounds, becomes more soluble as pH increases (Vlek and Lindsay, 1974), so that the mode of action of lime in increasing the availability of Mo seems straightforward.

Earlier, Amin and Joham (1958) proposed a Mo cycle within the soil in which it seemed possible that oxidation might contribute to Mo availability in the long term and that the immediate availability of the element might be governed by the state of equilibrium between MoO_3 and the salt form. Under acidic conditions, the balance would be in favor of MoO_3, thus reducing the amounts of molybdate salts present. That would account for the fact that Mo availability is lowered by increasing acidity.

The highest amounts of available Mo are found in soils derived from limestone and shale parent material (Stone and Jencks, 1963). In general, the need for Mo fertilization of most soils can be met by adequate liming of the soil. Ahlrichs, Hanson, and MacGregor (1963) reported that for Minnesota soils, the recommended liming practices substantially increased the Mo content of vernal alfalfa (*Medicago sativa* L.), and thus

Table 5.1. *Effects of soil pH on Mo concentrations in a few crops grown on two soils*

	Mo concentration (mg kg^{-1})					
	Cauliflower		Alfalfa		Bromegrass	
Soil pH[a]	No Mo	Mo (2.5 ppm)	No Mo	Mo (2.5 ppm)	No Mo	Mo (2.5 ppm)
Silty clay loam soil						
5.0	Trace	0.02	Trace	0.43	0.11	0.95
5.5	Trace	0.21	0.51	4.40	0.30	1.80
6.0	0.11	1.62	0.91	4.63	0.27	1.67
6.5	0.56	6.43	1.48	4.93	0.62	2.30
Culloden sandy loam soil						
5.0	Trace	0.39	Trace	0.11	0.02	0.35
5.5	Trace	1.34	Trace	2.04	0.02	1.09
6.0	Trace	3.15	Trace	2.01	0.04	3.59
6.5	Trace	3.58	Trace	3.32	0.05	3.77

[a] Soil:water ratio 1:2.

direct Mo fertilization for those soils was not essential. Likewise, Rolt (1968) reported that with 2 tons of lime plus Mo per acre, or 4 tons of lime alone, the maximum yield of white clover could be obtained. Studies from Alabama showed that soybeans did not respond to Mo when the soil pH was 5.8–6.5 (Odom et al., 1980).

Applications of Mo did not increase crop yields from acid soils in Brazil limed to pH > 6.0 (Franco and Day, 1980). Likewise, Boswell (1980) found that Mo application did not increase soybean yields from soils limed to values above pH 6.2.

In a number of crops there have been significant interaction effects between applied Mo and lime whereby the increases in Mo concentrations with added lime have been of much greater magnitude with applied Mo than without added Mo (James, Jackson, and Harward, 1968; Gupta, 1969; Gupta, Chipman, and MacKay, 1978). On the same soil, pH increases resulted in greater plant Mo contents in some crops, such as alfalfa and cauliflower (*Brassica oleracea* var. *botrytis* L.), than in bromegrass (*Bromus inermis* Leyss.) (Table 5.1). Unpublished data from the Agriculture and Agri-Food Canada Charlottetown Research Centre showed that increasing the soil pH from 4.9–5.6 to 7.0 increased plant-tissue Mo concentrations from 0.13–0.19 to 0.66 ppm, and further increases to pH 7.8–8.0 resulted in much higher tissue Mo concentrations: 1.69–3.42 ppm. Generally, Mo concentrations in plants grown on calcareous soils are higher than those for plants grown on acid soils (Elseewi and Page, 1984).

The principle of Mo release following liming is dependent on the total Mo content of the soil. For example, lime applications to pH 6.5 on two soils from Prince Edward Island did not produce maximum yields of alfalfa and cauliflower in a greenhouse experiment (Gupta, 1969); some addition of Mo was necessary to achieve maximum yields. Application of lime alone to an acid soil (pH 5.4) low in available Mo did not correct the Mo deficiency problem in burley tobacco (*Nicotiana tabacum* L.) grown on a silt loam soil (Sims and Atkinson, 1976). Recent studies by Khan, Mulchi, and McKee (1994) showed that leaf Mo in tobacco grown in Maryland increased with increasing Mo rates, especially at higher pH values. Likewise, on a sphagnum peat soil, some addition of Mo, besides liming, was necessary to obtain maximum yields of some vegetable crops (Gupta et al., 1978). Studies of forage legumes showed that application of Mo to limed soils occasionally resulted in plant Mo concentrations that were in the hazardous range (5–10 mg kg^{-1}) for livestock (Mortvedt and Anderson, 1982).

The observation of Anderson (1956) that plants grown on low-pH soil responded to wood ashes, but not to lime, indicated the need for something that was in the ashes that was not provided simply by raising the soil pH. Well-limed or naturally neutral soils can also be depleted of available Mo by cropping or by leaching (Johnson, Pearson, and Stout, 1952). Rubins (1956) noted that some Mo deficiencies have occurred on well-limed or naturally neutral soils and on acid soils derived from calcareous parent materials. On highly oxidized soils in the southeastern United States, a high initial application of lime at a rate of 11,200 kg ha^{-1} or the equivalent amount in annual applications did not eliminate the need for Mo application to maintain alfalfa stands on those soils (Giddens and Perkins, 1972).

Soil Type and Texture

Studies from Western Australia showed that approximately 10% of added Mo was removed by leaching from two gray sands (pH 5.8–6.1), whereas negligible quantities were removed from more acidic (pH 5.0–5.4) sandy soils (Riley et al., 1987). Leaching did not appear to be an important factor in the occurrences and recurrences of Mo deficiencies in the yellow-brown acidic sand-plain soils of the Western Australian wheat belt. Smith and Leeper (1969) concluded that leaching is not likely to move significant amounts of Mo through acidic profiles, except perhaps in very sandy soils. Liming of soils resulted in higher plant Mo concentrations in crops on fine-textured soils than in crops on coarse-textured soils (Gupta, 1969).

Calcareous soils along the Yellow River in China, derived from loess and the alluvium along the river, have been found to be deficient in total Mo and available Mo, as reviewed by Shorrocks (1989). It has also been pointed out that there is a second group including mainly acid soils in which the total Mo is high but the available Mo is low. The highest soil Mo contents have been found in the lateritic soils derived from granite and basalt. Shorrocks (1989) further stated that in Ireland, Mo deficiencies are found in the sandstone and granite areas and also in some of the limestone soils, but not in the muddy limestones nor in the carboniferous shales, which have a high Mo content. Coarse-textured loams and silty loams low in organic matter, as well as severely eroded and/or heavily weathered soils, responded to Mo applications (Boswell 1980). Application of molybdate fertilizer remains effective for a longer period of time on a soil that has a low ability to retain molybdate (Barrow et al., 1985).

Organic Matter

The amount of Mo adsorbed has been found to be closely related to the soil's organic-matter content (Karimian and Cox, 1978). Because Mo acts as an anion in the soil, it is difficult to explain its adsorption by organic matter. However, certain soils high in organic matter have been found to be deficient in Mo (Mulder, 1954; Davies, 1956). Mulder (1954) found pronounced responses to added Mo in several crops grown on soils with high amounts of organic matter.

Coarse-textured and silty loam soils relatively low in organic matter have been found to respond to Mo applications (Boswell, 1980). The presence of organic matter may also facilitate the availability of certain elements, presumably by supplying soluble complexing agents that interfere with the fixation of elements. For example, soils rich in organic matter contain the readily exchangeable mobile MoO_4^{2-} ion (Koval'skiy and Yarovaya, 1966). Laboratory experiments involving additions of various organic materials such as compost, farmyard manure, and peat to the soil showed increased extractions of Mo after 4–12 weeks of incubation (Gupta, 1971b). However, in the presence of added Mo, the amounts of extracted Mo were consistently lower when organic materials were added. Bloomfield and Kelso (1973) reported that the anionic form of Mo persisted in solution even after a 3-week incubation period with anaerobically decomposing plant material.

Drainage

There have been few studies on the chemistry of Mo in wet soils that produce high-Mo plants (Allaway, 1977), perhaps because the wet, poorly aerated conditions found in the field are difficult to replicate in a laboratory setting suitable for precise measurements of Mo solubility.

However, soil wetness is one of the chief factors affecting the availability of Mo. Poorly drained soils accumulate so much of the available MoO_4^{2-} that plants grown on them are toxic when fed to animals (Davies, 1956; Kubota et al., 1961). Peats and mucks are products of a wet environment and have been associated with Mo toxicity in the soils of the California delta, the Klamath area in Oregon, and the Everglades in Florida (Kubota, 1972). some Scottish soils with poor drainage produce plants containing large quantities of Mo (Mitchell, 1974). The long-standing use of the term "Peat scours" to describe Mo toxicity in cattle and sheep strongly suggests that the problem has occurred frequently on wet soils. Plant Mo concentrations sufficiently high to cause molybdenosis in cattle and sheep are generally confined to poorly drained sites where soils have been formed from high-Mo basalt materials, usually alluvium from igneous rocks or shales (Welch et al., 1991).

Well-drained soils (e.g., podzols and krasnozems) are likely to be low in Mo, and little enrichment has been found in any particular layer (Mitchell, 1974). Wet soils such as those in swamps and lakes that are not leached tend to be high in organic matter and to have large amounts of Mo that are highly available, low pH notwithstanding (Kubota et al., 1961). The soluble Mo fraction in submerged acid soils is generally high, presumably as a result of decreased adsorption of MoO_4^{2-} in the presence of high concentrations of ammonia and waterlogging (Ponnamperuma, 1985). It might be expected that Australian soils, being in areas that are now quite arid, would be well supplied with Mo. However, the deficient soils invariably show signs of having been formed in earlier geologic periods and having experienced heavy leaching. Thus, the current low contents of Mo reflect substantial earlier losses that were never replenished.

Plant Factors

Stage of Plant Growth and Plant Part Sampled

The Mo content in tobacco (*Nicotiana tabacum* L.) leaves has been found to be greater than that in other plant parts (Pal et al., 1976). Studies of tomatoes (*Lycopersicon esculentum* Mill.) by Stout et al. (1951), of alfalfa by Reisenauer (1956), and of soybeans by Singh and Kumar (1979) have shown that leaf tissues contain considerably higher amounts of Mo than stems. The leaves of forage legumes and vegetable crops have been found to contain higher Mo concentrations than the stems (Table 5.2). Among vegetable crops, the lower half of each plant contains more Mo than the upper half (Gupta, 1991). Such differences are greater in plants with higher Mo concentrations. Molybdenum, as an anion, has been found to be readily mobile when fed to the primary leaves of beans, as most of the Mo is transported to the stem and roots (Kannan and Ramani, 1978). That downward movement of Mo in those plants may at least partially explain the higher Mo concentrations in the lower halves of plants, as reported by Gupta (1991). Likewise, Boswell (1980) reported that seeds from the lower third of the soybean plant contained more Mo than did seeds from the top third or middle third of the plant. The interveinal areas of leaves have been found to preferentially accumulate Mo (Stout and Meagher, 1948). In the case of cereal crops grown on low-Mo soils, the Mo concentration in the grain was generally lower than that in the straw (Table 5.3); however, when Mo was applied at rates of 0.5 ppm or higher, the Mo concentration in grain was considerably lower than that in the boot-stage tissue (Gupta, 1971a). With increasing maturity, the Mo content in leaves and stems was found to decrease in soybeans (Singh and Kumar, 1979); it was also noted that the grain contained higher quantities of Mo than did the leaves, stems, or pod husks. In solution-culture studies of *Phaseolus vulgaris* L., the roots were found to contain higher quantities of Mo than the stems and leaves (Wallace and Romney, 1977).

Bhatt and Appukuttan (1982) showed that 10-day-old cotton (*Gossypium hirsutum* L.) seedlings had maximum amounts of Mo. After the seedling stage, the maximum amount of Mo was found in the leaves from the time of flower-bud initiation to peak flowering, and Mo amounts declined rapidly after boll bursting. In forage species, Mo concentrations were lower in later samplings than in earlier samplings (Gupta, 1977). Applications of fly ash high in Mo to acid soils resulted in higher plant

Table 5.2. *Mo concentrations in various plant parts in forage and vegetable crops*

| Crop | Mo concentration (mg kg^{-1}) | | | |
	Leaves only	Stem only	Upper halves of plants	Lower halves of plants
Alfalfa	0.28	0.15	0.24	0.22
Red clover	0.12	0.15	0.13	0.13
Cauliflower	1.65	0.98	1.37	1.66
Broccoli	3.76	1.76	2.33	3.88
Rutabaga	0.65	0.32	0.49	0.58

Tale 5.3. *Mo concentrations in cereal grain and straw with and without added Mo*

| Mo added to soil (g ha^{-1}) | Mo concentration (mg kg^{-1}) | | | | | |
| | Wheat | | Barley | | Oats | |
	Grain	Straw	Grain	Straw	Grain	Straw
None	0.18	0.14	0.29	0.15	0.41	0.10
125	0.88	2.07	0.72	1.11	2.18	0.91
250	1.04	4.27	2.34	2.46	3.32	2.35
500	1.52	8.07	3.51	4.02	4.55	4.92

Mo concentrations in the late clippings of forage legumes than in the early clippings. However, the reverse was the case for a calcareous soil (Elseewi and Page, 1984).

Plant Species and Varieties

Differences in plant species have been found to have some influence on the Mo concentrations in plant tissues. Kubota and Allaway (1972) showed that various forage legumes grown on the same soil have nearly the same amounts of Mo. They further stated that grasses generally contain less Mo than legumes. In the case of soybeans, the top leaves had considerably higher Mo concentrations than the lower leaves (Table 5.4).

Table 5.4. *Mo concentrations in plant parts for two soybean varieties at various stages of sampling*

| | Mo concentration (mg kg⁻¹) | | | |
| | Var. High Protein | | Maple Isle | |
Stage of growth and plant part sampled	Range	Mean ± SD[a]	Range	Mean ± SD
First (whole young plants, stage V5)[b]	0.19–1.54	0.53 ± 0.40	0.24–0.69	0.41 ± 0.16
Second (most recently expanded leaves, stage R1)	0.24–2.72	1.05 ± 0.84	0.13–1.15	0.54 ± 0.27
Second (leaves at the third node from bottom, stage R1)	0.06–0.33	0.11 ± 0.07	0.06–0.24	0.08 ± 0.04
Third (most recently expanded leaves near top, stage R4)	0.06–3.71	0.73 ± 1.14	0.15–1.40	0.50 ± 0.35
Third (leaves at third node from bottom, stage R4)	0.06–0.19	0.09 ± 0.03	0.08–0.24	0.10 ± 0.04
Seeds at harvest	0.8–10.14	5.07 ± 3.19	1.61–6.25	3.76 ± 1.79

[a] Mean Mo values represent average from 10 locations.
[b] Growth stages as described by Fehr et al. (1971).

Also, soybean seeds contained much higher Mo concentrations than other parts of the plants (Table 5.4). The Mo content in seeds varies from one plant species to another. Among seeds, soybeans (*Glycine max*) and peas (*Pisum sativum* L.) accumulate much higher amounts of Mo than wheat (*Triticum aestivum* L.), barley (*Hordeum vulgare* L.), and oats (*Avena sativa* L.) when compared at the same level of Mo fertilization (Tables 5.3 and 5.4). In a study conducted on Prince Edward Island (U. C. Gupta, unpublished data), samples collected from adjacent sites showed the Mo contents of the seeds of three crops to decrease in the following order: peas > barley > wheat. Molybdenum concentrations in barley and wheat grains were considerably less than those in soybean grain. With an Mo application of 0.44 kg ha⁻¹, the Mo concentration in soybean grain increased fourfold. However, the increases in Mo in barley and wheat grains were very small (Hawes, Sims, and Wells, 1976). Studies by Brogan, Fleming, and Byrne (1973) showed that Mo concentrations were higher in clover than in grass grown on soils high in Mo. Gupta (1979), however, working on podzol soils in eastern Canada, could not find consistent differences in the Mo concentrations in red clover (*Trifolium pratense* L.) and alfalfa as compared with that in timothy (*Phleum pratense* L.).

The cultivars of plant species have been found to differ in their ability to absorb Mo. For example, Chipman et al. (1970) found that among cauliflower cultivars, Pioneer was not as efficient as Snowball 84 in the uptake of Mo, especially at low levels of Mo supply. Unpublished findings at the Charlottetown Research Centre showed that a high-protein cultivar of soybean contained more Mo than did the cultivar Maple Isle (Table 5.4).

The seeds of the *Gossypium arboreum* cultivar and a hybrid contained less Mo than did other cultivars of cotton (Bhatt and Appukuttan, 1982). Recent studies by Shivashankar and Hagstrom (1991) showed that peas contained much higher Mo concentrations than cereals and corn. They also reported that the hybrid cotton–lucerne (alfalfa) combination resulted in 50% greater Mo uptake than when followed by the combination of groundnut (*Arachis hypogaea* L.) and wheat or by cluster beans. Cereal grains contained much less Mo in unfertilized plots than in Mo-fertilized plots (Table 5.3).

Sampling Season

In studies in Ireland, herbage Mo concentrations were greatly affected by the seasons, with Mo values of $2 \, \text{mg} \, \text{kg}^{-1}$ in the spring to as high as $11 \, \text{mg} \, \text{kg}^{-1}$ in the autumn (Shorrocks, 1989).

Effects of Phosphorus, Sulfur, Nitrogen, Manganese, and Chloride

Plant uptake of Mo is usually enhanced by the presence of soluble phosphorus (P) and decreased by the presence of available sulfur (S). Earlier studies by Stout et al. (1951) showed that higher amounts of P in culture solutions could increase Mo uptake as much as 10-fold, and therefore soil applications of Mo with P fertilizers may be effective. Phosphorus fertilization was found to increase Mo concentrations in hybrid corn (Hulagur and Dangarwala, 1982). Increasing the application rates of P reduced the overall Mo concentrations in herbage grown on peatland (Feely, 1990). Since then, several researchers have observed similar effects of P on Mo uptake by plants. Higher amounts of P have been found to result in relatively more Mo desorption (Xie and MacKenzie, 1991).

The effect of P to increase the Mo concentrations in plants has been reported to be associated with the stimulating effect of the PO_4^{3-} ion on the uptake of Mo because of formation of a complex phosphomolybdate anion that is absorbed more readily by the plants (Barshad, 1951). On the other hand, in tomato plants, P deficiency was found to enhance Mo uptake (Heuwinkel et al., 1992), from which it was concluded that Mo uptake occurs through P-binding/transporting sites at the plasma membranes of root cells.

In field experiments on groundnut (*Arachis hypogaea* L.) in western Africa, fertilization of acid sandy soils with single superphosphate (SSP) decreased the Mo concentrations, particularly in the nodules (Rebafka, Ndunguru, and Marschner, 1993). However, the same authors reported that soil application of triple superphosphate (TSP) enhanced Mo uptake. In view of the negative effects of sulfate on Mo uptake, application of TSP is clearly superior to the use of SSP for Mo-deficient soils. In the long run, however, TSP provides no real solution for legumes grown on these acid, sandy soils that are also low in S. Therefore, applications of S fertilizers and SSP as a source of P should be accompanied by Mo fertilization for Mo-deficient soils if efficient N_2 fixation in legumes is to be achieved.

The relationship between P and Mo in tobacco plants is also influenced by N, for application of NH_4-N can enhance the concentrations of both native and applied P, and the P can enhance Mo concentrations when Mo is applied, but not influence Mo nutrition in the absence of added Mo (Schwamberger and Sims, 1991). The same authors further showed that this enhanced Mo nutrition due to P application may be great enough to overcome the negative effects of NH_4 on Mo nutrition. Enhanced Mo concentrations due to P application may also be dependent on the presence of adequate N or addition of N.

Addition of S has been found to decrease the Mo content in crops such as Brussels sprouts (Gupta and Munro, 1969), raya (*Brassica juncea* L.) (Pasricha and Randhawa, 1972), berseem (*Trifolium alexandrinum* L.) (Sisodia, Sawarkar, and Rai, 1975), soybeans (Singh and Kumar, 1979), tobacco (Sims, Leggett, and Pal, 1979), and perennial ryegrass (*Lolium perenne* L.) and clover (Williams and Thornton, 1972). A study by Singh and Kumar (1979) showed that applications of S at 40 ppm significantly increased the total Mo uptake in 45- and 100-day-old soybean plants.

The interaction effects of Mo and S will be dealt with in detail in Chapter 14.

MacLachlan (1955) reported that deficiencies of P and S can limit the responses of plants to Mo unless P and S are applied. In each case there is deficiency of P or S rather than an indirect effect on Mo uptake. It was further stated that a deficiency of P, S, or Mo was no indication of deficiency or sufficiency of the other two.

It has been suggested that acidification of soils by N fertilizers and crop removal has been one of the causes for the decline in forage Mo concentrations in California (Phillips and Meyer, 1993). Molybdenum applied in combination with N increased wheat-grain Mo concentrations (Butorina, Yagodin, and Feofanov, 1991). Heavy rates of N application in Romania were found to be associated with Mo deficiency in corn and sunflower, as reviewed by Shorrocks (1987). Likewise, increasing N rates have been reported to reduce the Mo content in the midribs of stored cabbage (Berard, Senecal, and Vigier, 1990) and in cut herbage (Reith et al., 1984). Similarly, increasing applications of ammonium nitrate to substrates of acidic sandy and sandy loam soils decreased the Mo content in cooksfoot (*Dactylus glomerata* L.) (Gembarzewski and Sienkiewicz, 1989).

A decrease in soil pH due to high N application, especially NH_4NO_3 or urea together with K_2SO_4, was found to be the main cause of decreased Mo content in cured tobacco leaves (Sims, Atkinson, and Smitobol, 1975). Those authors further reported an inverse relationship between the Mo and Mn concentrations in the leaves. Because both Mn and Mo are inversely affected by soil pH, one cannot be certain whether poor growth of plants in soil with a low pH is caused by Mo deficiency or Mn toxicity or both. Nitrogen fertilization of tobacco has been found to increase the Mn concentration in burley tobacco leaves (Sims and Atkinson, 1974). The Mo concentration in burley tobacco leaves has been found to increase with increasing additions of chloride as $CaCl_2 \cdot 2H_2O$, and the effect has been much greater with high additions of Mo fertilizer than with low additions (Eivazi et al., 1983). The exact mechanism for the positive effect of chloride salts on Mo uptake by plants is unknown.

Summary

There are many soil and plant factors that can affect the availability of Mo to crop plants: the nature of the parent rock, the forms of Mo in the soil, soil pH, drainage, the presence of organic matter, the status of P, S, N, Mn, and Cl in the soil, variations in plant species, and the part of the plant sampled for analysis. However, the soil pH, the parent material from which the soil has been derived, and the drainage conditions probably are the most important factors. Molybdenum is the only trace mineral whose availability to plants increases with an increase in soil pH under most conditions.

References

Ahlrichs, L. E., Hanson, R. G., and MacGregor, J. M. (1963). Molybdenum effect on alfalfa grown on thirteen Minnesota soils in the greenhouse. *Agronomy J.* 55:484–6.
Allaway, W. H. (1977). Perspectives on molybdenum in soils and plants. In *Molybdenum in the Environment*, vol. 2, ed. W. R. Chappell and K. K. Petersen, pp. 317–39. New York: Marcel Dekker.
Amin, J. V., and Joham, H. E. (1958). A molybdenum cycle in the soil. *Soil Sci.* 85:156–60.
Anderson, A. J. (1956). Molybdenum as a fertilizer. *Adv. Agron.* 8:163–202.
Barrow, N. J., Leahy, P. J., Southey, I. N., and Purser, D. B. (1985). Initial and residual effectiveness of molybdate fertilizer in two areas of southwestern Australia. *Aust. J. Agric. Res.* 36:579–87.
Barshad, I. (1951). Factors affecting the molybdenum content of pasture plants. II. Effect of soluble phosphates, available nitrogen and soluble sulfates. *Soil Sci.* 71:387–98.
Berard, L. S., Senecal, M., and Vigier, B. (1990). Effects of nitrogen fertilization on stored cabbage. II. Mineral composition in mid-rib and head tissues of two cultivars. *J. Hort. Sci.* 65:409–16.
Berger, K. C., and Pratt, P. F. (1965). Advances in secondary and micronutrient fertilization. In *Fertilizer Technology and Usage*, ed. M. H. McVickar, G. L. Bridger, and L. B. Nelson, pp. 313–17. Madison, WI: Soil Science Society of America.
Bhatt, J. G., and Appukuttan, E. (1982). Manganese and molybdenum contents of the cotton plant at different stages of growth. *Commun. Soil Sci. Plant Anal.* 13:463–71.
Bloomfield, C., and Kelso, W. I. (1973). The mobilization and fixation of molybdenum, vanadium, and uranium by decomposing plant matter. *J. Soil Sci.* 24:368–79.
Boswell, F. C. (1980). Factors affecting the response of soybeans to molybdenum application. In *World Soybean Research Conference II*, ed. F. T. Corbin, pp. 417–32. Boulder, CO: Westview Press.
Brogan, J. C., Fleming, G. A., and Byrne, J. E. (1973). Molybdenum and copper in Irish pasture soils. *Irish J. Agric. Res.* 12:71–81.
Butorina, E. P., Yagodin, B, A., and Feofanov, S. N. (1991). Effect of late foliar application of urea and molybdenum on winter wheat grain yield and quality. *Agrokhimiya* 4:17–20. Reprinted 1993, *Soils Fert.* 56(12):1473.
Chipman, E. W., MacKay, D. C., Gupta, U. C., and Cannon, H. B. (1970). Response of cauliflower cultivars to molybdenum deficiency. *Can. J. Plant Sci.* 50:163–7.

Davies, E. B. (1956). Factors affecting molybdenum availability in soils. *Soil Sci.* 81:209–21.

Donald, C. M., and Williams, C. H. (1954). Fertility and productivity of a podzolic soil as influenced by subterranean clover (*Trifolium subterraneum* L.) and superphosphate. *Aust. J. Agric. Res.* 5:664–87.

Eivazi, F., Sims, J. L., Casey, M., Johnson, G. D., and Leggett, J. E. (1983). Growth and molybdenum concentration of burley tobacco as influenced by potassium, molybdenum, and chloride in transplant fertilizer solutions. *Can. J. Plant Sci.* 63:531–8.

Elseewi, A. A., and Page, A. L. (1984). Molybdenum enrichment of plants grown on fly-ash treated soils. *J. Environ. Qual.* 13:394–8.

Feely, L. (1990). Agronomic effectiveness of nitrogen, sulphur and phosphorus for reducing molybdenum uptake by herbage grown on peatland. *Irish J. Agric. Res.* 29:129–39.

Fehr, W. R., Caviness, C. E., Burmood, D. T., and Pennington, J. S. (1971). Stage of development descriptions for soybeans, *Glycine max* (L.) Merrill. *Crops Sci.* 11:929–31.

Franco, A. A., and Day, J. M. (1980). Effects of lime and molybdenum on nodulation and nitrogen fixation of *Phaseolus vulgaris* L. in acid soils of Brazil. *Turrialba* 30:99–105.

Gembarzewski, H., and Sienkiewicz, U. (1989). Lowering cooksfoot yields caused by manganese toxicity and molybdenum deficiency caused by acidifying effect of ammonium nitrate application. In *Soil and Forage Plant Nutrition: Proceedings of the XVI International Grasslands Congress*, ed. R. Desroches, pp. 7–8.

Giddens, J., and Perkins, H. F. (1972). Essentiality of molybdenum for alfalfa on highly oxidized piedmont soils. *Agronomy J.* 64:819–20.

Gupta, U. C. (1969). Effect and interaction of molybdenum and limestone on growth and molybdenum content of cauliflower, alfalfa, and bromegrass on acid soils. *Soil Sci. Soc. Am. Proc.* 33:929–32.

Gupta, U. C. (1971a). Boron and molybdenum nutrition of wheat, barley and oats grown in Prince Edward Island soils. *Can. J. Soil Sci.* 51:415–22.

Gupta, U. C. (1971b). Influence of various organic materials on the recovery of molybdenum and copper added to a sandy clay loam soil. *Plant Soil* 34:249–53.

Gupta, U. C. (1977). Effect of methods of molybdenum application on molybdenum uptake by crops on podzol soils of eastern Canada. In *Molybdenum in the Environment*, vol. 2, ed. W. R. Chappell and K. K. Petersen, pp. 651–63. New York: Marcel Dekker.

Gupta, U. C. (1979). Effect of methods of application and residual effect of molybdenum on the molybdenum concentration and yield of forages on podzol soils. *Can. J. Soil Sci.* 59:183–9.

Gupta, U. C. (1991). Boron, molydenum and selenium status in different plant parts in forage legumes and vegetable crops. *J. Plant Nutr.* 14:613–21.

Gupta, U. C., Chipman, E. W., and MacKay, D. C. (1978). Effect of molybdenum and lime on the yield and molybdenum concentration of crops grown on acid sphagnum peat soil. *Can. J. Plant Sci.* 58:983–92.

Gupta, U. C., and Munro, D. C. (1969). Influence of sulfur, molybdenum and phosphorus on chemical composition and yields of Brussels sprouts and of molybdenum on sulfur contents of several plant species grown in the greenhouse. *Soil Sci.* 107:114–18.

Hawes, R. L., Sims, J. L., and Wells, K. L. (1976). Molybdenum concentration of certain crop species as influenced by previous applications of molybdenum fertilizer. *Agronomy J.* 68:217–18.

Heuwinkel, H., Kirkby, E. A., Le Bot, J., and Marschner, H. (1992). Phosphorus deficiency enhances molybdenum uptake by tomato plants. *J. Plant Nutr.* 15:549–68.

Hulagur, B. F., and Dangarwala, R. T. (1982). Effect of zinc, copper and phosphorus fertilization on the uptake of iron, manganese and molybdenum by hybrid maize. *Madras Agric. J.* 69:11–16.

James, D. W., Jackson, T. L., and Harward, M. E. (1968). Effect of molybdenum and lime on the growth and molybdenum content of alfalfa grown on acid soils. *Soil Sci.* 105:397–402.

Jarrell, W. M., and Dawson, M. D. (1978). Sorption and availability of molybdenum in soils of western Oregon. *Soil Sci. Soc. Am. J.* 42:412–15.

Johnson, C. M., Pearson, G. A., and Stout, P. R. (1952). Molybdenum nutrition of crop plants. II. Plant and soil factors concerned with molybdenum deficiencies in crop plants. *Plant Soil* 4:178–96.

Jones, L. H. P. (1956). Interaction of molybdenum and iron in soils. *Science* 123:1116.

Jones, L. H. P. (1957). The solubility of molybdenum in simplified systems and aqueous soil suspensions. *J. Soil Sci.* 8:313–27.

Kannan, S., and Ramani, S. (1978). Studies on molybdenum absorption and transport in bean and rice. *Plant Physiol.* 62:179–81.

Karimian, N., and Cox, F. R. (1978). Adsorption and extra stability of molybdenum in relation to some chemical properties. *Soil Sci. Soc. Am. J.* 42:757–61.

Karimian, N., and Cox, F. R. (1979). Molybdenum availability as predicted from selected soil chemical properties. *Agronomy J.* 71:63–5.

Khan, N. A., Mulchi, C. L., and McKee, C. G. (1994). Influence of soil pH and molybdenum fertilization on the productivity of Maryland tobacco. I. Field investigations. *Commun. Soil Sci. Plant Anal.* 25:2103–16.

Koval'skiy, V. V., and Yarovaya, G. A. (1966). Molybdenum in filtrated biogeochemical provinces. *Agrokhimiya* 8:68–91.

Krauskopf, K. B. (1972). Geochemistry of micronutrients. In *Micronutrients in Agriculture*, ed. J. J. Mortvedt, P. M. Giordano, and W. L. Lindsay, pp. 7–40. Madison, WI: Soil Science Society of America.

Kubota, J. (1972). Sampling of soils for trace element studies. In *Geochemical Environment in Relation to Health and Disease*, ed. H. C. Hopps and H. L. Cannon, pp. 105–15. Boulder, CO: Geological Society of America.

Kubota, J., and Allaway, W. H. (1972). Geographic distribution of trace element problems. In *Micronutrients in Agriculture*, ed. J. J. Mortvedt, P. M. Giordano, and W. L. Lindsay, pp. 525–54. Madison, WI: Soil Science Society of America.

Kubota, J., Lazar, V. A., Langan, L. N., and Beeson, K. C. (1961). The relationship of soils to molybdenum toxicity in cattle in Nevada. *Proc. Soil Sci. Soc. Am.* 25:227–32.

Kubota, J., Lemon, E. R., and Allaway, W. H. (1963). The effect of soil moisture content upon the uptake of molybdenum, copper, and cobalt by Alsike clover. *Proc. Soil Sci. Soc. Am.* 27:679–83.

Lindsay, W. L. (1972). Inorganic phase equilibria of micronutrients in soils. In *Micronutrients in Agriculture*, ed. J. J. Mortvedt, P. M. Giordano, and W. L. Lindsay, pp. 41–57. Madison, WI: Soil Science Society of America.

MacLachlan, K. D. (1955). Phosphorus, sulphur and molybdenum deficiencies in soils from eastern Australia in relation to nutrient supply and some characteristics of soil and climate. *Aust. J. Agric. Res.* 6:673–84.

Manheim, F. T., and Landergren, S. (1978). Molybdenum: isotopes in nature. In *Handbook of Geochemistry*, vol. II-4, ed. K. H. Wedepohl, pp. 42-B-1–42-O-2. Berlin: Springer-Verlag.

Massey, H. F., and Lowe, R. H. (1961). High molybdenum content of certain Kentucky soils. *Soil Sci. Soc. Am. Proc.* 25:161–2.

Mitchell, R. L. (1964). Trace elements in soils. In *Chemistry of the Soil*, 2nd ed., ed. F. E. Bear, pp. 320–68. Princeton, NJ: Van Nostrand Reinhold.

Mitchell, R. L. (1974). Trace element problems on Scottish soils. *Neth. J. Agric. Sci.* 22:295–304.

Mortvedt, J. J., and Anderson, J. J. (1982). *Forage Legumes: Diagnosis and Correction of Molybdenum and Manganese Problems*. Southern Cooperative Series bulletin 278.

Mulder E. G. (1954). Molybdenum in relation to growth of higher plants and microorganisms. *Plant Soil* 5:368–415.

Norrish, K. (1975). Geochemistry and mineralogy of trace elements. In *Trace Elements in Soil-Plant-Animal Systems*, ed. J. D. Nicholas and A. R. Egan, pp. 55–81. Proc. Jubilee Symp. Waite Agric. Res. Inst. New York: Academic Press.

Odom, J. W., Thurlow, D. L., Eson, J. T., Starling, J. G., and Pitts, J. A. (1980). Molybdenum deficiency triggered by low soil pH. *Highlights Agric. Res.* 27(1):4 (Auburn University, Auburn, AL).

Pal, U. R., Gossett, D. R., Sims, J. L., and Leggett, J. E. (1976). Molybdenum and sulfur nutrition effects on nitrate reduction in burley tobacco. *Can. J. Bot.* 54:2014–22.

Pasricha, N. S., and Randhawa, N. S. (1972). Interaction effect of sulphur and molybdenum on the uptake and utilization of these elements by raya (*Brassica juncea* L.). *Plant Soil* 37:215–20.

Phillips, R. L., and Meyer, R. D. (1993). Molybdenum concentration of alfalfa in Kern County. California: 1950 versus 1985. *Commun. Soil Sci. Plant Anal.* 24:2725–31.

Ponnamperuma, F. N. (1985). Chemical kinetics of wetland rice soils relative to soil fertility. In *Wetland Soils: Characterization, Classification, and Utilization*. pp. 71–89. Los Banos, Philippines: International Rice Research Institute.

Rebafka, F.-P., Ndunguru, B. J., and Marschner, H. (1993). Single superphosphate depresses molybdenum uptake and limits yield response to phosphorus in groundnut (*Arachis hypogaea* L.) grown on an acid sandy soil in Niger, West Africa. *Fert. Res.* 34:233–42.

Reisenauer, H. M. (1956). Molybdenum content of alfalfa in relation to deficiency symptoms and response to molybdenum fertilization. *Soil Sci.* 81: 237–42.

Reith, J. W. S., Burridge, J. C., Berrow, M. L., and Caldwell, K. S. (1984). Effects of fertilizers on the contents of copper and molybdenum in herbage cut for conservation. *J. Sci. Food Agric.* 35:245–56.

Riley, M. M., Robson, A. D., Gartrell, J. W., and Jeffery, R. C. (1987). The absence of leaching of molybdenum in acid soils from western Australia. *Aust. J. Soil Res.* 25:179–84.

Rolt, W. F. (1968). Some effects of lime and molybdenum on the growth of white clover in Autea clay. *N.Z. J. Agric. Res.* 11:193–205.

Rubins, E. J. (1956). Molybdenum deficiencies in the United States. *Soil Sci.* 81:191–7.

Schwamberger, E. C., and Sims, J. L. (1991). Effects of soil pH, nitrogen source, phosphorus, and molybdenum on early growth and mineral nutrition of burley tobacco. *Commun. Soil Sci. Plant Anal.* 22:641–57.

Shivashankar, K., and Hagstrom, G. R. (1991). Molybdenum fertilizer sources and their use in crop production. In *International Symposium on the Role of Sulphur, Magnesium and Micronutrients in Balanced Plant Nutrition (1991: Sichuan, China)*, ed. S. Portch, pp. 297–305 Hong Kong: Potash and Phosphate Institute.

Shorrocks, V. (1987). *Micronutrient News* 7(3) (Micronutrient Bureau, M.B. House, Hertfordshire, U.K.).

Shorrocks, V. (1989). *Micronutrient News* 9(3) (Micronutrient Bureau, M.B. House, Hertfordshire, U.K.).

Sims, J. L., and Atkinson, W. O. (1974). Soil and plant factors influencing accumulation of dry matter in burley tobacco growing in soil made acid by fertilizer. *Agronomy J.* 66:775–8.

Sims, J. L., and Atkinson, W. O. (1976). Lime, molybdenum, and nitrogen source effects on yield and selected chemical components of burley tobacco. *Tobacco Sci.* 20:181–4.

Sims, J. L., Atkinson, W. O., and Smitobol, C. (1975). Mo and N effects on growth, yield and Mo concentration of burley tobacco. *Agronomy J.* 67:824–8.

Sims, J. L., Leggett, J. E., and Pal, U. R. (1979). Molybdenum and sulfur interaction effects on growth, yield, and selected chemical constituents of burley tobacco. *Agronomy J.* 71:75–8.

Singh, M., and Kumar, V. (1979). Sulfur, phosphorus, and molybdenum interactions on the concentration and uptake of molybdenum in soybean plants (*Glycine max*). *Soil Sci.* 127:307–12.

Sisodia, A. K., Sawarkar, N. J., and Rai, M. M. (1975). Effect of sulphur and molybdenum on yield and nutrient uptake in berseem (*Trifolium alexandrinum*). *J. Indian Soc. Soil Sci.* 23:96–102.

Smith, B. H., and Leeper, G. W. (1969). The fate of applied molybdate in acidic soils. *J. Soil Sci.* 20:246–54.

Stone, K. L., and Jencks, E. M. (1963). *The Available Molybdenum Status of Some West Virginia Soils*. West Virginia University, Agricultural Experiment Station Bulletin 484.

Stout, P. R., and Meagher, W. R. (1948). Studies of the molybdenum nutrition of plants with radioactive molybdenum. *Science* 108:471–3.

Stout, P. R., Meagher, W. R., Pearson, G. A., and Johnson, C. M. (1951). Molybdenum nutrition of crop plants. I. The influence of phosphate and sulfate on the absorption of molybdenum from soils and solution cultures. *Plant Soil* 3:51–87.

Thornton, I., and Webb, J. S. (1973). *Environmental Geochemistry: Some Recent Studies in the United Kingdom*, pp. 89–98. Proc. 7th Conf. on Trace Substances in Environmental Health. Columbia: University of Missouri.

Vlek, P. L. G., and Lindsay, W. L. (1974). Molybdenum solubility relationship in soils irrigated with high Mo water. *Agron. Abstr.*, p. 126 (Madison, WI: American Society of Agronomy).

Wallace, A., and Romney, E. M. (1977). *Biological Implications of Metals in the Environment*, ed. A. Wallace and E. M. Romney, pp. 370–9. ERDA symposium series 42.

Welch, R. M., Allaway, W. H., House, W. A., and Kubota, J. (1991). Geographic distribution of trace element problems. In *Micronutrients in Agriculture*, ed. J. J. Mortvedt, F. R. Cox, L. M. Shuman, and R. M. Welch, pp. 31–57. Madison, WI: Soil Science Society of America.

Williams, C., and Thornton, I. (1972). The effect of soil additives on the uptake of molybdenum and selenium from soils from different environments. *Plant Soil* 36:395–406.

Xie, R. J., and MacKenzie, A. F. (1991). Molybdate sorption–desorption in soils treated with phosphate. *Geoderma* 48:321–33.

6

Analytical Techniques for Molybdenum Determination in Plants and Soils

FRIEDA EIVAZI and JOHN L. SIMS

Introduction

Molybdenum (Mo) has been the subject of many investigations since Arnon and Stout (1939) first showed that Mo is essential for higher plants. The biological importance of Mo was reported to lie in the processes of nitrogen (N) fixation and nitrate reduction (Bortels, 1936; Mulder, 1948). The literature on Mo in soils, plants, and animals has been reviewed by Gupta and Lipsett (1981). Because Mo plays important roles in plant and animal nutrition and is an environmental concern (Council for Agricultural Science and Technology, 1976; U.S. Environmental Protection Agency, 1993), a critical study of the analytical procedures available for Mo determinations is warranted.

In general, selection of a particular analytical method or technique to measure an element of interest is based on the availability of the proper instrumentation and facilities. The characteristics of the analytical instrument used and the acceptable limits of variability, with emphasis on sensitivity, reproducibility, and specificity, are determining factors in selecting a technique. These issues are particularly important for determinations of micronutrients, because the critical concentrations of these elements for plants can be close to the limits of detection for most analytical procedures. Also, one's ability to determine true values using any method is dependent on the use of reliable standards (Alvarez, 1980; Taylor, 1985), the precision of the analytical procedure (Horwitz, 1982; Dux, 1986), and careful evaluation at each step in the analytical process (Sterrett et al., 1987; Munter, Halverson, and Anderson, 1984).

The investigation reported in this chapter (A6-102-96) was conducted in connection with a project of the Division of Food and Agricultural Sciences, Cooperative Research, Lincoln University, and the Kentucky Agricultural Experiment Station and is published with approval of the directors.

Many methods have been proposed for routine determination of Mo in plant materials. The most widely used colorimetric procedures can be divided into two groups, mainly on the basis of the reagents used to produce the colored Mo complex. One procedure is based on formation of the colored complex $Fe[MoO(SCN)_5]$, produced by reaction of Mo with an alkali thiocyanate and excess iron (Fe) in the presence of a reducing agent, stannous chloride. Isoamyl alcohol dissolved in carbon tetrachloride (Chapman and Pratt, 1961) is used to extract the complex from the aqueous phase. The Mo content is determined by comparing the absorbance of the sample with appropriate standards (Evans, Purvis, and Bear, 1950; Johnson and Arkley, 1954). The dithiol procedure developed by Piper and Beckworth (1948) and modified and used by Clark and Axley (1955), Bingley (1959), and Gupta and MacKay (1965) for plants and soils is based on precipitation and extraction of a green-colored Mo-dithiol complex after removal of interfering ions from the test solution. This procedure, though too tedious and time-consuming for routine work, has been found to be more sensitive and more precise than the thiocyanate method. The green-colored complex formed between Mo and dithiol is stable for at least 24 hours (Gupta and MacKay, 1965; Gupta, 1993). Most of the chemical methods previously described involve solvent extraction, and both the accuracy and precision of the methods deteriorate at Mo concentrations approaching $0.1 \, mg \, kg^{-1}$, the critical concentration for many plants. As for the thiocyanate spectrophotometric procedure (Johnson and Arkley, 1954), difficulties have been encountered with the use of stannous chloride ($SnCl_2$) as the reducing agent, for it varies in purity from batch to batch of commercially available product and is not always free from traces of Mo. Also, it tends to form a persistent film on the cuvettes, another potential source of error. Interference from Fe is another difficulty with the thiocyanate method. The compound $Fe[MoO(SCN)_5]$ is formed when sufficient Fe is present, and it can be a part of the colored complex.

Yatsimirskii (1964) proposed a kinetic method for determination of Mo on the basis of its catalytic action on the potassium iodide–hydrogen peroxide (KI-H_2O_2) reaction. Fuge (1970) adapted the method for use with an autoanalyzer in order to overcome the difficulties of performing the method manually. The reaction is time-dependent and is sensitive to minor changes in reagent concentrations. Fuge (1970) used the method to determine Mo concentrations in geological and biological samples. Bradfield and Stickland (1975) and Quin and Woods (1979) adapted the method for Mo determinations in plant materials. Several simple but significant changes in the method were made by Quin and Woods (1979) to give a rapid and sensitive automated procedure for determination of Mo in plant materials (Eivazi, Sims, and Crutchfield, 1982; Sims and Crutchfield, 1992). This procedure will be discussed in detail later in this chapter.

Procedures such as direct atomic-absorption spectrophotometry (AAS), both flameless (Henning and Jackson, 1973; Wilson 1979; Curtis and Grusovin, 1985), and flame (Butler and Mathews, 1996; Khan, Cloutier, and Hidiroglou, 1979), provide adequate sensitivity for Mo and are relatively rapid, but are subject to matrix interferences. The use of spectrographic methods is restricted because of inadequate limits of detection caused by the refractory nature of Mo oxides. Recent modifications and improvements in flameless atomic-absorption procedures (Wilson, 1979, 1992) show promise and may offer an alternative method suitable for some laboratories. Methods involving inductively coupled plasma (ICP) atomic-emission spectrometry have been developed using both wet digestion procedures (Havlin and Soltanpour, 1980) and plant-tissue ash (Jones, 1977). Direct-current plasma spectrometry (DCP) for determination of Mo in plant tissue was developed by Pierzynski, Crouch, and Jacobs (1986), but the procedure showed that DCP results were enhanced at low Mo concentrations, as compared with AAS, when high amounts of calcium (Ca) and magnesium (Mg) were present in the sample matrix.

The total Mo content in soils usually ranges between 0.5 and 3.0 mg kg^{-1} and is dependent on the Mo content of the parent rock (Massey, Lowe, and Bailey, 1967). However, some soils have been found to contain as little as 0.1 mgkg^{-1} total Mo (Williams, 1971) or as much as 30.0 mgkg^{-1} (Kubota, 1976). Davis (1956) classified the forms of soil Mo into the following fractions: (1) unavailable Mo (held within the crystal lattices of primary and secondary minerals), (2) conditionally available Mo (retained as the MoO_4^{2-} anion by clay minerals, and available to a greater or lesser degree depending on pH and probably on the phosphate status of the soil), (3) organic-matter-chelated Mo, and (4) water-soluble Mo. The available Mo is not closely related to the total Mo content of a soil, because the availability of Mo is dependent on pH and on other chemical and biological factors in the soil (Stone and Jencks, 1963). Where the availability of Mo for uptake by plants is important, a method for determination of the mobile or readily extractable Mo is required. However, for pollution monitoring, a method for determining the total content of Mo in the soil is necessary.

Although the acid ammonium oxalate procedure (Grigg, 1953) is the method most commonly used to determine the available Mo in soils, the findings have not been consistent (Grigg, 1960; Lombin, 1985). Other extractants have also been used to determine the available Mo: water (Lavy and Barber, 1964; Gupta and MacKay, 1965), hot water (Lowe and Massey, 1965; Zbiral, 1992), anion-exchange resin (Bhella and Dawson, 1972; Jackson and Meglen, 1975; Karimian and Cox, 1978; Ritchie, 1988; Sherrell, 1989), ammonium bicarbonate–diethylenetriamine pentaacetic acid (AB-DTPA) (Boon, 1984; Soltanpour, 1985). Soltanpour (1991) used the AB-DTPA extractant and ICP detection to determine the bioavailability of several potentially toxic elements in various soils, mine spoils, and soils treated with sewage sludge: lead (Pb), cadmium (Cd), nickel (Ni), selenium (Se), arsenic (As), boron (B), sulfur (S), and Mo. A method was developed by Rowbottom (1991) for direct electrothermal atomic-absorption spectrometric (ET-AAS) determinations of Mo extractable with ammonium acetate and Mo soluble in aqua regia (HCl + HNO_3, $3:1$) in soils and sewage sludge. Mortvedt and Anderson (1982) used graphite-furnace atomic-absorption spectrometry (GF-AAS) to measure extracted Mo. Several authors have also used GF-AAS to measure Mo in soil extracts (Little and Kerridge, 1978; Lombin, 1985; Zbiral, 1992; Wang, Reddy, and Munn, 1994). A detailed review of the literature on extractants and selected procedures for determining soil concentrations of Mo available to plants is presented in Chapter 7 of this book.

Determination of Molybdenum in Plants by Automated Catalytic Procedures

Principles

Yatsimirskii (1964) studied many catalytic reactions involving hydrogen peroxide in acid media. One of those reactions was the oxidation of potassium iodide by hydrogen peroxide in the presence of a catalyst according to the equation

$$H_2O_2 + 2KI + 2H^+ \rightarrow 2H_2O + I_2 + 2K^+$$

The reaction is catalyzed by compounds of Fe(III), by elements of the middle group in the periodic system such as zirconium (Zr), niobium (Nb), tantalum (Ta), tungsten (W), and Mo, and also, apparently, by hafnium (Hf) compounds (Yatsimirskii 1964). He found that the rate of evolution of iodine was a function of the concentration of the catalyst if the reaction time was held constant and that the rate of reaction could be measured by following the changes in the optical density of an iodine-starch solution. Yatsimirskii (1964) used that dependence to develop very sensitive procedures for determinations of Mo, W, Fe, and Zr in soils. However, it is difficult to perform such procedures manually, as the reaction rate is time-dependent and sensitive to minor changes in reagent concentrations. The procedure was used by Bradfield and Stickland (1975) and Quin and Woods (1979) to measure Mo in several plant materials: wheat (*Triticum aestivum*) leaves, clover (*Trifolium repens*), and lucerne (*Medicago sativa*). Significant improvements in the sensitivity and reproducibility of the method have been reported by Eivazi et al. (1982) and Sims and Crutchfield (1992). The method has been used successfully in our laboratories for determination of Mo in tobacco (*Nicotiana tabacum* L.), soybean [*Glycine max* (L.) Merr.], and corn (*Zea mays* L.). The method we describe in detail here incorporates the improvements made by Eivazi et al. (1982) and Sims and Crutchfield (1992), and permission to publish has been granted by Marcel Dekker Inc.

Apparatus

1. Erlenmeyer flasks, 25 mL
2. Muffle furnace
3. Electric or steam hot plate
4. Autoanalyzer System II or equivalent (see Figure 6.1 for modules and configuration details)
5. Analytical balance
6. Reagent dispensers, watch glasses, volumetric flasks, disposable sample cups, and assorted common laboratory glassware

Reagents and Solutions

1. Hydrochloric acid (HCl), 2-M
2. Hydrogen peroxide (H_2O_2), 30%
3. Hydrochloric acid, 0.125-M [a large quantity of this reagent will be consumed with each batch of samples; add Triton X-100 (wetting agent) at $0.5 \, mL \, L^{-1}$]
4. Potassium iodide (KI), 0.5% (w/v); add Triton X-100 at $0.5 \, mL \, L^{-1}$
5. Hydrogen peroxide solution; dilute 0.65 mL of 30% H_2O_2 to 1 L
6. Ammonium fluoride solution (NH_4F), 0.25% (w/v)
7. Molybdenum stock solution, $1,000 \, mg \, L^{-1}$ (commercially available)

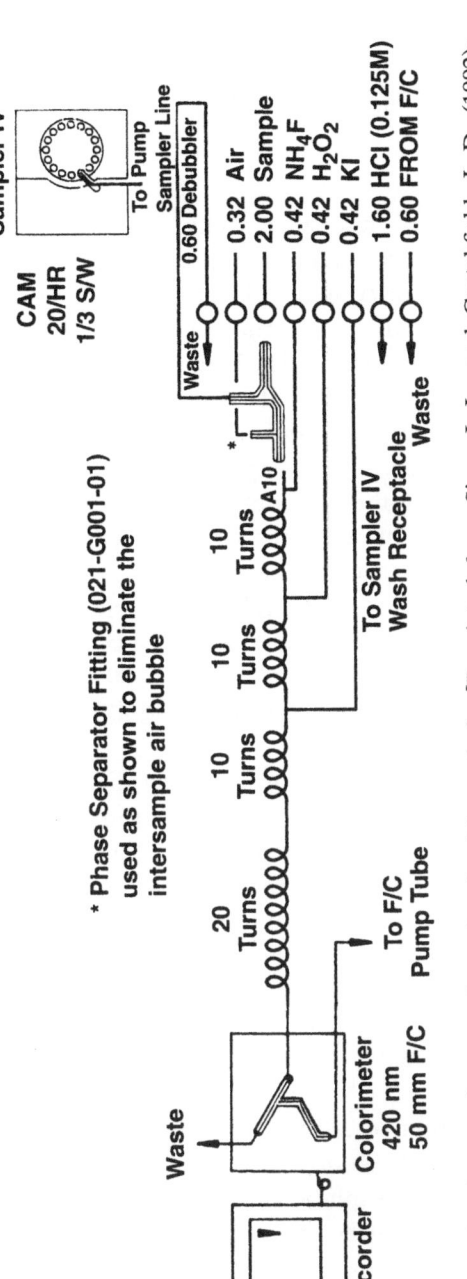

Figure 6.1. Autoanalyzer configuration for Mo analysis. [Reprinted from Sims, J. L., and Crutchfield, J. D. (1992): Determination of molybdenum in plants. In *Plant Analysis Reference Procedures for the Southern Region of the United States*, ed. C. O. Plank, pp. 61–6, University of Georgia Southern Cooperative Series bulletin 368.]

These reagents should be prepared shortly before use, with deionized water, and should be kept tightly sealed at all times.

Procedure

Weigh 0.5 g of dried plant material into a carefully cleaned 25-mL Erlenmeyer flask or a silica crucible, place it in a muffle furnace, and heat it at 500°C for at least 4 hours. After cooling, add 1 mL of 30% H_2O_2, 4 mL of 2-M HCl, and two glass beads (antibumping granule). Heat it, uncovered, on a hot plate at approximately 80°C until dry, and then allow it to cool. Add 10 mL of 0.125-M HCl to the residue; cover it and heat for 20–30 minutes at 80°C; then swirl and mix the samples thoroughly to ensure complete dissolution of the sample. Allow the samples to settle overnight before measuring the Mo content. Carefully decant sufficient liquid into disposable sample cups for analysis on the auto-analyzer. Prepare blanks with each set of samples.

To begin the analysis, turn on the colorimeter, recorder, and proportioning pump for 10–15 minutes, until the baseline on the recorder is stable. Adjust the peak height of highest standard to 75–80% of full scale; then proceed with the analysis. Usually a blank and a full set of standards are run with each tray of samples.

A sampler/wash ratio of 1:2 was originally proposed by Eivazi et al. (1982). To be able to use a computer program (Sigmacan) to read the charts, the sampler/wash ratio of 1:2 was changed to 1:3 timing (Sims and Crutchfield, 1992). Although that reduces the daily output by 50%, a better baseline resolution between peaks on the recorder is achieved, which allows more accurate interpolation of the data. Analysis of 100 digested samples per day is possible even at the slower rate.

Calibration

To prepare a standard Mo solution of 10 mg L^{-1}, pipette 5 mL of the commercial stock solution (1,000 mg L^{-1}) into a 500-mL volumetric flask, and dilute to volume with 0.125-M HCl solution. The working standards are prepared daily from the standard Mo solution (10 mg L^{-1}), containing Mo at 0.04–0.20 mg L^{-1}, using the 0.125-M HCl solution. It is important that the acid concentrations of the samples, standards, and intersample wash solution be exactly the same to avoid the possibility of baseline drifting. Also, the sampler wash receptacle should be rinsed out with fresh 0.125-M HCl daily.

Calculations

A plot of peak height versus Mo concentration of the standard yields a sigmoidal curve that is nearly linear in the middle range. Very low Mo concentrations in samples can be approximated by dissolving the original dry sample in 10 mL of a middle-range standard and then subtracting the value of that standard from the sample value read from the standard curve. High samples should be diluted to fall in the optimum range. If 0.5 g of tissue and 10 mL of acid are used, the milligrams-per-liter readings from the standard curve should be multiplied by 20 (to determine the Mo concentration as milligrams per kilogram).

Comments

Range and sensitivity. The baseline rate of formation of iodine is a function of the concentration of $KI-H_2O_2$ and of the pH. The reagent concentrations and autoanalyzer configuration (Figure 6.1) used in this procedure are optimized to produce maximum linearity for the concentration range between 0.04 and 0.20 mg L^{-1}. The concentrations of Mo in plant materials will range from less than 0.1 mg kg^{-1} in plants grown on acidic soils heavily fertilized with ammonium-N to 25 mg kg^{-1} in plants grown on neutral or alkaline soils fertilized with Mo. The concentrations of plant Mo can vary greatly depending on whether the Mo fertilizer was applied broadcast (low), was banded with the seed (high), or was placed in transplant water near the plant root (high). The sensitivity of this procedure for Mo is less than 0.1 mg kg^{-1} in plant material using a 0.5-g plant sample.

Interferences. Iron(III) and various heavy metals, such as titanium (Ti), vanadium (V), chromium (Cr), Zr, Nb, Ta, Hf, and W, have been shown to catalyze the oxidation of iodide by hydrogen peroxide. However, these metals are present in plant materials in very small concentrations, and Mo has a greater catalytic effect on the reaction. The NH_4F added to complex Fe(III) reduces interference from all other metals (Quin and Woods, 1979). Wet ashing using perchloric acid should not be used, because perchlorate salts may also interfere with the iodide-H_2O_2 reaction. Our studies (Eivazi et al., 1982) have shown that concentrations of 5% or more of sodium (Na), potassium (K), and Mg as their chlorides do not have any effect on the rate of reaction, but higher concentrations of

Ca (4% or more) and Mg (1.5% or more) will slightly depress the rate of reaction, causing a drop in peak height. If excessive matrix variation between samples and standards is suspected, the magnitude of the associated error can be estimated by using the method of "single-standard addition" described in the next section. Anions such as phosphate, arsenate, oxalate, and tartrate form complex compounds with molybdic acid and can cause the catalytic action of Mo to disappear completely. Adding the H_2O_2 prior to the KI in the automated system precludes that effect (Eivazi et al., 1982), and there is no interference from phosphate concentrations in the range common in most plant tissues. The accuracy of the method depends on the extent to which the standard and sample matrices are matched. This can be estimated by using the single-standard technique:

Extract and analyze several test samples using the current procedure. Note the shape of the standard curve, and select those samples that have absorbance readings in the middle of the linear portion of the curve. Take two aliquots of each of these samples. To one, add a quantity of Mo standard that will give a combined absorbance reading not above the linear portion of the curve. To the other aliquot, add the same volume of blank solution. Analyze the samples, and calculate the corrected concentration (C_s) for each according to the equation

$$C_s = A_b / (A_a - A_b) \times (C_b \times V_b / V_s)$$

where C_s is the original concentration of the sample, A_a is the adsorbance of the sample with standard added, A_b is the adsorbance of the sample with blank added, C_b is the concentration of the standard added, V_b is the volume of the standard or blank added, and V_s is the volume of the sample aliquot. A comparison of the corrected values with those derived from the original standard curve should provide a measure of error from the matrix effect. Repeated analyses of the same plant sample for Mo should yield coefficients of variability of about 5%.

The quantitative accuracy of the system was measured by adding Mo at $0.4\,mg\,kg^{-1}$ (on a plant dry-weight basis) to tobacco samples containing a range of Mo concentrations. Green burley tobacco was used, and the lamina sections of the leaves from the whole plant were taken for measurement. The coefficient of determination (R^2) for the relationship was 0.98 and was significant at a probability level of less than 0.01%, with the slope of the equation approaching 1.0 (Eivazi et al., 1982). Thus, the data show that reliable results can be achieved with the procedure because of the sensitivity of the reaction and the precision of the autoanalyzer.

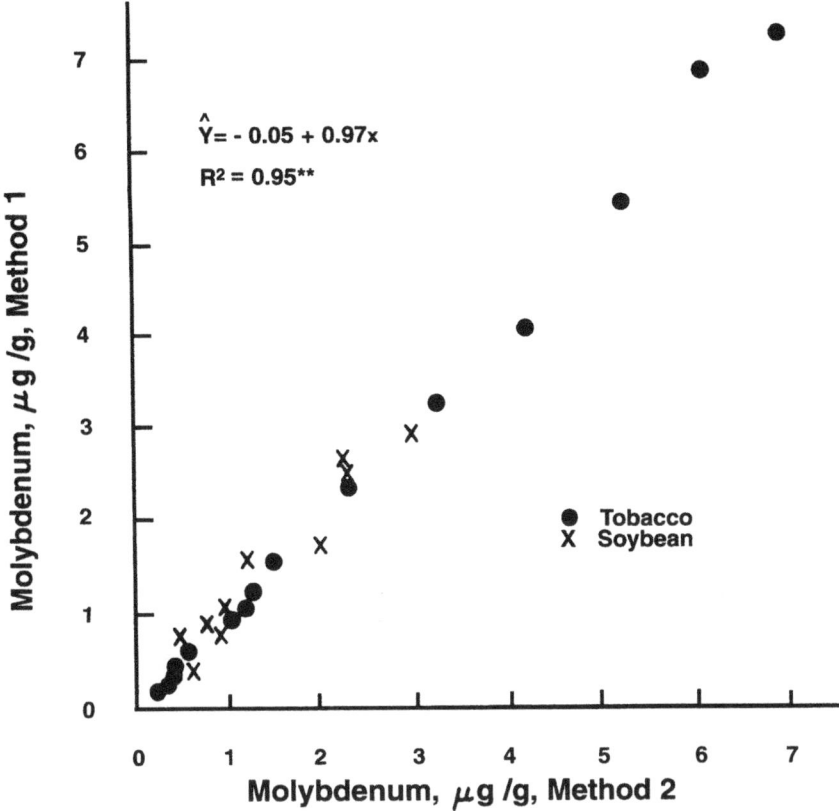

Figure 6.2. Molybdenum concentrations in tobacco and soybean leaves. Method 1 is by a spectrophotometric procedure, and method 2 by the KI-H_2O_2 procedure. (Reprinted from Eivazi et al., 1982, with permission of Marcel Dekker, Inc.)

Representative results. To test the performance of the method, the findings obtained with the automated method were compared with those of the thiocyanate spectrophotometric procedure, as modified from the work of Johnson and Arkley (1954). Tobacco and soybean samples with varying amounts of Mo were analyzed with both methods (Figure 6.2). There was good correlation ($R^2 = 0.95^{**}$) between the values obtained with the KI-H_2O_2 method and the manual thiocyanate procedure. As compared with the manual thiocyanate method, the advantages of the automated procedure are increased productivity (20 determinations of

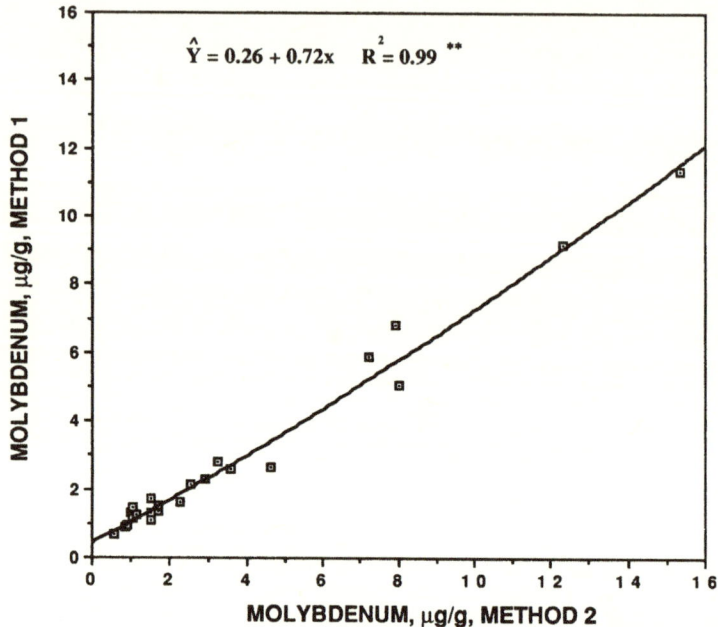

Figure 6.3. Molybdenum concentrations in soybean samples. Method 1 is by the KI-H_2O_2 procedure, and method 2 by the graphite-furnace procedure. [Reprinted from Wilson, O. D. (1992): Determination of molybdenum in plant tissue using the graphite furnace. In *Plant Analysis Reference Procedures for the Southern Region of the United States*, ed. C. O. Plank, pp. 61–6. University of Georgia Southern Cooperative Series bulletin 368.]

digested samples per hour) and lower skill and experience requirements. In another experiment, the Mo contents of soybeans grown on 12 soils in a greenhouse and treated with Mo at either 0.3 or 0.6 mg kg^{-1} were determined with the KI-H_2O_2 procedure, and the findings were compared with those from the graphite-furnace method described by Wilson (1992). Excellent correlation ($R^2 = 0.99**$) was found between the two procedures (Figure 6.3).

Determination of Total Molybdenum in Soils

This determination is as described by Little and Kerridge (1978), with modifications.

Principles

The colorimetric procedures, though tedious and time-consuming, remain useful when the Mo concentrations in soils and plats are low. The stannous chloride–thiocyanate procedure, originally described by Marmoy (1939), and revised by Evans et al. (1950), Purvis and Peterson (1956), and Johnson and Arkley (1954), is one of the most widely used methods for determination of total Mo in soils. This procedure is based on formation of the colored complex $Fe[MoO(SCN)_5]$ produced by the reaction of Mo with an alkali thiocyanate and excess Fe in the presence of a reducing agent, stannous chloride. Isoamyl alcohol dissolved in carbon tetrachloride (Chapman and Pratt, 1961) is used to extract the complex from the aqueous phase. The Mo content is determined by comparison of the absorbance of the sample with appropriate standards.

The total amount of Mo in a soil is commonly determined by extraction using perchloric acid digestion (Reisenauer, 1965). A colorimetric determination of Mo in soils and sediments using zinc dithiol was described by Stanton and Hardwick (1967). Dry ashing of soil and the extraction of ash using concentrated acids was employed by Perrin (1946) and Grigg (1953). Sodium carbonate fusion has also been used for extraction of total Mo (Purvis and Peterson, 1956; Jackson, 1958). Little and Kerridge (1978) introduced a method for determination of total Mo using atomic-absorption spectrophotometry (AAS). In this procedure, for Mo concentrations between 0.1 and $1 \, mg \, kg^{-1}$, methyl isobutyl ketone is used to extract the Mo as the thiocyanate complex, and the extract is then analyzed using a nitrous oxide/acetylene flame. This procedure has been modified and used in our laboratory. The modified method is described here:

Apparatus

1. Volumetric flasks, 10 mL and 25 mL
2. Platinum crucible, 30-mL capacity, with lid
3. Meker burner
4. Centrifuge
5. Atomic-absorption spectrophotometer (AAS)

Reagents

1. Sodium carbonate (Na_2CO_3), anhydrous
2. Hydrochloric acid (HCl), 6-M

3. Hydrochloric acid, 2-M
4. Methyl isobutyl ketone (MIBK)
5. Potassium thiocyanate (KCNS), 5-M
6. Stannous chloride ($SnCl_2 \cdot 2H_2O$), 10% (prepared in 10-M HCl)
7. Standard Mo solution, $1,000\,mg\,L^{-1}$ (commercially available)

Procedure

The sodium carbonate fusion method of Jackson (1958) is used to prepare soil samples. Take 1 g of oven-dried and fine-ground soil sample, place it in a platinum (Pt) crucible, and heat it to 900°C. Add 5 g of Na_2CO_3 to the Pt crucible containing the ignited samples, and mix it thoroughly using a glass rod. Place 1 g of Na_2CO_3 on top of the mixture. Place the covered crucible at a slight angle on a triangle, and warm it over a Meker burner. Gradually increase the heat to use almost the full flame of the burner (900°C), until the bottom of the crucible is a cherry red. Maintain this temperature for 15–20 minutes. Lift the lid periodically to provide an oxidizing environment. During the last stages of fusion, adjust the cover so that the crucible top is about one-fourth open. Increase the heat to about 1,000°C, and continue heating for 5–10 minutes. Remove the cover and heat the crucible a few minutes more to finish the fusion. Remove the flame and rotate the crucible in such a way that the fusion will solidify along the sides. This will facilitate removal of the solidified melt. Transfer the solidified material into a beaker, and rinse the crucible and lid using 5 mL of 6-M HCl. To the contents of the beaker add 40–50 mL of water, and heat the beaker gently on an electric hot plate, stirring it until the contents are dissolved. This process takes 2–3 hours. The contents of the beaker should then be transferred to 50-mL plastic tubes and centrifuged at 3,500 rpm for 5 minutes. Transfer the clear solution to a 100-mL beaker, add HCl (2 mL of 12-N HCl per 1 g of Na_2CO_3 contained), and evaporate to dryness. Add 30–40 mL of 1-M HCl to the residue, and heat the contents on gentle heat while stirring. Transfer the contents of the beaker into 50-mL tubes, and centrifuge at 3,500 rpm for 5 minutes. Bring the volume of the clear supernatant solution to 50 mL using 1-M HCl. Prepare the blank solutions in a similar way as the samples.

Take an aliquot (1–23 mL) of soil solution into a 10- or 25-mL volumetric flask (depending on extract volume). Add 0.2 mL of 5-M KCNS and mix the contents of the flask; add 0.5 mL of 10% $SnCl_2$ and again mix; add 1 mL of MIBK and bring to volume using the blank solution. Shake the flask vigorously for 50–60 seconds to ensure thorough mixing of the two phases. When the phases have separated, take the aliquot from the MIBK organic phase, and measure Mo with a nitrous oxide/acetylene flame using the AAS.

Calibration Standards

Prepare a standard Mo solution ($1 mg L^{-1}$) by pipetting 1 mL of the commercial stock Mo solution ($1,000 mg L^{-1}$) into a 1,000-mL volumetric flask, and dilute the volume with 1-M HCl solution. Prepare working standards containing 0, 0.5, 1.0, 1.5, 2.0, and 3.0 µg of Mo from the standard solution (1 mg Mo per liter). Add 0.2 mL of KCNS, 0.5 mL of $SnCl_2$ solution, and 1 mL of MIBK, and bring the volume to 10 mL using blank solution. The procedure outlined earlier for the preparation of samples should be followed in preparing the working standards.

Comments

Molybdenum in soils and plant extracts has been determined successfully by colorimetric methods (Johnson and Arkley, 1954; Kubota and Cary, 1982; Gupta and MacKay, 1965). However, colorimetric procedures are time-consuming. Recently, advances in AAS, DCP, and automated instrumentation have made it much easier to measure Mo concentrations much more rapidly (Jones, 1977). In our laboratory, simple but significant modifications have been added to the procedure of Little and Kerridge (1978). In the method described by Little and Kerridge (1978), the oven-dried samples (1 g) were digested with $HF/HClO_4$. The digest was taken up in 5 mL of concentrated HCl and made to a total volume of 50 mL. An aliquot of that solution was used for separation and extraction of Mo into MIBK. In our laboratory, an Na_2CO_3 fusion procedure is used to prepare the soil solution; a 5-M KCNS solution is applied instead of 10-M KCNS (easier to monitor); the volume of soil solution is increased to 23 mL, which is then treated with KCNS, $SnCl_2$, and MIBK. This allows more sensitive measurements for soil samples with low Mo concentrations. Little and Kerridge (1978) did not attempt to correlate their Mo concentrations determined with AAS to those determined using the thiocyanate spectrophotometric method. The comparison for 15 soil extracts was made only between the carbon-rod analyzer (CRA) and the spectrophotometric procedure.

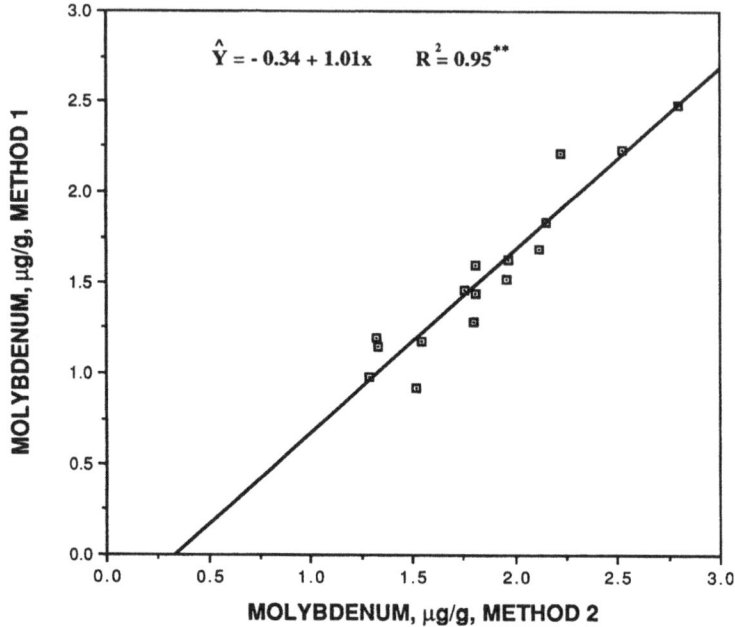

Figure 6.4. Total Mo in soil samples. Method 1 is by the colorimetric procedure (Johnson and Arkley, 1954), and method 2 by modified AAS.

Representative Results

Soil samples with a wide range of total Mo concentrations were tested using the standard colorimetric procedure of Johnson and Arkley (1954) and the modified procedure described here. For 16 soil samples with Mo concentrations between 0.9 and 2.8 mg kg^{-1}, there was good correlation ($R^2 = 0.95**$) between the values obtained using the modified Little and Kerridge (1978) method and those obtained with the method of Johnson and Arkley (1954). Figure 6.4 shows a plot of the data from the two methods. Despite the good agreement, the modified AAS procedure is less time-consuming and features increased sensitivity and greater productivity, as compared with the colorimetric method. Also, because of the increased number of operations required in the colorimetric procedure, the chances for error are greatly increased.

References

Alvarez, R. (1980). NBS plant tissue standard reference materials. *J. Assoc. Off. Anal. Chem.* 63:806–8.

Arnon, D. I., and Stout, P. R. (1939). Molybdenum as an essential element for higher plants. *Plant Physiol.* 14:599–602.

Bhella, H. S., and Dawson, M. D. (1972). The use of anion exchange resin for determining available soil molybdenum. *Soil Sci. Soc. Am. Proc.* 36:177–8.

Bingley, J. B. (1959). Simplified determination of molybdenum in plant material by 4-methyl-1,2-dimercaptobenzene, dithiol. *J. Agric. Food Chem.* 7:269–70.

Boon, D. Y. (1984). *The Ammonium Bicarbonate–DTPA Soil Test (AB-DTPA) for Determination of Plant Available Pb, Cd, Ni, and Mo in Mine Tailings and Contaminated Soils.* Sixth High Altitude Revegetation Workshop information series no. 53. Fort Collins: Colorado State University.

Bortels, H. (1936). Further studies on the significance of Mo, V, and W, and other soil materials on nitrogen fixing and other microorganisms. *Zentr. Bakeriol. Parasilenk. Abt. II* 95:193–218.

Bradfield, E. G., and Stickland, J. F. (1975). The determination of molybdenum in plants by an automated catalytic method. *Analyst* 100:1–6.

Butler, L. R. P., and Mathews, P. M. (1966). The determination of trace quantities of molybdenum by atomic absorption spectrophotometer. *Anal. Chem. Acta* 36:310–27.

Chapman, H. D., and Pratt, P. F. (1961). *Methods of Analysis for Soils, Plants, and Waters*; pp. 145–9. Davis: University of California, Division of Agricultural Sciences.

Clark, L. J., and Axley, J. H. (1955). Molybdenum determination in soils and rocks with dithiol. *Anal. Chem.* 27:2000–3.

Council for Agricultural Science and Technology (1976). *Application of Sewage Sludge to Cropland: Appraisal of Potential Hazards of the Heavy Metals to Plants and Animals.* Report no. 64. Ames, IA: CAST.

Curtis, P. R., and Grusovin, J. (1985). Determination of molybdenum in plant tissue by graphite furnace atomic absorption spectrophotometry (GFAAS). *Commun. Soil Sci. Plant Anal.* 16:1279–91.

Davis, E. B. (1956). Factors affecting molybdenum availability in soils. *Soil Sci.* 81:209–21.

Dux, J. P. (1986). *Handbook of Quality Assurance for the Analytical Chemistry Laboratory.* New York: Van Nostrand Reinhold.

Eivazi, F., Sims, J. L., and Crutchfield, J. (1982). Determination of molybdenum in plant materials using a rapid, automated method. *Commun, Soil Sci. Plant Anal.* 13:135–50.

Evans, H. J., Purvis, E. R., and Bear, F. E. (1950). Colorimetric determination of molybdenum by means of nitric and perchloric acids. *Anal. Chem.* 22:1568–9.

Fuge, R. (1970). An automated method for the determination of molybdenum in geological and biological samples. *Analyst* 95:171–6.

Grigg, J. L. (1953). Determination of the available molybdenum of soils. *N.Z. J. Sci. Technol., Sect. A* 34:405–14.

Grigg, J. L. (1960). The distribution of molybdenum in the soils of New Zealand. I. Soils of the North Island. *N.Z. J. Agric. Res.* 3:69–86.

Gupta, U. C. (1993). Boron, molybdenum, and selenium. In *Soil Sampling and Methods of Analysis*, ed. M. R. Carter, pp. 91–9. Chelsea, MI: Lewis Publishers.

Gupta, U. C., and Lipsett, J. (1981). Molybdenum in soils, plants, and animals. *Adv. Agron.* 34:73–115.

Gupta, U. C., and MacKay, D. C. (1965). Determination of molybdenum in plant materials using 4-methyl-1,2-dimercaptobenzene (dithiol). *Soil Sci.* 99:414–15.

Havlin, J. L., and Soltanpour, P. N. (1980). A nitric acid plant tissue digest method for use with inductively coupled plasma spectrometry. *Commun. Soil Sci. Plant Anal.* 11:969–80.

Henning, S., and Jackson, T. L. (1973). Determination of molybdenum in plant tissue using flameless atomic absorption. *Atomic Absorption Newsletter* 12:100–1.

Horwitz, W. (1982). Evaluation of analytical methods used for regulation of foods and drugs. *Anal. Chem.* 54:67A–76A.

Jackson, D. R., and Meglen, R. R. (1975). A procedure for extraction of molybdenum from soil with anion-exchange resin. *Soil Sci. Soc. Am. J.* 39:373–4.

Jackson, M. L. (1958). *Soil Chemical Analysis.* Englewood Cliffs, NJ: Prentice-Hall.

Johnson, C. M., and Arkley, T. H. (1954). Determination of molybdenum in plant tissue. *Anal. Chem.* 26:572–4.

Jones, J. B. (1977). Elemental analysis of soil extracts and plant tissue ash by plasma emission spectroscopy. *Commun. Soil Sci. Plant Anal.* 8:349–65.

Khan, S. U., Cloutier, R. O., and Hidiroglou, M. (1979). Atomic absorption spectroscopic determination of molybdenum in plant tissue and blood plasma, *J. Assoc. Off. Anal. Chem.* 62:1062–964.

Karimian, N., and Cox, F. R. (1978). Adsorption and extraction of molybdenum in relation to some chemical properties of soil. *Soil Soc. Sci. Am. J.* 43:757–61.

Kubota, J. (1976). Molybdenum status of United States soils and plants. In *Molybdenum in the Environment, Geochemistry, Cycling, and Industrial Uses of Molybdenum*, vol. 2, ed. W. B. Chappel and K. K. Petersen, pp. 555–81. New York: Marcel Dekker.

Kubota, J., and Cary, E. E. (1982). In *Methods of Soil Analysis*, part 2, ed. A. L. Page et al., pp. 485–500. Madison, WI: American Society of Agronomy.

Lavy, T. L., and Barber, S. A. (1964). Movement of molybdenum in the soils and its effect on availability to the plant. *Proc. Soil Sci. Soc. Am.* 28:93–7.

Little, I. P., and Kerridge, P. C. (1978). A laboratory assessment of the molybdenum status of nine Queensland soils. *Soil Sci.* 125:102–6.

Lombin, G. (1985). Micronutrient soil test for the semi-arid savannah of Nigeria: boron and molybdenum. *Soil Sci. Plant Nutr.* 31:1–11.

Lowe, R. H., and Massey, H. F. (1965). Hot water extraction for available soil molybdenum. *Soil Sci.* 100:238–43.

Marmoy, F. B. (1939). Determination of molybdenum in plant materials. *J. Soc. Chem. Ind. (London)* 58:275.

Massey, H. F., Lowe, R. H., and Bailey, H. H. (1967). Relation of extractable molybdenum to soil series and parent rock in Kentucky. *Soil Sci. Soc. Am. Proc.* 31:200–2.

Mortvedt, J. J., and Anderson, O. E. (1982). *Forage Legumes: Diagnosis and Correction of Molybdenum and Manganese Problems.* University of Georgia Southern Cooperative Series bulletin 278.

Mulder, E. C. (1948). The importance of molybdenum in the nitrogen metabolism of microorganisms and higher plants. *Plant Soil* 1:94–119.

Munter, R. C., Halverson, T. L., and Anderson, R. D. (1984). Quality assurance of plant tissue analysis by ICP-AES. *Commun. Soil Sci. Plant Anal.* 15:1285–322.

Perrin, D. D. (1946). *N.Z. J. Sci. Technol.* 28:183–7.

Pierzynski, G. M., Crouch, S. R., and Jacobs, L. W. (1986). Use of direct-current plasma spectrometry for the determination of molybdenum in plant tissue digests and soil extracts. *Commun. Soil. Sci. Plant Anal.* 17:419–28.

Piper, C. S., and Beckworth, R. S. (1948). A new method for determination of molybdenum in plants. *J. Soc. Chem. Ind.* 67:374–8.

Plank, C. O. (1992). *Plant Analysis Reference Procedures for the Southern Region of the United States.* University of Georgia Southern Cooperative Series bulletin 368.

Purvis, E. R., and Peterson, N. K. (1956). Methods of soil and plant analysis for molybdenum. *Soil Sci.* 81:223–8.

Quin, B. F., and Woods, P. H. (1979). Automated catalytic method for the routine determination of molybdenum in plant materials. *Analyst* 104:552–9.

Reisenauer, H. M. (1965). Molybdenum. In *Methods of Soil Analysis: Chemical and Microbiological Properties,* part 2, ed. C. A. Black, pp. 1050–8. Madison, WI: American Society of Agronomy.

Ritchie, G. S. P. (1988). A preliminary evaluation of resin extractable molybdenum as a soil test. *Commun. Soil Sci. Plant Anal.* 19:507–16.

Rowbottom, W. H. (1991). Determination of ammonium acetate extractable molybdenum in soil, and aqua regia (hydrochloric acid and nitric acid, 3 + 1) soluble molybdenum in soil and sewage sludge by electrothermal atomic absorption spectrometry. *J. Anal. Atomic Spectrometry* 6:123–7.

Sherrell, C. G. (1989). Evaluation of an anion resin method for plant available molybdenum in New Zealand soils. *Commun. Soil Sci. Plant Anal.* 20:61–74.

Sims, J. L., and Crutchfield, J. D. (1992). Determination of molybdenum in plants. In *Plant Analysis Reference Procedures for the Southern Region of the United States,* ed. C. O. Plank, pp. 61–6. University of Georgia Southern Cooperative Series bulletin 368.

Soltanpour, P. N. (1985). Use of ammonium bicarbonate DTPA to evaluate elemental availability and toxicity. *Commun. Soil Sci. Plant Anal.* 16:323–38.

Soltanpour, P. N. (1991). Determination of nutrient availability and elemental toxicity by AB-DTPA soil test and ICPS. *Adv. Agron.* 16:165–87.

Stanton, R. E., and Hardwick, A. J. (1967). The colorimetric determination of molybdenum in soils and sediments by zinc dithiol. *Analyst* 92:387–90.

Sterrett, S. B., Smith, C. B., Masianica, M. P., and Demchek, K. T. (1987). Comparison of analytical results from plant analysis laboratories. *Commun. Soil Sci. Plant Anal.* 18:287–99.

Stone, K. L., and Jencks, E. M. (1963). *The Available Molybdenum Status of Some West Virginia Soils.* West Virginia University Agricultural Experiment Station bulletin no. 484.

Taylor, J. K. (1985). *Handbook for SRM Users.* NBS special publication 260-100. Gaithersburg, MD: U.S. Department of Commerce, National Bureau of Standards.

U.S. Environmental Protection Agency (1993). Standards for the use or disposal of sewage sludge. *Federal Register* 58:9248–415.

Wang, L., Reddy, K. J., and Munn, L. C. (1994). Comparison of ammonium bicarbonate-DTPA, ammonium carbonate, and ammonium oxalate to assess the availability of molybdenum in mine spoils and soils. *Commun. Soil. Sci. Plant Anal.* 25:523–36.

Williams, J. H. (1971). *Trace Elements in Soils and Crops.* Ministry of Agriculture, Fisheries and Food (U.K.) technical bulletin no. 21, pp. 119–36.

Wilson, O. D. (1979). Determination of molybdenum in wet-ashed digests of plant material using flameless atomic absorption. *Commun. Soil Sci. Plant Anal.* 10:1319–30.

Wilson, O. D. (1992). Determination of molybdenum in plant tissue using the graphite furnace. In *Plant Analysis Reference Procedures for the Southern Region of the United States*, ed. C. O. Plank, pp. 67–70. University of Georgia Southern Cooperative Series bulletin 368.

Yatsimirskii, K. B. (1964). The use of catalytic reactions involving hydrogen peroxide in the study of the formation of complexes and in the development of very sensitive analytical methods. In *Catalysis and Chemical Kinetics*, ed. A. A. Balandin et al., pp. 201–6. New York: Academic Press.

Zbiral, J. (1992). Determination of molybdenum in hot water soil extracts: influence of pH and available iron on the molybdenum content. *Commun. Soil Sci. Plant Anal.* 23:817–25.

7

Testing for Molybdenum Availability in Soils

JOHN L. SIMS and FRIEDA EIVAZI

Introduction

The testing of soils for molybdenum (Mo) and other micronutrients has been reviewed extensively in recent publications (Gupta and Lipsett, 1981; Anderson and Mortvedt, 1982; Cox, 1987; Johnson and Fixen, 1990; Sims and Johnson, 1991; Sims, 1996). The general objectives for testing soils for any nutrient have been to assess the soil's capacity to supply plant-available nutrient during the growth of crops and to gather data that can guide producers in obtaining the best economic response to fertilizer application. Fitts and Nelson (1956) suggested that soil testing can be divided into four phases: (1) sampling the soil, (2) conducting tests to determine nutrient availability, (3) calibrating test findings with crop responses, and (4) interpreting the findings and making recommendations.

Because of the relatively small amounts of Mo in soils (0.1–30 mg kg^{-1}) (Kubota, 1976), the importance of seed Mo reserves in supplying crop needs (Peterson and Purvis, 1961; Harris, Parker, and Johnson, 1965; Gurley and Giddens, 1969), the importance of soil properties that affect Mo availability (Lowe and Massey, 1965; Massey, Lowe, and Bailey, 1967; Karimian and Cox, 1978, 1979; Burmester, Adams, and Odom, 1988), and the low requirements of most crops for Mo (0.1–0.5 mg kg^{-1} tissue), the testing of soils for Mo in the classic sense is rendered difficult. In one study alone, using the method of Grigg, Massey et al. (1967) found that Mo concentrations varied widely among 73 soils in Kentucky derived from various parent materials, ranging from 0.08 mg kg^{-1} for soils of

The investigation reported in this chapter (95-06-038) was in connection with a project of the Kentucky Agricultural Experiment Station and the Division of Food and Agricultural Sciences, Cooperative Research, Lincoln University, and is published with approval of the directors.

Pennsylvanian sandstone and shale origin to 9.4 mg kg⁻¹ for soils from Devonian black fissle shale. A more realistic goal may be to use analytical tests that (1) will be relatively rapid, reproducible, and accurate, (2) can categorize fields into deficient and nonresponsive groups from a plant-growth standpoint, and (3) can identify contaminated soils and situations of soil additives (e.g., sewage sludge) that can produce forages that are toxic to animals or can be a threat to the environment (Miltmore and Mason, 1971; Anonymous, 1976; Allaway, 1977; Elseewi and Page, 1984; Thom, 1987; U.S. Environmental Protection Agency, 1993).

Progress in testing for Mo in the past was hampered by a lack of understanding of the sites or "pools" of available Mo in soils (Viets, 1962). According to Reisenauer, Walsh, and Hoeft (1973), the available Mo includes the Mo in solution, that found in labile form adsorbed on sesquioxide surfaces, that chelated by soil organic matter, and that at the surfaces of slightly soluble crystalline compounds. The relative amounts of Mo found among these fractions vary widely among soils and have not been extensively studied, particularly the Mo associated with soil organic matter.

The lack of research into ways of concentrating Mo during analytical analysis has also hampered progress in testing soils for Mo. The colorimetric methods used with success in the past (Johnson and Arkley, 1954; Purvis and Peterson, 1956; Gupta and MacKay, 1965a) are tedious and time-consuming. These disadvantages have been overcome partially by recent advances in graphite-furnace atomic-absorption spectroscopy (GF-AAS), plasma-emission spectrometry, and automated spectrophotometric instrumentation (Jones, 1991). However, some of the instruments used in those procedures are not sensitive enough to measure Mo in plant (or soil) materials containing Mo at less than 0.5 mg kg⁻¹, and few have been correlated or calibrated to the crop needs for Mo.

The purposes of this chapter are to review the literature on extraction and analytical procedures for determining plant-available Mo in soils and to outline selected methods for determining the plant-available Mo in soils.

Extraction Procedures for Available Molybdenum

Ammonium Oxalate

One of the earliest and most widely used extractants for determining the available Mo in soils was acid ammoniun oxalate (pH 3.3), as proposed by Grigg (1953a). Earlier, Kurtz, DeTurk, and Bray (1946) and Low and

Black (1950) had shown that the anion oxalate could be used to extract more PO_4 from soil than could citrate, acetate, or benzoate and that oxalate was second only to fluoride and hydroxyl ions in its strength of PO_4 replacement. Grigg (1953a) stated that oxalate not only had the advantage of determining the exchangeable Mo but also, once the Mo was extracted, formed strong complexes with oxalic acid, rendering the exchange irreversible.

An extractant pH of 3.3 was used to buffer and to prevent any material change in extractant pH when shaken with soil, and also because differences in Mo values between soils were greater at pH 3.3 than with an oxalate solution of neutral pH (Grigg, 1953b). Solutions at pH 3.3 also extracted about threefold more Mo from soils than did solutions at pH 6.5. Using NH_4 oxalate at pH 3.3, Grigg categorized responsive soils as those containing Mo at less than $0.20 \, mg \, kg^{-1}$ at pH < 6.3. When Grigg (1953a) tested other anion extractants, the use of PO_4 as the replacing ion led to analytical difficulties and erroneous results. Extraction with 2% NaOH gave increased amounts of Mo, and the findings suggested that a fraction of soil Mo was associated with organic matter. In later work, Grigg (1960) decided that the method was unreliable for diagnosing Mo deficiencies, because oxalate extracts a portion of Fe-bound Mo that is unavailable to plants. He concluded that soil type was a better guide to the available Mo than was the chemical test.

Since the work by Grigg, acid ammonium oxalate has been tested extensively on a range of soils, with variable results. Haley and Melsted (1957) studied the available Mo extracted by different extractants using Illinois soils at various pH values (5.2–6.8), various NH_4-oxalate-extractable Mo contents (0.05–$0.53 \, mg \, kg^{-1}$), and plant Mo concentrations of 0.27–$10.65 \, mg \, kg^{-1}$. Those authors found the following order for the amounts of Mo extracted: 0.1-N NaOH > acid ammonium oxalate > 0.03-N NH_4F > 0.1-N HCl + 0.03-N NH_4F. The amounts of soil Mo extracted by both NaOH and NH_4 oxalate extractants were significantly correlated with the amounts of Mo removed by plants ($r = .85^{**}$ and $.80^{**}$, respectively). Similarly, Gupta and MacKay (1966) compared NH_4 oxalate (pH 3.0) with 2% citric acid, 0.1-N HCl, and 0.1-N NaOH for extracting exchangeable Mo and Cu from podzol soils in the maritime provinces of Canada. Sixteen hours of shaking time was necessary for maximum release of both elements. Generally, the amounts of soil Mo extracted followed the order NH_4 oxalate > citric acid > 0.1-N NaOH > 0.1-N HCl.

Soil Mo extractable with ammonium oxalate was found to be significantly correlated ($r = .70^{**}$) with Mo uptake by peanuts (*Arachis* *hypogaea* L.) in a group of soils in Nigeria (NH_4-oxalate-extractable Mo of 0.07–0.14 mg kg^{-1}), but the best prediction of Mo uptake was obtained when the percentage of soil organic matter was included with the NH_4-oxalate-extractable Mo as independent variables (Lombin, 1985). Six field experiments were conducted in four states in the southern United States over 3 years to determine the responses of soybeans (*Glycine max* L.) to Mo applications (Anderson and Mortvedt, 1982). The amounts of Mo extractable with acid ammonium oxalate were significantly related to the leaf and seed Mo concentrations and to relative yields in some locations and years. But yield responses often were more closely related to soil pH or leaf nitrogen (N) concentration than to soil or plant Mo. Because of the great variations in soil Mo where the responses to Mo could be determined, a critical soil Mo level could not be established. Generally, yields increased with additions of Mo or lime to soils of pH < 5.7, suggesting that this may be a critical pH for obtaining responses to Mo or lime in the region.

In contrast to the foregoing studies, acid-NH_4-oxalate-extractable soil Mo was not found to be related to plant Mo in a number of other studies (Lowe and Massey, 1965; Pathak, Shankar, and Misra, 1969; Little and Kerridge, 1978; Karimian and Cox, 1979; Mortvedt and Anderson, 1982; Cox, 1987; Burmester et al., 1988). Generally, in those studies the amount of plant Mo was more closely related to some soil property other than Mo content. Also, predictions of plant Mo concentrations were greatly improved by including an additional soil property, such as pH (Lowe and Massey, 1965; Pathak et al., 1969; Mortvedt and Anderson, 1982; Cox, 1987) or soil organic matter (Lombin, 1985; also see Karimian and Cox, 1978), in the prediction. The best predictions for plant Mo concentrations and relative yields of cauliflower (*Brassica oleracea* var. *botrytis* L.) were obtained from the use of soil pH and the ratio of amorphous iron (Fe) to free Fe (Karimian and Cox, 1979), and the relative yields of soybeans were highly related to the ratio between Fe and the Mo extracted by acid NH_4 oxalate (Burmester et al., 1988). The amount of Mo extracted with ammonium oxalate (pH 3.3) was correlated with total Fe ($r = .75$), but was not related to plant responses to added Mo (Little and Kerridge, 1978).

Recently, Liu (1995) studied how the pH values of NH_4 oxalate extractants affected the amounts of Mo extracted from Kentucky soils and the relationships between soil Mo and the Mo taken up by tobacco (*Nicotiana tabacum* L.) and soybean plants grown in a greenhouse. The rationale for the study was that in humid regions, plants absorb Mo from soils with pH values ranging from 5 to 7, and the amounts of Mo extracted by extractants at similar pH values should be more closely related to the amounts of plant Mo uptake. Six NH_4 oxalate solutions (0.3-M) were prepared, with pH values ranging from 3.3 to 6.4. Molybdenum in the extracts was measured after 18 hours of shaking by the automated colorimetric method of Eivazi and Sims (Chapter 6, this volume). Prior to cropping, surface samples from 12 soils (pH range 5.2–7.46) had been left in the greenhouse without any change or with two rates of Na_2MoO_4 addition (0.3 and 0.6 mg Mo per 1 kg of soil). The data showed that the amounts of Mo extracted from the soils decreased markedly with increasing pH of the extractant. However, correlation coefficients for the relationship between soil Mo and Mo uptake by tobacco plants increased from $r = .02$ for pH-3.3 extracts (Grigg's reagent) to $r = .81^{**}$ for pH 6.0, and then decreased again at pH 6.4. Coefficients for similar correlations at an extractant pH of 6.0 for soybeans were significant, but lower. The findings from that study show promise of predictive value from such tests in acid soils.

Water

There has been some success in extracting the available soil Mo with water when soil Mo concentrations have been sufficiently high (Barshad, 1951; Kubota, Lemon, and Allaway, 1963). But in another study, water-soluble Mo was not detected by a colorimetric procedure in half of the samples investigated (Gupta and MacKay, 1965b). Barshad (1951) measured water-soluble Mo in soils with a wide range of pH values (4.8–7.8) that produced clover (*Trifolium repens* L.) plants with a wide range of Mo concentrations (5–118 $\mu g\,g^{-1}$). Soil Mo was found to be closely related to plant Mo concentration. Plant tissues containing Mo at more than 15 $\mu g\,g^{-1}$ were associated with soil concentrations of water-soluble Mo greater than 0.2 $\mu g\,g^{-1}$, suggesting that water extractions can be used to distinguish soils likely to produce forages toxic to animals.

Lowe and Massey (1965) continuously leached soils (10 hours) with hot water in a Soxhlet procedure to study the relation of water-soluble Mo to Mo uptake by alfalfa (*Medicago sativa*) grown in a greenhouse. Molybdenum uptake by alfalfa was significantly related to soil Mo ($r = .83^{**}$) for a group of cultivated soils. Among woodland soils, similar correlations for soil Mo were significant, but lower ($r = .66^{**}$), and for plant Mo versus soil pH ($r = .66^{**}$). Those findings for Mo extracted by hot water were confirmed by Pathak et al. (1969) for a number of soils of India.

Recently, Zbiral (1992) used hot water to extract plant-available Mo from 138 soil samples in the former Czechoslovakia. Molybdenum from water extracts, after boiling (10 minutes) and filtering, was concentrated as a complex with 8-hydroxyquinoline into chloroform. Determination of Mo was by flameless (graphite-furance) atomic-absorption spectroscopy (GF-AAS). The procedure is rapid and avoids many of the difficulties in concentrating Mo that are encountered with the older colorimetric procedures. Zbiral (1992) used the Mo data to derive an equation describing soil Mo as a function of exchangeable soil acidity and Fe extracted with diethylenetriamine pentaacetic acid (DTPA) (Lindsay and Norvell, 1978). The calculated and measured soil Mo values were found to be in good agreement, except for 6 of the 138 soil samples. The equation was found to be valid only for soils in equilibrium with respect to Mo additions. Recent additions of Mo to soils in the greenhouse increased the measured values markedly, as compared with the calculated values.

Anion-Exchange Resin

Anion-exchange resins have been used to extract Mo in a number of studies (Bhella and Dawson, 1972; Dawson and Bhella, 1972; Jackson and Meglen, 1975; Jarrell and Dawson, 1978; Karimian and Cox, 1978; Ritchie, 1988; Sherrell, 1989). Bhella and Dawson (1972) limed a group of soils with $Ca(OH)_2$ to produce soils with pH values of 4.95–7.10 for greenhouse studies. Anion-exchangeable Mo was highly correlated with soil pH ($r = .80$) in soils without added Mo and with Mo and N concentrations in clover (*Trifolium subterraneum*) plants ($r = .86$ and $r = .88$, respectively). Those findings suggested that anion-exchange resin could detect increases in available soil Mo brought about by increases in soil pH in the pH range of approximately 5.0–7.0. Jackson and Meglen (1975) cited the earlier work of S. R. Olsen (Colorado State University, USDA-ARS) showing that the amount of Mo extracted by the resin method increased linearly as a function of soil pH, whereas the amount extracted by acid ammonium oxalate was constant and was not related to pH.

Other researchers have noted significant correlations between the amount of Mo extracted with anion resin and the plant content of Mo (Jarrell and Dawson, 1978; Lombin, 1985; Ritchie, 1988). In contrast,

Karimian and Cox (1978) found no correlation between plant Mo uptake and soil Mo extracted by anion resin or acid ammonium oxalate. Lombin (1985) found better correlations for either 0.1-N NaOH or acid ammonium oxalate extraction and Mo uptake by peanuts than for anion-resin extraction and Mo uptake. Sherrell (1989) concluded that the resin method was not accurate enough for diagnosis of Mo deficiency in the pastoral soils of New Zealand, although the resin method was more useful than the ammonium oxalate method.

Ammonium Carbonate and Ammonium Bicarbonate–DTPA Extractants

Vlek and Lindsay (1977) extracted Mo from soils for 12 hours with 1-M $(NH_4)_2CO_3$ at pH 9.0 to release the more labile fraction of soil Mo. The values with that extractant were shown to correlate closely ($r = .977$) with Mo uptake by alfalfa when data for soils of pH greater than 7.0 were used. The correlation coefficients decreased markedly when data for soils of pH less than 7.0 were included in the correlations, because the $(NH_4)_2CO_3$ extracted more Mo than was available to plants.

Boon (1984) cited earlier work showing that the $(NH_4)_2CO_3$ and ammonium oxalate procedures extracted similar amounts of Mo from regraded mine spoils in Montana (Neuman and Munshower (1983) and that values for extracted Mo by the $(NH_4)_2CO_3$ and the ammonium bicarbonate – DTPA (AB-DTPA) procedures were highly correlated ($r = .98$).

The AB-DTPA method was developed for simultaneous extractions of phosphorus (P), potassium (K), iron (Fe), manganese (Mn), copper (Cu), and zinc (Zn) in calcareous soils (Soltanpour and Schwab, 1977). The advantages and theoretical basis of the method were reported by Havlin and Soltanpour (1981). The thrust of the effort was to develop a method to extract multiple macronutrients and micronutrients at the same time, a single method that would replace three earlier methods: ammonium acetate, $NaHCO_3$ (Olsen et al., 1954), and DTPA (Lindsay and Norvell, 1978). As noted earlier and in other literature (Pierzynski and Jacobs, 1986a,b; Soltanpour, 1991), the AB-DTPA test has been extended to include extraction of Mo. Because the extractant can be used in conjunction with inductively coupled plasma (ICP) emission analysis, it offers the added potential for measuring Mo during routine soil analysis of multiple nutrients when the concentrations of Mo arc within the detection limits of ICP analysis.

Molybdenum, both the inorganic salt and that contained in Mo-enriched sludge, and lime (CaO) were applied to a sandy loam soil in a greenhouse to evaluate the influence of soil pH on plant uptake and extractability of Mo (Pierzynski and Jacobs, 1986a). Molybdenum in corn (*Zea mays* L.), soybean, and alfalfa increased with increases in soil pH. Molybdenum extracted with AB-DTPA also increased with increasing soil pH, but values for Mo extracted with acid ammonium oxalate were not influenced by soil pH. Plant Mo was more closely related to values obtained by the AB-DTPA method than those by the ammonium oxalate method. Including soil pH in the multiple-regression analyses of plant Mo or extractable soil Mo improved predictions in 17 of 24 prediction equations.

Wang, Reddy, and Munn (1994) compared the $(NH_4)_2CO_3$, AB-DTPA, and ammonium oxalate methods to assess the availability of Mo in mine spoils and soils. Crested wheatgrass (*Agropyron cristatum*) and alfalfa were grown in a greenhouse on soils with added Mo, with pH values ranging from 5.3 to 8.4, to provide data for correlation purposes. Of the three extractants, ammonium oxalate (pH 3.3) removed the greatest amounts of Mo. The amounts of Mo extracted by all extractants were equally and significantly correlated with plant Mo uptake. All three methods were considered useful for assessing Mo availability and the potential toxicity from plant uptake of Mo from reclaimed spoils.

Analytical Procedures for Plant-available Molybdenum in Soils

Colorimetric Methods

Analytical procedures used for measuring the available Mo in soils and plants were reviewed in Chapter 6 of this volume. The procedure of choice may depend on which instruments are available, but also on the intended use of the data. Growing plants on alkaline or contaminated soils high in Mo is likely to result in plant and soil extracts containing Mo at more than 1.0 mg kg^{-1}, whereas soils deficient in Mo will result in extracts with Mo concentrations ranging from 0.5 mg kg^{-1} to as low as undetectable.

Among the most commonly used chemical reagents for extractions from biological materials containing low amounts of Mo are dithiol and thiocyanate (Gupta and Lipsett, 1981). In these procedures, Mo forms a complex with dithiol or thiocyanate in aqueous solution and then is concentrated in an organic solvent prior to spectrophotometric analysis.

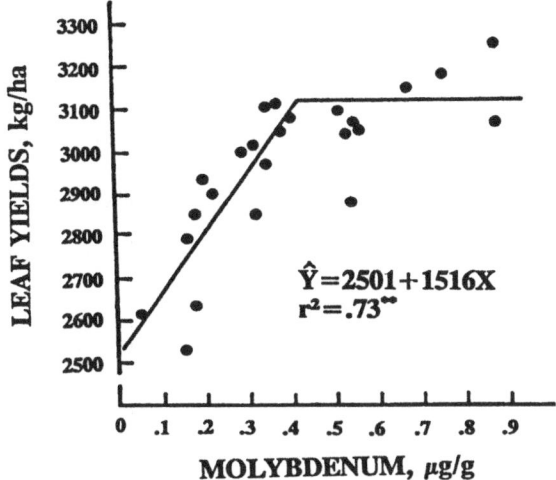

Figure 7.1. Relationship of leaf Mo concentrations and cured-leaf yields of burley tobacco. The fitted equation is of the form $\hat{y} = b_0 + b_1[\min(x, A)]$, where b_0 is the intercept and b_1 is the slope of the line up to the point $X = A$. [Reproduced from Sims and Atkinson (1976): Lime, molybdenum, and nitrogen source effects on yield and selected chemical components of burley tobacco, *Tobacco Science* 20:181–4, courtesy of Lockwood Trade Journal Co., Inc., New York, NY, and *Tobacco Science*.]

The dithiol procedure, developed by Piper and Beckworth (1948), and modified later (Williams, 1955; Bingley, 1963; Gupta and MacKay, 1965a), has been considered to be more sensitive and more precise than the thiocynate method. However, Purvis and Peterson (1956) found the thiocynate method to be more adaptable to routine work.

Although colorimetric methods are tedious and require much time, they remain useful for measurements in materials with low Mo concentrations. A modified stannous chloride–thiocyanate procedure described by Marmoy (1939) and Johnson and Arkley (1954) has been used extensively in our laboratory for many years and has yielded useful results (Sims and Atkinson, 1974; Pal et al., 1976; Hawes, Sims, and Wells, 1976; Sims, Leggett, and Pal, 1979; Sims, 1996). An example of such data is shown in Figure 7.1 (Sims and Atkinson, 1976). In that study, an acid soil (pH 5.4) had been treated with various amounts of agricultural limestone and N, P, and K fertilizers in the form of $NaNO_3$ or urea, concentrated superphosphate, and K_2SO_4, respectively. Maximum cured yields of burley tobacco were associated with leaf-tissue Mo concentrations of about $0.42\,mg\,kg^{-1}$ in this largely acid soil environment.

An automated colorimetric method that has been described in detail by Eivazi and Sims (Chapter 6, this volume) is also useful for materials of low Mo concentrations. This method is sensitive for Mo concentrations below $0.1\,mg\,kg^{-1}$ and is rapid and precise, and the values obtained with this method have been found to be closely correlated to the Mo values determined by the stannous chloride–thiocyanate method (Eivazi, Sims, and Crutchfield, 1982) and those determined by GF-AAS (Chapter 6, this volume).

Atomic-Absorption and Plasma-Emission Spectroscopy

GF-AAS has been used successfully to measure Mo in plant digests in the range of $0.05-2.0\,mg\,kg^{-1}$ (Henning and Jackson, 1973; Wilson, 1979, 1992; Soon and Bates, 1985). Several authors have used GF-AAS to determine low concentrations of Mo in soil extracts (Little and Kerridge, 1978; Lombin, 1985; Zbiral, 1992; Wang et al., 1994). GF-AAS is particularly useful when Mo concentrations in materials are below $0.5\,mg\,kg^{-1}$. For materials with Mo concentrations greater than about $1.0\,mg\,kg^{-1}$, flame atomic-absorption and ICP emission spectrometry have been used (Soltanpour, Jones, and Workman, 1982; Boon and Soltanpour, 1983; Pierzynski and Jacobs, 1986a,b; Soltanpour, 1991).

Selected Methods for Determination of Available Molybdenum in Soils

Among the attributes of a good soil test for the available soil nutrients are close correlation and calibration of the amounts of nutrients extracted by the test and the amounts taken up by plants, as well as easy adaptation of the test to routine laboratory analysis. In addition, the test should be accurate, economical, and reproducible. The tests to be outlined here have been selected with these factors in mind, keeping in mind the importance of how soil pH affects the availability of Mo to plants. For each unit increase in soil pH above pH 5.0, the soluble Mo concentration can increase 100-fold (Lindsay, 1972). The absorption of Mo by plants increases rapidly and logarithmically with increasing pH (Hewitt, 1958). Lowe and Massey (1965) and Bhella and Dawson (1972) showed that 35–65% of the variation in plant Mo uptake could be accounted for by regression analysis of plant Mo on soil pH. In many other studies, prediction of plant Mo content was improved markedly by including both pH and extractable soil Mo as independent variables in the regressions (Pathak et al., 1969; Mortvedt and Anderson, 1982; Pierzynski and Jacobs, 1986a; Cox, 1987; Liu, 1995).

The anion-resin and the AB-DTPA extractants have been shown to be responsive to changes in soil pH (Bhella and Dawson, 1972; Jackson and Meglen, 1975; Pierzynski and Jacods, 1986a), whereas the strongly buffered ammonium oxalate extractant (pH 3.3) was designed to be unresponsive to differences in soil pH. That may partially explain why the values reported with the acid ammonium oxalate extractant have not correlated with plant Mo uptake in many studies; plant uptake of Mo from soil is responsive to differences in soil pH and occurs at pH values much higher than 3.3.

The anion-resin and AB-DTPA methods to be outlined next were reported previously (Soltanpour et al., 1982; Sims, 1996) and are published here with permission of the American Society of Agronomy, Inc., and the Soil Science Society of America, Inc., Madison, WI.

Anion-Resin Method

Principles

In the method outlined here, aqueous resin–soil suspensions are equilibrated for 18 hours, the soil is washed from the resin, the sorbed Mo is displaced from the resin with 2-M $NaNO_3$, and the Mo in the leachate is measured. The extraction procedure is a modification of that of Jarrell and Dawson (1978), with the procedure for organic-matter destruction as described by Jackson and Meglen (1975). Measurement of the Mo obtained after extraction is carried out using the automated $KI\text{-}H_2O_2$ procedure (Chapter 6, this volume).

Apparatus

1. Reciprocating shaker, top-loading
2. Sieve, 60 mesh (0.25 mm), brass
3. Peristaltic pump
4. Hot plate

Reagents

1. AG1-X4 anion-exchange resin (Bio-Rad Laboratories, Melville, NY), Cl-saturated
2. Sodium chloride (NaCl), reagent-grade

3. Sodium hydroxide (NaOH), reagent-grade
4. Sodium nitrate (NaNO$_3$), reagent-grade
5. Hydrogen peroxide (H$_2$O$_2$), 30%, reagent-grade
6. Sulfuric acid (H$_2$SO$_4$), reagent-grade
7. Hydrochloric acid (HCl), reagent-grade

Procedure

AG1-X4 anion-exchange resin (chloride-saturated) is used to extract Mo from soil suspensions. The resin should be preconditioned to avoid decreasing the pH of soil suspensions. Alternatively, rinse the resin with 2-M NaOH and 2-M NaCl solutions, followed by distilled water. After sieving, the resin fraction greater than 60-mesh size is stored in moist form until used. Grind 10 g of air-dried soil to pass a 60-mesh sieve, and place the soil in a glass tube, 2.0 cm (inside diameter) × 20 cm. Add 5 g of AG1-X4 resin and 20 mL of double-distilled water to each tube. Insert the stopper and shake the soil–resin mixture for 18 hours at 25°C. Separate the resin from the soil by washing with distilled water on a 60-mesh brass screen. Transfer the resin to a glass funnel with a 1.5-cm (outside diameter) stem inserted in a 10-mL disposable pipette tip. Prior to addition of resin, glass wool should be placed near the tip of each pipette. Connect the tip of the pipette to a peristaltic pump with small tubing, and add 30 mL of 2-M NaNO$_3$ to the funnel. Allow the resin and NaNO$_3$ to equilibrate for 45 minutes prior to displacing the NaNO$_3$ solution and Mo from the resin with the pump over an additional 45-minute period. Evaporate the extractant NaNO$_3$ solution from a 50-mL Erlenmeyer flask to dryness on a hot plate at low heat (80°C), to avoid loss of sample by splattering. After cooling, add 2 mL of 30% H$_2$O$_2$, 8 mL of 2-M HCl, and two glass beads to each flask. Heat, uncovered, on a hot plate at 80°C until dry. If necessary, repeat the H$_2$O$_2$ portion of the organic-matter-destruction step to render the residue colorless. Allow the samples to cool and add 10 mL of 0.125-M HCl; cover and heat for 20 minutes at 80°C. Swirl the flasks to ensure dissolution of the samples. Allow the samples to settle overnight before determining the Mo content. Carefully decant sufficient liquid into disposable sample cups for analysis on the autoanalyzer (Chapter 6, this volume).

Interpretation

Modifications made in the procedure of Jarrell and Dawson (1978) include the use of a reciprocating shaker, a funnel, larger-diameter leaching tubes, and a peristaltic pump, as well as $NaNO_3$ to replace Mo from the resin. Several tubes contained in Styrofoam or another shaking rack can be shaken on their sides at one time. The funnel and tubes of slightly larger diameter permit easier transfer of resin from shaking tube to leaching tube, and the peristaltic pump ensures measured and more even dispensing of the $NaNO_3$ extracting solution of multiple samples. Because Lombin (1985) reported that high chloride concentrations tended to depress, and sometimes enhance, the Mo signal in the GF-AAS technique, $NaNO_3$ was used in place of NaCl. The resin extracts lend themselves to analysis of Mo by methods other than the automated $KI-H_2O_2$ method when Mo concentrations fall within the detection limits of the instrument.

Representative Findings

Recently, this procedure was used in preliminary tests in our laboratory with the Mo-uptake data of Liu (1995) for tobacco to correlate with soil Mo values, as described in the first section of this chapter. Soil Mo values obtained by the anion-resin method were significantly related to plant uptake of Mo ($r^2 = .86**$).

AB-DTPA Method

Principles

In this procedure (Soltanpour et al., 1982) the NH_4HCO_3 acts as both an extractant of Mo and a buffer, in a manner similar to the extraction of PO_4. During extraction, the pH of the soil–AB-DTPA mixture increases to pH 8.5, dissolving calcium (Ca), aluminum (Al), and Fe molybdates (Soltanpour, 1991). Also, the HCO_3^- ions desorb molybdates from the surfaces of soil colloids. The DTPA may aid the Mo extraction process, but is included primarily for the extraction of multiple cations. Molybdenum determination is accomplished through use of inductively coupled plasma spectrometry (ICPS). The procedure has not been tested extensively on acid soils, but is outlined for use on alkaline or Mo-contaminated soils or spoils.

Apparatus

1. Reciprocal shaker
2. Inductively coupled plasma (ICP) instrument

Reagents

1. Ammonium bicarbonate (NH_4HCO_3)
2. Diethylenetriamine pentaacetic acid (DTPA)
3. Ammonium hydroxide (NH_4OH)
4. Nitric acid (HNO_3)
5. AB-DTPA

A solution of 1-M NH_4HCO_3 and 0.005-M DTPA can be made by adding 1.97 g of DTPA to 800 mL of distilled deionized water (DDW). To aid dissolution and to prevent effervescence, add approximately 2 mL of 1:1 NH_4OH. After most of the DTPA is dissolved, add 79.06 g of NH_4HCO_3 and stir gently until dissolved. Adjust the pH to 7.6 with either HCl or NH_4OH, and dilute the solution to 1.0 L with DDW. The pH of the extracting solution may change unless it is stored under about 3 cm of mineral oil. However, use of a fresh solution is preferable.

Standards

The standards must be made in the AB-DTPA extracting solution. Because of the linearity of ICP standard curves usually only two standards are used to standardize an element. The low standard for Mo contains only AB-DTPA. To prepare the high standard, take 1 mL of a certified reference standard containing 1 mg of Mo, and make the volume up to 500 mL with AB-DTPA solution.

Procedure

To 10 g of a 2-mm-sieved soil in a 125-mL conical flask add 20 mL of AB-DTPA solution. Shake on a reciprocal shaker for 15 minutes at 180 cycles per minute with flasks kept open. Use Whatman no. 42 filter paper or its equivalent to filter the extracts. Place 0.25 mL of concentrated HNO_3 in a 10-mL beaker, and carefully add 2.5 mL of AB-DTPA extract or standard solution to the beaker. Mix on a rotary shaker for 15 minutes to eliminate carbonate species. Determine Mo on an ICP spectrometer.

Interpretation

Nebulizers, devices used for aspiration of the sample into plasma, often clog when high-salt solutions are aspirated; the HNO_3 acid step is designed to eliminate that problem. The HNO_3 acid treatment can be eliminated if the ICP spectrometer is equipped with a high-salt nebulizer (Soltanpour, 1991).

Summary

Because of the relatively small amounts of Mo in soil, the importance of seed Mo reserves in supplying crop needs, the importance of soil properties that affect Mo availability, and the low requirements of most crops for Mo, the testing of soils for Mo in the classic sense is rendered difficult. A more realistic goal may be to use analytical tests that (1) will be relatively rapid, reproducible, and accurate, (2) can categorize fields into deficient and nonresponsive groups from a plant-growth standpoint, and (3) can identify contaminated soils and situations of soil additives (e.g., sewage sludge) that can result in forages that are toxic to animals or can be a threat to the environment. Progress in testing for Mo in the past was hampered by a lack of understanding of the sites or 'pools' of available Mo in soils and by lack of easy and adequate methods for concentrating Mo to measurable amounts during analytical analysis. Instrumentation for Mo analysis has been greatly advanced relative to those earlier inadequacies and now is much better for correlation and calibration of tests in the field.

For Mo extraction, unbuffered NH_4 oxalate solutions of pH 6.0 show promise for acid soils and warrant further research. Currently, anion-exchange resins are being used in many different geographic areas, with generally good success in determining the Mo available to plants. The procedure for use of anion-exchange resin on acid soils that has been outlined here results in extracts that can be analyzed by a number of analytical methods, and the AB-DTPA procedure of Soltanpour et al. (1982) has been outlined for use on alkaline or contaminated soils.

References

Allaway, W. H. (1977). Perspectives on molybdenum in soils and plants. In *Molybdenum in the Environment*, vol. 2, ed. W.B. Chappell and K. K. Petersen, pp. 317–39. New York: Marcel Dekker.

Anderson, O. E., and Mortvedt, J. J. (1982). *Soybeans: Diagnosis and Correction of Manganese and Molybdenum Problems*. University of Georgia Southern Cooperative Series bulletin 281.

Anonymous (1976). *Application of Sewage Sludge to Cropland: Appraisal of Potential Hazards of the Heavy Metals to Plants and Animals*. CAST report no. 64. Ames, IA: CAST.

Barshad, I. (1951). Factors affecting the Mo content of pasture plants. I. Nature of soil molybdenum, growth of plants, and soil pH. *Soil Science* 71:297–313.

Bhella, H. S., and Dawson, M. D. (1972). The use of anion exchange resin for determining available soil molybdenum. *Soil Sci. Soc. Am. Proc.* 36:177–8.

Bingley, J. B. (1963). Molybdenum in plants and animals. Determination of molybdenum in biological materials with dithiol-control of copper interference. *J. Agric. Food Chem.* 11:130–1.

Boon, D. Y. (1984). *The Ammonium Bicarbonate–DTPA Soil Test (AB-DTPA) for Determination of Plant Available Pb, Cd, Ni, and Mo in Mine Tailings and Contaminated Soils.* Sixth High Altitude Revegetation Workshop information series no. 53. Fort Collins: Colorado State University.

Boon, D. Y., and Soltanpour, P. N. (1983). The ammonium bicarbonate–DTPA soil test for determination of plant available lead, cadmium, and molybdenum in mine tailings and contaminated soils. Presented at the 75th annual meeting of the American Society of Agronomy, Washington, DC.

Burmester, C. H., Adams, J. F., and Odom, J. W. (1988). Response of soybean to lime and molybdenum on ultisols in northern Alabama. *Soil Sci. Soc. Am. J.* 52:1391–4.

Cox, F. R. (1987). Micronutrient soil tests: correlation and calibration. In *Soil Testing: Sampling. Correlation. Calibration, and Interpretation*, ed. J. R. Brown, pp. 97–117. Madison, WI: Soil Science Society of America.

Dawson, M. D., and Bhella, H. S. (1972). Subterranean clover yield and nutrient content as influenced by molybdenum status. *Agronomy J.* 64:308–11.

Eivazi, F., Sims, J. L., and Crutchfield, J. (1982). Determination of molydenum in plant materials using a rapid, automated method. *Commun. Soil Sci. Plant Anal.* 13:135–50.

Elseewi, A. A., and Page, A. L. (1984). Molybdenum enrichment of plants grown on fly-ash-treated soils. *J. Environ. Qual.* 13:394–8.

Fitts, J. W., and Nelson, W. L. (1956). The determination of lime and fertilizer requirements of soils through chemical tests. In *Advances in Agronomy*, vol. 8, ed. A. G. Norman, pp. 241–82. New York: Academic Press.

Grigg, J. L. (1953a). Determination of available molybdenum of soils. *N.Z. J. Sci. Technol.* 34:405–14.

Grigg, J. L. (1953b). Determination of available soil molybdenum. *N.Z. Soil News* 3:37–40.

Grigg, J. L. (1960). The distribution of molybdenum in soils of New Zealand. I. Soils of the North Island. *N.Z. J. Agric. Res.* 3:69–86.

Gupta, U. C., and Lipsett, J. (1981). Molybdenum in soils, plants, and animals. *Adv. Agron.* 34:73–115.

Gupta, U. C., and MacKay, D. C. (1965a). Determination of Mo in plant materials using 4-methyl-1,2-dimercaptobenzene (dithiol). *Soil Science* 99:414–15.

Gupta, U. C., and MacKay, D. C. (1965b). Extraction of water soluble copper and molybdenum in podzol soils. *Soil Sci. Soc. Am. Proc.* 29:323.

Gupta, U. C., and MacKay, D. C. (1966). Procedure for the determination of exchangeable copper and molybdenum in podzol soils. *Soil Science* 101:93–7.

Gurley, W. H., and Giddens, J. (1969). Factors affecting uptake, yield response, and carryover of molybdenum in soybean seed. *Agronomy J.* 61:7–9.

Haley, L. E., and Melsted, S. W. (1957). Preliminary studies of molybdenum in Illinois soils. *Soil Sci. Soc. Am. Proc.* 21:316–19.

Harris, H. B., Parker, M. B., and Jonhnson, B. J. (1965). The influence of molybdenum content of soybean seed and other factors associated with seed source on progeny response to applied molybdenum. *Agronomy J.* 57:397–9.

Havlin, J. L., and Soltanpour. P. N. (1981). Evaluation of the NH₄HCO₃-DTPA soil test for iron and zinc. *Soil Sci. Soc. Am. J.* 45:70–5.

Hawes, R. L., Sims, J. L., and Wells, K. L. (1976). Molybdenum concentration of certain crop species as influenced by previous applications of fertilizer. *Agronomy J.* 68:217–18.

Henning, S., and Jackson, T. L. (1973). Determination of molybdenum in plant tissue using flameless atomic absorption. *Atomic Absorption Newsletter* 12:100–1.

Hewitt, E. J. (1958). Some aspects of mineral nutrition in legumes. In *Nutrition of the Legumes*, ed. E. G. Hallsworth, pp. 15–41. New York: Academic Press.

Jackson, D. R., and Meglen, R. R. (1975). A procedure for extraction of molybdenum from soil with anion exchange resin. *Soil Sci. Soc. Am. Proc.* 39:373–4.

Jarrell, W. M., and Dawson, M. D. (1978). Sorption and availability of molybdenum in soils of western Oregon. *Soil Sci. Soc. Am. Proc.* 42:412–15.

Johnson, C. M., and Arkley, T. H. (1954). Determination of molybdenum in plant tissue. *Anal. Chem.* 26:572–4.

Johnson, G. V., and Fixen, P. E. (1990). Testing soils for sulfur, boron, molybdenum, and chlorine. In *Soil Testing and Plant Analysis*, 3rd ed., ed. R. L. Westerman, pp. 265–73. Madison, WI: Soil Science Society of America.

Jones, J. B., Jr. (1991). Plant tissue analysis in micronutrients. In *Micronutrients in Agriculture*, 2nd ed., ed. J. J. Mortvedt, pp. 477–521. Madison, WI: Soil Science Society of America.

Karimian, N., and Cox, F. R. (1978). Adsorption and extractability of molybdenum in relation to some chemical properties of soil. *Soil Sci. Soc. Am. J.* 43:757–61.

Karimian, N., and Cox, F. R. (1979). Molybdenum availability as predicted from selected soil chemical properties. *Agronomy J.* 71:63–5.

Kubota, J. (1976). Molybdenum status of United States soils and plants. In *Molybdenum in the Environment. Geochemistry, Cycling, and Industrial Uses of Molybdenum*, vol. 2, ed. W. B. Chappell and K. K. Petersen, pp. 555–81. New York: Marcel Dekker.

Kubota, J., Lemon, E. R., and Allaway, W. H. (1963). The effect of soil moisture content upon the uptake of molybdenum, copper, and cobalt by alsike clover. *Soil Sci. Soc. Am. Proc.* 27:679–83.

Kurtz, T., DeTurk, E. E., and Bray, R. H. (1946). Phosphate adsorption by Illinois soils. *Soil Science* 61:111–24.

Lindsay, W. L. (1972). Inorganic phase equilibria of micronutrients in soil. In *Micronutrients in Agriculture*, ed. J. J. Mortvedt et al., pp. 41–57. Madison, WI: Soil Science Society of America.

Lindsay, W. L., and Norvell, W. A. (1978). Development of a DTPA soil
    test for zinc, iron, manganese, and copper. *Soil Sci. Soc. Am. Proc.*
    42:421–8.
Little, I. P., and Kerridge, P. C. (1978). A laboratory assessment of the
    molybdenum status of nine Queensland soils. *Soil Science* 125:102–6.
Liu, Dacheng. (1995). An index for plant available molybdenum in Kentucky
    soils. M.S. thesis, University of Kentucky, Lexington.
Lombin, G. (1985). Micronutrient soil tests for the semi-arid savannah of
    Nigeria: boron and molybdenum. *Soil Sci. Plant Nutr.* 31:1–11.
Low, P. F., and Black, C. A. (1950). Reactions of phosphate with kaolinite.
    Soil Science 70:272–90.
Lowe, R. H., and Massey, H. F. (1965). Hot water extraction for available soil
    molybdenum. *Soil Science* 100:238–43.
Marmoy, F. B. (1939). Determination of molybdenum in plant materials.
    J. Soc. Chem. Ind. (London) 58:275.
Massey, H. F., Lowe, R. H., and Bailey, H. H. (1967). Relation of extractable
    molybdenum to soil series and parent rock in Kentucky. *Soil Sci. Soc.
    Am. Proc.* 31:200–2.
Miltmore, J. E., and Mason, J. L. (1971). Copper to molybdenum ratio and
    molybdenum and copper concentrations in ruminant feeds. *Can. J.
    Animal Sci.* 51:193–200.
Mortvedt, J. J., and Anderson, O. E. (1982). *Forage Legumes: Diagnosis and
    Correction of Molybdenum and Manganese Problems.* University of
    Georgia Southern Cooperative Series bulletin 278.
Neuman, D. R., and Munshower, F. F. (1983). Copper and molybdenum
    uptake by legumes growing on topsoiled mine spoils. Paper prepared for
    Peabody Coal Company, Rocky Mountain Div., Denver, Co.
Olsen, S. R., Cole, C. V., Watanabe, F. S., and Dean, L. A. (1954). *Estimation
    of Available P in Soils by Extraction with NaHCO₃.* USDA circular 939.
    Washington, DC: U.S. Government Printing Office.
Pal, U. R., Gossett, D. R., Sims, J. L., and Leggett, J. E. (1976). Molybdenum
    and sulfur nutrition effects on nitrate reduction in burley tobacco. *Can.
    J. Bot.* 54:2014–22.
Pathak, A. N., Shankar, H., and Misra, R. V. (1969). Correlation of available
    molybdenum values obtained by different methods to molybdenum
    uptake by alfalfa. *J. Indian Soc. Soil Sci.* 17:151–3.
Peterson, N. K., and Purvis, E. R. (1961). Development of molybdenum
    deficiency symptoms in certain crops. *Soil Sci. Soc. Am. Proc.* 25:111–16.
Pierzynski, G. M., and Jacobs, L. W. (1986a). Extractability and plant
    availability of molybdenum from inorganic and sewage sludge sources.
    J. Environ. Qual. 15:323–6.
Pierzynski, G. M., and Jacobs, L. W. (1986b). Molybdenum accumulation by
    corn and soybeans from a molybdenum-rich sewage sludge. *J. Environ.
    Qual.* 15:394–8.
Piper, C. S., and Beckworth, R. S. (1948). A new method for determination of
    molybdenum in plants. *J. Soc. Chem. Ind.* 67:374–8.
Purvis, E. R., and Peterson, N. K. (1956). Methods of soil and plant analysis
    for molybdenum. *Soil Science* 81:223–8.
Reisenauer, H. M., Walsh, L. M., and Hoeft, R. G. (1973). Testing soils for
    sulfur, boron, molybdenum, and chlorine. In *Soil Testing and Plant
    Analysis*, ed. L. M. Walsh and J. D. Beaton, pp. 173–200. Madison, WI:
    Soil Science Society of America.

Ritchie, G. S. P. (1988). A preliminary evaluation of resin extractable molybdenum as a soil test. *Commun. Soil Sci. Plant Anal.* 19:507–16.

Sherrell, C. G. (1989). Evaluation of an anion exchange resin method for plant available molybdenum in New Zealand soils. *Commun. Soil Sci. Plant Anal.* 20:61–74.

Sims, J. L. (1996). Molybdenum and cobalt. In *Methods of Soil Analysis: Chemical Methods, Part 3*, 3rd ed., ed. D. L. Sparks, pp. 723–37. Madison, WI: American Society of Agronomy.

Sims, J. L., and Atkinson, W. O. (1974). Soil and plant factors influencing accumulation of dry matter in burley tobacco growing in soil made acid by fertilizer. *Agronomy J.* 66:775–8.

Sims, J. L., and Atkinson, W. O. (1976). Lime, molybdenum, and nitrogen source effects on yield and selected chemical components of burley tobacco. *Tobacco Sci.* 20:181–4.

Sims, J. T., and Johnson, G. V. (1991). Micronutrient soil tests. In *Micronutrients in Agriculture,* 2nd ed., ed. J. J. Mortvedt, pp. 427–76. Madison, WI: Soil Science Society of America.

Sims, J. L., Leggett, J. E., and Pal, U. R. (1979). Molybdenum and sulfur interaction effects on growth, yield, and selected chemical constituents of burley tobacco. *Agronomy J.* 71:75–8.

Soltanpour, P. N. (1991). Determination of nutrient availability and elemental toxicity by AB-DTPA soil test and ICPS. In *Advances in Soil Science,* vol. 16, ed. B. A. Stewart, pp. 165–90. Berlin: Springer-Verlag.

Soltanpour, P. N., Jones, J. B., Jr., and Workman, S. M. (1982). Optical emission spectrometry. In *Methods of Soil Analysis. Agronomy 9, Part 2,* 2nd ed., ed. A. L. Page, pp. 29–65. Madison, WI: American Society of Agronomy.

Soltanpour, P. N., and Schwab, A. P. (1977). A new soil test for simultaneous extraction of macro and micro-nutrients in alkaline soils. *Commun. Soil Sci. Plant Anal.* 8:195–207.

Soon, Y. K., and Bates, T. E. (1985). Molybdenum, cobalt, and boron uptake from sewage-sludge-amended soils. *Can. J. Soil Sci.* 65:507–17.

Thom, W. O. (1987). *Land Application of Sewage Sludge.* AGR-1, Department of Agronomy, University of Kentucky, Lexington.

U.S. Environmental Protection Agency. (1993). Standards for the use or disposal of sewage sludge. *Federal Register* 58:9248–415.

Viets, F. E. (1962). Chemistry and availability of micronutrients. *J. Agric. Food Chem.* 10:174.

Vlek, P. L. G. (1975). The chemistry, availability, and mobility of molybdenum in Colorado soils. Ph.D. dissertation, Colorado State University, Fort Collins.

Vlek, P. L. G., and Lindsay, W. L. (1977). Molybdenum contamination in Colorado pasture soils. In *Molybdenum in the Environment,* vol. 2, ed. W. B. Chappell and K. K. Petersen, pp. 619–50. New York: Marcel Dekker.

Wang, L., Reddy, K. J., and Munn, L. C. (1994). Comparison of ammonium bicarbonate–DTPA, ammonium carbonate, and ammonium oxalate to assess the availability of molybdenum in mine spoils and soils. *Commun. Soil Sci. Plant Anal.* 25:523–36.

Williams, C. H. (1955). A colorimetric method for the determination of molybdenum in soils. *J. Sci. Food Agric.* 6:104–10.

John L. Sims and Frieda Eivazi

Wilson, D. O. (1979). Determination of molybdenum in wet-ashed digests of plant material using flameless atomic absorption. *Commun. Soil Sci. Plant Anal.* 10:1319–30.

Wilson, D. O. (1992). Determination of molybdenum in plant tissue using the graphite furnace. In *Plant Analysis Reference Procedures for the Southern Region of the United States*, ed. C. O. Plank, pp. 67–70. University of Georgia Southern Cooperative Series bulletin 368.

Zbiral, J. (1992). Determination of molybdenum in hot-water soil extracts: influence of pH and available iron on the molybdenum content. *Commun. Soil Sci. Plant Anal.* 23:817–25.

8

Molybdenum Availability in Alkaline Soils

C. P. SHARMA and C. CHATTERJEE

Introduction

Molybdenum (Mo) is essential for higher plants (Arnon and Stout, 1939). As a constituent of nitrate reductase, sulfite reductase, and, in nodulated legumes, nitrogenase and xanthine oxidase, it is involved in electron-transfer reactions. Nitrate reductase, which catalyzes the conversion of nitrate to nitrite by transfer of electrons from Mo to nitrate, is an inducible enzyme that depends on Mo for its synthesis. There is a close relationship involving Mo supply, nitrate reductase activity, and plant growth. In the nitrogenase of legumes and other plants that fix atmospheric nitrogen, Mo is directly involved in the reduction of N_2 (Witt and Jungk, 1977). As a constituent of xanthine oxidase it catalyzes the oxidation of xanthine to uric acid, which is a precursor of allantoin and allantoic acid in most tropical and subtropical legumes. Besides playing a catalytic role, Mo is required for plant reproductive development (Agarwala et al., 1979) and nodulation of leguminous and nonleguminous N-fixing plants (Becking, 1961).

Compared with those for other micronutrients, the physiological requirements for Mo are very low (Stout and Meagher, 1948; Hewitt, 1963). The adequate concentrations of Mo in most crops range between 0.1 and 1.0 mg kg^{-1} (Bergmann, 1988), but the threshold for Mo toxicity in plants is much higher than that for any other micronutrient. Some plants can accumulate 1,000 times their adequate concentrations of Mo without exhibiting any visible signs of toxicity: cotton (*Gossypium* spp.), tomato (*Lycopersicon esculentum* Mill.), and *Phaseolus* beans (Joham, 1953; Johnson, 1966; Widdowson, 1966). In general, legumes will accumulate more Mo than grasses (Allaway, 1977; Singh, Kumar, and Lakhan, 1995).

131

In spite of the fact that the Mo requirements of plants are low and their thresholds for toxicity are high, Mo deficiencies and toxicities in plants are not uncommon. Whereas plants grown on well-drained acid soils tend to suffer from Mo deficiency, forage crops grown on alkaline soils tend to accumulate Mo in such large concentrations that it may be toxic to grazing cattle. To a large extent, those problems are attributed to the effects of soil reactions on the Mo availability in soils. This chapter reviews the published reports on the availability of Mo in alkaline soils and plants grown thereon. The different methods for determining the available Mo are compared in terms of their correlation with plant Mo uptake. The contributions of different soil factors that influence the availability of Mo in alkaline soils are also discussed.

Molybdenum Uptake by Plants and Its Availability in Soils

Although Mo is a metal, it is unlike the other metallic micronutrients [iron (Fe), manganese (Mn), copper (Cu), and zinc (Zn)] in that plants absorb Mo from the soil as an anion (MoO_4^{2-}). The primary source of Mo in soils is molybdenite (MoS_2). Other Mo-bearing minerals that contribute to soil Mo are powellite ($CaMoO_4$), wulfenite ($PbMoO_4$), and ferrimolybdite [$Fe_2(MoO_4)_3 \cdot 8H_2O$] (Goldschmidt, 1954). Molybdenum bound to organic matter may also contribute to the available Mo in soils.

In soils rich in iron, the most common molybdate is $Fe_2(MoO_4)_3$, and in soils rich in aluminum it is $Al_2(MoO_4)_3$. The presence of calcium, lead, copper, or iron cations (Ca^{2+}, Pb^{2+}, Cu^{2+}, Fe^{2+}) in soils in appreciable amounts may lead to precipitation of the molybdates.

Because oxides and hydroxides of iron and aluminum exist in diverse forms, Mo can be adsorbed to them in various ways. This may involve covalent bonding or an exchange with the surface hydroxyls (Ellis and Knezek, 1973). Compared with adsorption on hydrous Fe oxides, the strength of adsorption to Al oxide is much weaker (Jones, 1956, 1957).

Availability of Molybdenum in Alkaline Soils

The availability of Mo to plants depends on the amounts of Mo present in three forms: (1) soluble Mo, (2) adsorbed Mo, and (3) firmly held Mo (Barrow, 1977). Molybdenum in solution is in equilibrium with adsorbed Mo. Firmly held Mo is not in direct equilibrium with the molybdate in solution; it can undergo a change to the adsorbed form (and can change back again) and thus can indirectly affect the available Mo concentration.

Adsorption of Mo to hydrous Fe and Al oxides is pH-dependent (Reisenauer, Tabikh, and Stout, 1962), and the rate of adsorption is highest at acidic pH. It decreases with increases in pH from 4.45 to 7.5. Compared with acid soils, alkaline soils are high in soluble or available Mo (Davis, 1956). As in any other soil, in alkaline soil the availability of Mo is also influenced by several other factors. The relative Mo content of the parent rock, the process of soil evolution, and the physicochemical attributes of the soil (pH, calcareousness, organic-matter content, cation-exchange capacity, texture, moisture, relative concentrations of other mineral elements) all influence the availability of Mo.

In the semiarid tropics, where alkaline soils cover large areas, soil alkalinity is largely a function of high base saturation, particularly with sodium. These soils are therefore essentially sodic. Alkalinity is mainly due to the presence of the carbonates of sodium and calcium and bicarbonate of sodium. The sodium in the exchange complex is positively correlated to pH (Agarwala et al., 1964, 1978; Agarwal, Yadav, and Gupta, 1982). The sodic soils of the arid and semiarid regions of the world often contain high amounts of neutral soluble salts, dominated by chlorides and sulfates (saline-sodic soils), or high amounts of calcium carbonate (calcareous soils). The alkaline soils thus present wide ranges of ionic compositions and exchange capacities that influence the availability of Mo and modify the alkalinity (high-pH) effect.

Evaluation of Mo availability in alkaline soils has not received as much attention as in acid soils, wherein Mo deficiency can limit crop production. Because most plants have a high threshold for Mo toxicity, relatively large accumulations of Mo in plants go unnoticed. The concern about the available Mo in alkaline soils arises primarily because of its toxicity to ruminants feeding on plants that accumulate excessive concentrations of Mo. Cattle grazing on pastures or fodder containing excessive ($>5\,mg\,kg^{-1}$) concentrations of Mo can develop the clinical symptoms of Mo toxicity (molybdenosis), also referred to as "teart" disease (Dick, 1956). An intake of high concentrations of Mo interferes with assimilation and utilization of copper (Cu) (Thomson, Thornton, and Webb, 1972). The antagonism between Mo and Cu can have serious consequences in alkaline soils, whose high pH values lead to high availability of Mo and low availability of Cu.

Table 8.1. *Methods for assessing the availability of Mo in soils*

Method/extractant	Reference
Bioassay using *A. niger* (M.) strain	Mulder (1948)
1-M neutral normal ammonium acetate	Fujimoto and Sherman (1951)
Acid ammonium oxalate (AAO), pH 3.3	Grigg (1953)
Water	Lavy and Barber (1964),
	Gupta and MacKay (1965)
Hot water (Soxhlet)	Lowe and Massey (1965)
Anion-exchange resin	Bhella and Dawson (1972)
1-M ammonium carbonate, $(NH_4)_2CO_3$	Vlek and Lindsay (1977)
Ammonium acetate–EDTA	Williams and Thornton (1972)
Ammonium bicarbonate–DTPA	Soltanpour (1985)
(AB-DTPA)	

Assessment of Molybdenum Availability Using Different Extractants

There is no specific method that is best for assessing Mo availability in alkaline soils. In practically all of the reports on the status of the available Mo in alkaline soils, use has been made of one of the several methods essentially developed for determining available Mo in non-alkaline soils (Table 8.1). The principles and detailed procedures for such tests have been described in recent works on soil testing (Cox and Kamprath, 1972; Cox, 1987; Johnson and Fixen, 1990; Sims and Johnson, 1991) and have been described in relation to acid soils in Chapter 7 of this volume.

The bioassay for Mo initially developed by Mulder (1948) has been tested and found to be a suitable quantitative method for determining the Mo available in soils in nanogram quantities (Nicholas and Fielding, 1951; Thornton, 1953). The technique for culturing *Aspergillus niger* for Mo studies has been described by Agarwala et al. (1986). For Mo bioassay, the *Aspergillus niger* (M.) strain is grown in 500-mL Erlenmeyer flasks containing 50 mL of culture solution composed of glucose ($50 g L^{-1}$), KNO_3 ($5 g L^{-1}$), $MgSO_4 \cdot 7H_2O$ ($1 g L^{-1}$), $Ca(NO_3)_2$ ($0.6 g L^{-1}$), NaH_2PO_4 ($0.6 g L^{-1}$), Fe ($1.00 mg L^{-1}$), Mn ($0.07 mg L^{-1}$), Cu ($0.15 mg L^{-1}$), Zn ($0.45 mg L^{-1}$), Co ($0.02 mg L^{-1}$), and Ni ($0.02 mg L^{-1}$), adjusted to pH 2.0.

A known amount of the test soil, not exceeding 1 g, is added to the culture medium, and the medium is inoculated with a thick drop of spore suspension of *A. niger* with a platinum wire. The cultures are incubated

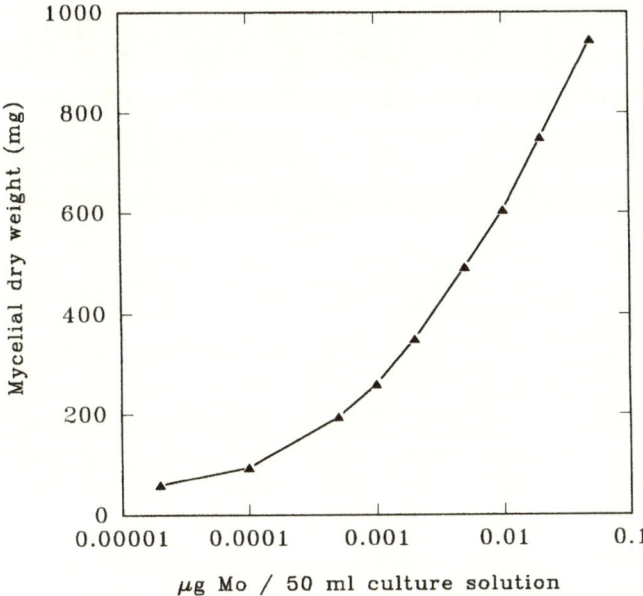

Figure 8.1. Relationship between molybdenum supply and mycelial dry weight of *A. niger* (M.) strain.

at 27°C for 5 days. At the end of 5 days, the mycelium is carefully washed free from the medium and is dried at 70°C for 48 hours. The mycelial yield is a function of the Mo it derives from the soil. The amount of Mo the fungus obtains (i.e., the available Mo) is ascertained with reference to a standard growth series run simultaneously with the test series. The standard series is prepared by culturing the fungus at graded levels of Mo (ranging from 0.0001 to 0.05 μg per 50 mL of culture solution) (Figure 8.1). Visual comparison of the growth of mycelium and the development of color of the spores between the standard and test samples gives a good estimate of the available Mo in the soil (Figure 8.2).

For assessing the availability of Mo in alkaline soils, various extractants have been used, including water-saturation extract, hot water (Lowe and Massey, 1965), acid ammonium oxalate (Grigg, 1953), ammonium carbonate (Vlek and Lindsay, 1977), and ammonium bicarbonate–DTPA (Soltanpour, 1985), as well as the bioassay procedure using *A. niger* (Mulder, 1948).

Barshad (1951a) found a good relationship between the Mo content of pasture plants and water-soluble Mo in soils with pH values ranging from

Figure 8.2. Molybdenum bioassay standard series. Arranged left to right are *A. niger* (M) cultures grown with nil, 0.0001, 0.0002, 0.001, 0.002, 0.005, 0.01, 0.02, and 0.05 µg Mo per 50 mL of culture solution.

4.7 to 7.5. In more alkaline soils (pH > 7.5), the water-soluble Mo content was even higher, but plant uptake of Mo was retarded. Agarwala et al. (1964, 1978) investigated the relationship between Mo in water-saturation extracts of naturally occurring alkaline (sodic) soils in Uttar Pradesh (India) and plant uptake of Mo. They failed to find any significant correlation between Mo content in water-saturation extracts and Mo uptake by rice (*Oryza sativa* L.) (Agarwala et al., 1964), but in the alkaline soils undergoing sullage reclamation, Mo uptake by barley (*Hordeum vulgare* L.) was positively correlated ($r = .43$) with Mo content in water-saturation extracts (Agarwala et al., 1978).

Ever since Grigg (1953) described a method for determining the available Mo in soils by extraction with 0.275-M acid ammonium oxalate (AAO) at pH 3.3, that method has continued to be widely used for determining the availability of Mo in alkaline soils. Agarwala et al. (1964, 1978) studied the relationships between AAO-extractable Mo and Mo uptake by rice (*Oryza sativa* L.) and barley (Agarwala et al., 1978) grown on alkaline soils in Uttar Pradesh. In those soils the AAO-extractable Mo did not show any significant correlation with plant Mo. Mortvedt and Anderson (1982) compared the amounts of AAO-extractable Mo and the Mo concentrations in forage legumes grown on 11 sites in five states in the southern United States. They also failed to find any significant correlation between AAO-extractable Mo and plant Mo when the latter was positively correlated with soil pH. On the other hand, some workers have found good correlations between AAO-extractable Mo and plant uptake of Mo. Pasricha and Randhawa (1971) found that the AAO-extractable Mo in recently reclaimed saline-sodic soils of the Sangrur district of Punjab (India) was significantly positively correlated ($r = .687$) with the Mo content in the top parts of berseem

(*Trifolium alexandrium* L.) grown thereon. Plants grown on poorly drained alkaline calcareous soils having AAO-extractable Mo concentrations of 0.15 mg kg^{-1} or more were found to accumulate Mo at concentrations of more than 10 mg kg^{-1}, concentrations reported to be toxic to ruminants (Dick, 1956).

Vlek and Lindsay (1977) examined the suitability of 1-M ammonium carbonate as an extractant for available Mo and found it to be a good indicator of Mo availability in neutral and alkaline soils.

Several workers have used more than one method of extracting soil Mo to determine the correlation between available soil Mo and plant Mo (Roschach, 1961; Agarwala et al., 1964, 1978; Lowe and Massey, 1965; Pathak, Shankar, and Misra, 1969; Verma and Jha, 1970; Karimian and Cox, 1979; Lombin, 1985; Pierzynski and Jacobs, 1986; Wang, Reddy, and Munn, 1994).

Lowe and Massey (1965) compared the continuous-leaching hot-water-extraction procedure developed by them with the AAO extraction procedure of Grigg (1953). They found the Mo concentration in alfalfa (*Medicago sativa* L.) to be better correlated with hot-water-extractable Mo than with AAO-extractable Mo. Pathak et al. (1969) used both of those methods for evaluating the availability of Mo in predominantly alkaline soils in the Uttar Pradesh province of India. Their findings were in agreement with those of Lowe and Massey (1965). Whereas the Mo concentration in alfalfa grown on those soils was significantly correlated with hot-water-extractable soil Mo ($r = .62**$), correlation between AAO-extractable Mo and plant Mo was very poor (Table 8.2). It has been suggested that hot-water extraction solubilizes as much Mo as is absorbed by the plants. On the other hand, the use of AAO, being a stronger extractant of Mo, leads to overextraction of soil Mo (Pathak et al., 1969).

Roschach (1961) reported good correlation between AAO-extractable Mo and Mo determined by the bioassay method using *A. niger*. Verma and Jha (1970) used the AAO extraction method and the bioassay method to assess the Mo status of 15 soils in Bihar (India) that varied from acidic (pH < 6.5) to alkaline (pH > 7.5). The available Mo, as determined by the two methods, was compared with the plant-tissue Mo concentrations in soybean [*Glycine max* (L.) Merr.] and groundnut (*Arachis hypogaea* L.) grown on those soils (Table 8.2). The bioassayed soil Mo was strongly correlated with plant-tissue Mo concentration in groundnut. The AAO-extractable Mo was correlated significantly with Mo concentration in groundnut, but not in soybean. Of the two methods

Table 8.2. *Correlations between plant uptake of Mo and available soil Mo using different methods*

Crop	Extractants	Correlation coefficient	Reference
Rice (*Oryza sativa* L.)	Water-saturation extract	−0.11	Agarwala et al. (1964)
	AAO	0.20	
Barley (*Hordeum vulgare* L.)	Water-saturation extract	0.43	Agarwala et al. (1978)
	AAO	0.20	
Alfalfa (*Medicago sativa* L.)	Hot water	0.62	Pathak et al. (1968)
	AAO	0.20	
Soybean [*Glycine max* (L.) Merr.]	AAO	NS	Verma and Jha (1970)
	Bioassay	0.77	
Groundnut (*Arachis hypogaea* L.)	AAO	0.54	
	Bioassay	0.69	
Alfalfa (*Medicago sativa* L.)	AAO	0.98	Wang et al. (1994)
	$(NH_4)_2CO_3$	0.98	
	AB-DTPA	0.98	
Crested wheatgrass (*Agropyron cristatum* L.)	AAO	0.77	
	$(NH_4)_2CO_3$	0.78	
	AB-DTPA	0.78	

for evaluating available Mo, bioassay with *A. niger* was found to be better correlated with plant Mo uptake.

In a comprehensive report on worldwide studies of micronutrients and the nutrient status of soils, prepared for the Food and Agriculture Organization (FAO) (Rome, Italy) and the Institute of Soil Science (Finland), Sillanpaa (1982) described the status of AAO-extractable Mo determinations. On a worldwide basis, determinations of AAO-extractable Mo have not been found to give a sufficiently reliable index of Mo availability to plants.

Pierzynski and Jacobs (1986) tested the suitability of AAO extraction and ammonium bicarbonate–diethylenetriamine pentaacetic acid (AB-DTPA) extraction for assessing available soil Mo. The procedure of AB-DTPA-extractable Mo was found to be more suitable for predicting Mo responses than was the AAO extraction procedure.

Wang et al. (1994) carried out pot experiments to evaluate the AB-DTPA, ammonium carbonate, and AAO extraction methods for assessing the availability of Mo in mine spoils and soils of high pH. The concentrations of available soil Mo as determined by the three extractants were significantly correlated with Mo uptake by alfalfa and crested wheatgrass (*Agropyron cristatum* L.) (Table 8.2). Although the three extractants were equally effective for assessing Mo availability in soils, the values obtained with AAO extraction were higher than the values from the two other extractants, possibly because oxalate, being a strong extractant of Mo, may solubilize some Mo adsorbed to the oxides of Fe and Al.

In determining the critical limits on predictions of field responses to Mo applications, Grigg (1953) had taken into account the pH factor, which gave the AAO extraction procedure a high rating in regard to the predictability of Mo responses. However, that was limited to soils with pH values ranging from acid to neutral. In the pH range encountered in alkaline soils, particularly where sodium dominates the adsorbed cations (sodic soils), causing soil pH values as high as 10 or more, AAO extraction may not be as suitable as it is in acid soils.

The procedure of resin extraction of Mo proposed by Bhella and Dawson (1972) has not been well tested for evaluating Mo availability in alkaline soils. A comparison between that method and the AAO extraction (Karimian and Cox, 1979) did not show any positive correlation between the Mo extracted by either of the methods and plant Mo uptake. Lombin (1985) tried the resin extraction procedure for assessing the availability of Mo in semiarid savannah soils of Nigeria, but did not find it suitable. Compared with resin extraction, AAO extraction showed good correlation with Mo uptake by peanuts, provided that soil organic-matter content was used as an independent variable.

Factors Affecting the Available Molybdenum in Alkaline Soils

Although pH is the predominant factor affecting the availability of Mo in alkaline soils, there are several other factors that influence its availability. Soil calcareousness, cation-exchange capacity (CEC), organic-matter content, and soil moisture are all known to affect the availability of Mo in alkaline soils.

Soil Reaction

Because Mo is weakly adsorbed on soils and hydrous oxides of Fe (Karimian and Cox, 1979) at alkaline pH, these soils have a relatively large proportion of Mo in the solution phase (Reisenauer et al., 1962).

Information about the relationship between Mo distribution and pH, in the range encountered in alkaline soils, is very limited. Prasad and Pagel (1976) reported that the arid and subarid soils of the tropics are high in available Mo, as determined by extraction with ammonium acetate at pH 4.0 (Mo at $0.216\,mg\,kg^{-1}$), and that the amount of Mo in those soils extractable with ammonium acetate was positively correlated with soil pH.

In India, several groups of workers have studied the relationship between pH or soil sodicity, expressed in terms of exchangeable sodium percentage (ESP), and Mo availability in soils or plant uptake of Mo. The high pH of the alkaline soils on the northern plains of India is largely a function of sodicity, salinity, and calcareousness. In those soils, pH and ESP show highly significant positive correlations (Agarwala et al., 1964, 1978; Agarwal et al., 1982).

In the alkali soils of Uttar Pradesh, Mo availability is positively correlated with soil pH (Singh and Singh, 1966; Pathak et al., 1968; Gupta and Dabas, 1980). Grewal, Bhumbla, and Randhawa (1969) also reported positive correlations between pH and available soil Mo in the alkaline soils of Punjab (region now in Pakistan). Rai et al. (1970) found a significant positive correlation ($r = .465***$) between pH and AAO-extractable Mo in the predominantly alkaline and calcareous black soils of Madhya Pradesh. Agarwala et al. (1978) found Mo concentrations in barley plants grown on alkaline soils in Uttar Pradesh undergoing sullage reclamation to be significantly correlated with pH and with ESP ($r = .42*$).

Wu, Son, and Chen (1986) studied the relationship between pH and adsorption capacity of soils in the Jilin province in China and found it to vary with soil type. At alkaline pH, Mo was most strongly adsorbed in chernozems, paddy soils, and dark brunisolic soils.

Even though a high pH favors an increase in Mo availability, plants grown on alkaline soils may respond to Mo fertilization because of the low content of Mo in the parent rocks and the adverse effects of the physicochemical attributes of such soils on the availability of Mo (Sharma et al., 1978).

Soil Calcareousness

The beneficial effect of liming on the availability of Mo is well established. Kavimandan, Badhe, and Ballal (1964) observed a positive correlation, approaching significance, between the available Mo and the

Table 8.3. *Effects of alkalinization of irrigation waters through NaHCO$_3$ addition (at constant CaCl$_2$ and MgSO$_4$) on Mo uptake by 4-week-old rice plants grown on soils of varying CaCO$_3$ contents and CEC values*

CEC	CaCO$_3$ (%)	Mo content (mg per plant) for the following alkalinities (SAR values):					
		<1	1	5	10	20	40
5.6	0.1	0.67	1.28	2.70	2.81	3.70	4.49
	2.1	4.48	3.55	3.35	3.45	3.45	4.39
13.5	1.8	2.36	2.00	1.99	1.42	1.04	1.78

Source: Adapted from Mehrotra and Agarwala (1979).

content of CaCO$_3$ in the alkaline soils of Vidarbha, India. In a pot culture study, Mehrotra and Agarwala (1979) investigated the effects of alkalinity, induced through the use of irrigation waters with a high sodium-adsorption ratio (SAR), on Mo uptake by rice plants grown on soils with varying CaCO$_3$ contents and CEC values. Their data (Table 8.3) show that in moderately alkaline soils (SAR < 5), Mo uptake from calcareous soils (CaCO$_3$, 2.14%) was much higher than that from noncalcareous soils (CaCO$_3$, 0.13%). Increases in alkalinity caused increases in Mo uptake from the noncalcareous soils (CaCO$_3$, 0.13%), but not from the calcareous soils (CaCO$_3$, 2.14%), thus indicating a strong interaction between alkalinity and calcareousness.

Rai et al. (1970) studied the AAO-extractable Mo content in deep black soils in Madhya Pradesh (pH ranging from 6.6 to 8.6, averaging 7.6, and CaCO$_3$ ranging from 1.25% to 10.77%, averaging 3.21%). The AAO-extractable Mo in those soils showed a significant positive correlation with CaCO$_3$ content ($r = .435$***). Verma, Mehta, and Singh (1982) and Singh et al. (1995) reported positive but nonsignificant correlations between the AAO-extractable Mo and the CaCO$_3$ content of those alkaline soils.

Cation-Exchange Capacity

The CEC is a major factor affecting Mo availability in alkaline soils. Prasad and Pagel (1976) reported a positive correlation between CEC and the amount of Mo extractable with ammonium acetate, at pH 4.0, for the arid and semiarid soils of the tropics. Mehrotra and Agarwala (1979) reported that the Mo-uptake responses to alkalinity by rice plants varied with the CEC of the soils (Table 8.3). An increase in alkalinity increased the Mo uptake by rice plants in low-CEC (4.8 mEq per 100 g) soils, but decreased it in high-CEC (13.5 mEq per 100 g) soils.

Organic Matter

There is some evidence to suggest that Mo may accumulate in the organic matter in a soil (Karimian and Cox, 1978). This is particularly so if soil drainage is impeded (Kubota et al., 1961). The availability of Mo from an Mo–organic-matter complex may depend on soil pH. The relationship between organic matter and available Mo in alkaline soils has not been well investigated. Kavimandan et al. (1964) reported a highly significant positive correlation between available Mo and organic-matter content in the alkaline soils of Vidarbha. Singh et al. (1995) found a significant positive correlation ($r = .38***$) between AAO-extractable Mo and organic-carbon content in the inceptisols of the Agra district of Uttar Pradesh, where pH values ranged from 7 to 9. Contrary to that, Pathak et al. (1968) observed a negative correlation between organic-matter content and available Mo in the saline-sodic soils of the northern plains of India.

Soil Moisture

High moisture contents in alkaline soils have been shown to increase the availability of Mo (Nayyar, 1972). Poonamperuma (1972) observed an increase in soluble-Mo concentration under wetland rice conditions. Flooding of soils increases the availability of Mo. Soils in the tropics that remain flooded for considerable periods have relatively high contents of Mo (Lopes, 1980), because a high moisture content in the soil reduces ferric iron to the ferrous form, which favors fixation of Mo (Jones, 1956).

Soil Texture

Molybdenum is adsorbed as an anion in the clay fraction, and that reduces its availability in fine-textured soils. Singh and Singh (1966) observed a significant negative correlation between available Mo and the silt or clay content in alkali soils. Prasad and Pagel (1976) also found a negative correlation between available Mo and the clay content of arid and subarid soils in the tropics. Other workers, however, have failed to find a consistent relationship or significant correlation between available Mo and soil texture (Barshad, 1951a; Kavimandan et al., 1964; Pathak et al., 1969).

There is some evidence showing differences in the available Mo contents of surface and subsurface soils, but there is little agreement about the relative concentrations of Mo in the two horizons. Pasricha and Randhawa (1971) and Agarwala et al. (1978) reported higher concentrations of AAO-extractable Mo in the subsurface horizon (15–30 cm) than in the surface horizon (0–15 cm) in the sodic soils of northern India. Contrary to that, Singh and Singh (1966), Grewal et al. (1969), and Gupta and Dabas (1980) found higher concentrations of AAO-extractable Mo in surface soils than in subsurface soils. The solubilization of organically bound Mo at high pH and the drainage conditions of the soil may account for the differences in the depthwise distributions of Mo in soils.

Nutrient Interactions

The availability of Mo to plants is influenced by the relative concentrations of other plant nutrients, which in turn can be affected by the several soil factors discussed earlier. Sulfate decreases plant uptake of Mo (Stout et al., 1951; Reisenauer, 1963). Sulfur fertilization of crops leads to decreases in Mo accumulation in plants (Barshad, 1951b; Stout et al., 1951; Reisenauer, 1963; Gupta and Cutcliffe, 1968; Gupta and Munro, 1969; Pasricha and Randhawa, 1972). There is antagonism between Mo and Cu (MacKay, Chipman, and Gupta, 1966; Butorec, 1981). Many alkaline soils are low in sulfur and copper, and application of each of these will reduce excessive accumulation of Mo (Pasricha and Randhawa, 1971; Gupta and Ram, 1980). In general, the nutrient interactions involving Mo under field conditions have received little attention (Chatterjee, 1992).

Sillanpaa (1982) presented a worldwide analysis of the contributions of six major soil factors that affect Mo availability: pH, texture, organic-carbon content, CEC, electrical conductivity, and $CaCO_3$ content. That study showed that all of those factors (except CEC) affect AAO-extractable soil Mo and plant Mo content to varying extents. Increasing the soil pH increased the AAO-extractable Mo, up to pH 5–6. That was followed by a decrease, reaching a minimum at pH 7.5–7.7, which was then followed by another pH-dependent increase. In contrast, the Mo contents of plants increased up to pH 7.5–7.7 (i.e., the pH level at which soil Mo reaches a minimum). Thus we see that the different soil factors affect the relationship between available soil Mo and plant uptake of Mo, illustrating the need for in-depth study of the correlations between the individual factors in determining the availability of Mo in alkaline soils.

Summary

Several chemical extraction procedures have been developed to determine the availability of Mo in soils. These procedures are intended to facilitate evaluation of soils and to determine the critical limits of deficiency and sufficiency of Mo and toxicity from Mo. The AAO extraction procedure for determining the available Mo, developed by Grigg (1953) for acid soils, has found wide use for assessing Mo availability in alkaline soils. This method has an advantage in that oxalate has a high buffering capacity and forms a strong complex with molybdate. Because of this, there is little change in the natural pH during extraction with AAO, and a single equilibrium extraction suffices for determination of plant-available Mo. The strong coordination complex between oxalic acid and Mo, however, can lead to overestimation of the plant-available Mo in the pH range encountered in alkaline soils, where the high alkalinity is a function of the high amount of sodium in the exchange complex. This test, therefore, needs to be calibrated for alkaline soils, taking into consideration the pH factor, as has been done for acid soils (Grigg, 1953). The hot-water-extraction method and the bioassay method using *A. niger* give good correlations with plant Mo concentrations in alkaline soils, but these have not been used extensively for determining the critical limits for Mo deficiency and toxicity from Mo as has been done for the ammonium carbonate, ammonium acetate, and AB-DTPA extractions.

Because of its good buffering capacity and the high pH (9.0) at which hydroxyl (OH^-) and carbonate (CO_3^-) ions are exchanged for adsorbed Mo, the method of ammonium carbonate extraction should be suitable for alkaline soils, but this needs to be tested under field conditions.

The three universal extractants, ammonium carbonate, ammonium acetate, and AB-DTPA, have an advantage in that the extracts can be read directly for their Mo concentrations by graphite-furnace atomic-absorption spectrometry, direct-current plasma-emission spectrometry, or inductively coupled plasma (ICP) spectroscopy. The ICP method has a distinct advantage in that it allows rapid multielement analysis of a single soil extract.

The chemistry of Mo makes a larger proportion of adsorbed molybdates soluble and available at alkaline pH than at acid pH. However, pH is only one of several factors that affect the availability of Mo. Assessment of Mo availability in an alkaline soil, as in any other soil, necessitates determination of the critical limits for Mo in soils and in plants based on specific soil–plant systems. The determination of the critical limits should make allowance for the dominant soil factors that affect availability and for the plant efficiency of absorption and utilization of Mo. Use of multiple-regression equations to account for the contributions of the individual factors (Sillanpaa, 1982) will make the critical limits more predictable.

References

Agarwal, R. R., Yadav, J. S. P., and Gupta, R. N. (1982). *Saline and Alkali Soils of India*. New Delhi: Indian Council of Agricultural Research.

Agarwala, S. C., Chatterjee, C., Nautiyal, N., and Sharma, C. P. (1986). Molybdenum nutrition of isolates of four *Aspergillus* species. *Can. J. Microbiol.* 32:557–61.

Agarwala, S. C., Chatterjee, C., Sharma, P. N., Sharma, C. P., and Nautiyal, N. (1979). Pollen development in maize plants subjected to Mo deficiency. *Can. J. Bot.* 57:1946–50.

Agarwala, S. C., Mehrotra, N. K., Sharma, C. P., and Sinha, B. K. (1978). Plant nutrient behavior with reference to barley growing (salt affected) soils under sullage reclamation. *J. Indian Soc. Soil Sci.* 26:372–9.

Agarwala, S. C., Sharma, C. P., Sinha, B. K., and Mehrotra, N. K. (1964). Soil–plant relationship with particular reference to trace elements in *usar* soils of Uttar Pradesh. *J. Indian Soc. Soil Sci.* 12:343–54.

Allaway, W. H. (1977). Perspectives on Mo in soils and plants. In *Molybdenum in the Environment*, vol. 1, ed. W. R. Chappell and K. K. Petersen, pp. 317–39. New York: Marcel Dekker.

Arnon, D. I., and Stout, P. R. (1939). Molybdenum as an essential element for higher plants. *Plant Physiol.* 14:599–602.

Barrow, N. J. (1977). Factors affecting the molybdenum status of soils. In *Molybdenum in the Environment*, vol. 2, ed. W. R. Chappell and K. K. Petersen, pp. 583–95. New York: Marcel Dekker.

Barshad, I. (1951a). Factors affecting the molybdenum content of pasture plants. 1. Nature of molybdenum, growth of plants, and soil pH. *Soil Sci.* 71:297–313.

Barshad, I. (1951b). Factors affecting the molybdenum content of pasture plants 2. Effect of soluble phosphates, available nitrogen and soluble sulfates. *Soil Sci.* 71:387–98.

Becking, J. H. (1961). A requirement of molybdenum for the symbiotic nitrogen fixation in alder. *Plant Soil* 15:217–27.

Bergmann, W. (1988). *Ernährungsstorungen bei Kulturpflanzen, Entstehung, vissuelle und analytische Diagnose.* Jena: Gustav Fisher Verlag.

Bhella, H. S., and Dawson, M. D. (1972). The use of anion exchange resin for determining available soil molybdenum. *Proc. Soil Sci. Soc. Am.* 36:177–8.

Butorec, K. (1981). Response of leucerne (*Medicago sativa* L.) to nitrogen molybdenum and copper fertilization. *Poljopr. Znan. Smotra* 55:269–87.

Chatterjee, C. (1992). Interactions of copper or molybdenum with other nutrients. In *Management of Nutrient Interactions in Agriculture*, ed. H. L. S. Tandon, pp. 78–95. New Delhi: Fertilizer Development and Consultation Organization.

Cox, F. R. (1987). *Micronutrient Soil Test: Correlation and Calibration Soil Testing: Sample Correlation, Calibration and Interpretation*, pp. 97–117. SSSA special publication 21. Madison, WI: Soil Science Society of America.

Cox, F. R., and Kamprath, E. J. (1972). Micronutrient soil tests. In *Micronutrients in Agriculture*, 2nd ed., ed. J. J. Mortvedt et al., pp. 289–317. Madison WI: Soil Science Society of America.

Davis, E. B. (1956). Factors affecting molybdenum availability in soils. *Soil Sci.* 81:209–22.

Dick, A. T. (1956). Molybdenum in animal nutrition. *Soil Sci.* 81:229–36.

Ellis, B.G., and Knezek, B. D. (1973). Adsorption reactions of micronutrients in soils. In *Micronutrients in Agriculture*, ed. J. J. Mortvedt et al., pp. 59–78. Madison, WI: Soil Science Society of America.

Fujimoto, G., and Sherman, G. D. (1951). Molybdenum content of typical soils and plants of the Hawaiian island. *Agronomy J.* 43:424–9.

Goldschmidt, V. M. (1954). *Geochemistry.* Oxford: Clarendon Press.

Grewal, J. S., Bhumbla, D. R., and Randhawa, N. S. (1969). Available micronutrient status of Punjab, Haryana and Himachal soils. *J. Indian Soc. Soil Sci.* 17:27–31.

Grigg, J. L. (1953). Determination of the available molybdenum of soils. *N.Z. J. Sci. Technol.* A34:405–14.

Gupta, U. C., and Cutcliffe, J. A. (1968). Influence of phosphorus on molybdenum content of Brussels sprouts under field and greenhouse conditions and on recovery of added molybdenum in soil. *Can. J. Soil Sci.* 48:117–23.

Gupta, U. C., and MacKay, D. C. (1965). Extraction of water-soluble copper and molybdenum from podzol soils. *Proc. Soil Sci. Soc. Am.* 29:323.

Gupta, U. C., and Munro, D. C. (1969). Influence of sulfur, molybdenum and phosphorus on chemical composition and yields of Brussels sprouts and of molybdenum on sulfur contents of several plant species grown in the green house. *Soil Sci.* 107:114–18.

Gupta, V. K., and Dabas, D. S. (1980). Distribution of molybdenum in some saline-sodic soils from Haryana. *J. Indian Soc. Soil Sci.* 28:28–30.

Gupta, V. K., and Ram, K. (1980). Effect of copper and molybdenum on the copper/molybdenum ratio and their concentration in different organs of cowpea (*Vigna sinensis* L.). *Plant Soil* 56:235–41.

Hewitt, E. J. (1963). Mineral nutrition of plants in culture media. In *Plant Physiology. Vol. III: Inorganic Nutrition of Plants*, ed. F. C. Steward, pp. 97–133. New York: Academic Press.

Joham, H. J. (1953). Accumulation and distribution of molybdenum in the cotton plant. *Plant Physiol.* 28:275–80.

Johnson, C. M. (1966). Molybdenum. In *Diagnostic Criteria for Plants and Soils*, ed. H. D. Chapman, pp. 286–306. University of California, Division of Agricultural Sciences.

Johnson, G. V., and Fixen, P. E. (1990). Testing soils for sulfur, boron, molybdenum, and chlorine. In *Soil Testing and Plant Analysis*, 3rd ed., ed. R. L. Westerman et al., pp. 266–71. Madison WI: Soil Science Society of America.

Jones, L. H. P. (1956). Interaction of molybdenum and iron in soil. *Science* 123:1116.

Jones, L. H. P. (1957). Solubility of molybdenum in simplified systems and aqueous soil suspensions. *J. Soil Sci.* 8:313–27.

Karimian, N., and Cox, F. R. (1978). Absorption and extractability of molybdenum in relation to some chemical properties of soil. *Soil Sci. Soc. Am. J.* 42:757–61.

Karimian, N., and Cox, F. R. (1979). Molybdenum availability as predicted from selected chemical properties. *Agronomy J.* 71:63–5.

Kavimandan, S. K., Badhe, N. N., and Ballal, D. K. (1964). Available copper and molybdenum in Vidarbha soils. *J. Indian Soc. Soil Sci.* 12:281–8.

Kubota, J., Lazar, V. A., Langan, L. N., and Beeson, K. C. (1961). The relationship of soils to molybdenum toxicity in cattle in Nevada. *Proc. Soil Sci. Soc. Am.* 25:227–32.

Lavy, T. L., and Barber, S. A. (1964). Movement of molybdenum in the soil and its effect on availability to plants. *Proc. Soil Sci. Soc. Am.* 28:93–7.

Lombin, G. (1985). Micronutrient soil tests for the semi-arid savannah of Nigeria: boron and molybdenum. *Soil Sci. Plant Nutr. (Tokyo)* 31:1–11.

Lopes, A. S. (1980). Micronutrients in the soils of the tropics as constraints to food production. In *Soil Related Constraints to Food Production in the Tropics*, pp. 277–98. Manila: International Rice Research Institute.

Lowe, R. H., and Massey, H. F. (1965). Hot water extraction for available soil molybdenum. *Soil Sci.* 100:238–43.

MacKay, D. C., Chipman, E. W., and Gupta, U. C. (1966). Copper and molybdenum nutrition of crops grown on acid sphagnum peat soil. *Proc. Soil Sci. Soc. Am.* 30:755–9.

Mehrotra, N. K., and Agarwala, S. C. (1979). Nutrient composition of rice plants subjected to high alkalinity (SAR) in irrigation waters and soil calcareousness. In *Micronutrients in Agriculture*, ed. S. C. Agarwala and C. P. Sharma, pp. 95–102. Lucknow, India: University of Lucknow.

Mortvedt, J. J., and Anderson, O. E. (1982). *Forage legumes: Diagnosis and Correction of Molybdenum and Manganese Problems*. University of Georgia Southern Cooperative Series bulletin 278.

Mulder, E. G. (1948). Importance of molybdenum in the nitrogen metabolism of microorganisms and higher plants. *Plant Soil* 1:94–119.

Nayyar, V. K. (1972). Studies on Mo and Cu in calcareous flood-plain soils of Punjab. Ph.D. thesis, Punjab Agricultural University, Ludhiana, India.

Nicholas, D. J. D., and Fielding, A. H. (1951). The use of *Aspergillus niger* (M) for determination of magnesium, zinc, copper and molybdenum available in soil to crop plants. *J. Hort. Sci.* 26:125–47.

Pasricha, N. S., and Randhawa, N. S. (1971). Available molybdenum status of some recently reclaimed saline-sodic soils and its effect on concentration of molybdenum, copper, sulphur and nitrogen in berseem (*Trifolium alexandrinum*) grown on these soils. In *Proceedings of the International*

Symposium on Soil Fertility Evaluation, vol. 1, pp. 1017–25. New Delhi: International Society of Soil Science.

Pasricha, N. S., and Randhawa, N. S. (1972). Interaction effect of sulphur and molybdenum on the uptake and utilization of these elements by raya (*Brassica juncea* L.). *Plant Soil* 37:215–20.

Pathak, A. N., Shankar, H., and Misra, R. V. (1968). Molybdenum status of certain Uttar Pradesh soils. *J. Indian Soc. Soil Sci.* 16:399–404.

Pathak, A. N., Shankar, H., and Misra, R. V. (1969). Correlation of available molybdenum values obtained by different methods of molybdenum uptake by alfalfa. *J. Indian Soc. Soil Sci.* 17:151–3.

Pierzynski, G. M., and Jacobs, L. W. (1986). Extractability and plant availability from inorganic and sewage sludge sources. *J. Environ. Qual.* 15:323–6.

Poonamperuma, F. N. (1972). The chemistry of submerged soils. *Adv. Agron.* 24:29–96.

Prasad, R. N., and Pagel, H. (1976). Comparative investigations into the content of available molybdenum in important soils of the arid and semi arid tropics. *Beitr. Trop. Landwirtsch. Veterinarmed.* 14:79–87.

Rai, M. M., Sitholey, D. B., Pal, A. R., Vakil, P., and Gupta, S. K. (1970). Available micronutrient status of deep black soils of Madhya Pradesh. *J. Indian Soc. Soil Sci.* 18:383–9.

Reisenauer, H. M. (1963). The effect of sulphur on the absorption and utilization of molybdenum by peas. *Proc. Soil. Soc. Am.* 27:553–5.

Reisenauer, H. M., Tabikh, A. A., and Stout, P. R. (1962). Molybdenum reactions with soils and hydrous oxides of iron, aluminium and titanium. *Proc. Soil Sci. Soc. Am.* 26:23–7.

Roschach, H. (1961). Experiences with the *Aspergillus* method in determining plant-available molybdenum. *Z. Pflanzenernaehr. Dueng.* 94:134–40.

Sharma, P. N., Chatterjee, C., Sharma, C. P., Mehrotra, N. K., and Agarwala, S. C. (1978). Availability of molybdenum for plant growth in major soil types of Uttar Pradesh. *Geophytology* 8:1–9.

Sillanpaa, M. (1982). *Micronutrients and the Nutrient Status of Soil: A Global Study*. FAO soils bulletin 48. Rome: Food and Agriculture Organization.

Sims, J. T., and Johnson, G. V. (1991). Micronutrient soil tests. In *Micronutrients in Agriculture*, 2nd ed., ed. J. J. Mortvedt et al., pp. 427–76. Madison, WI: Soil Science Society of America.

Singh, S., and Singh, B. (1966). Trace element studies on some alkali and adjoining soils of Uttar Pradesh. I. Profile distribution of molybdenum. *J. Indian Soc. Soil Sci.* 14:19–25.

Singh, V., Kumar, R., and Lakhan, R. (1995). Molybdenum status in soils and common fodders. *J. Indian Soc. Soil Sci.* 43:135–6.

Soltanpour, P. N. (1985). Use of ammonium bicarbonate–DTPA soil test to evaluate elemental availability and toxicity. *Commun. Soil Sci. Plant Anal.* 16:323–38.

Stout, P. R., and Meagher, W. R. (1948). Studies of the molybdenum nutrition of plants with radioactive molybdenum. *Science* 108:471–3.

Stout, P. R., Meagher, W. R., Pearson, G. A., and Johnson, C. M. (1951). Molybdenum nutrition of crop plants. I. The influence of phosphate on the absorption of molybdenum from soils and solution cultures. *Plant Soil* 68:287–91.

Thomson, I., Thornton, I., and Webb, J. S. (1972). Molybdenum in black shales and the bovine hypocuprosis. *J. Sci. Food Agric.* 23:879–91.

Thornton, R. H. (1953). *Aspergillus niger* and the estimation of molybdenum available in soil. *N.Z. Soil News* 3:42–4.
Verma, K. P., and Jha, K. K. (1970). Studies on soil molybdenum of Bihar. *J. Indian Soc. Soil Sci.* 18:37–9.
Verma, S. R., Mehta, V. S., and Singh, V. (1982). Molybdenum in soil and plant samples of Agra district. *Indian J. Agric. Sci.* 52:845–7.
Vlek, P. L. G., and Lindsay, W. L. (1977). Molybdenum contamination in Colorado pasture soils. In *Molybdenum in the Environment*, vol. 2, ed. W. R. Chappell and K. K. Petersen, pp. 619–50. New York: Marcel Dekker.
Wang, L., Reddy, K. J., and Munn, L. C. (1994). Comparison of ammonium bicarbonate–DTPA, ammonium carbonate and ammonium oxalate to assess the availability of molybdenum in mine spoils and soils. *Commun. Soil Sci. Plant Anal.* 25:523–36.
Widdowson, J. P. (1966). Molybdenum uptake by french beans on two recent soils. *N.Z. J. Agric. Res.* 9:59–67.
Williams, C., and Thornton, I. (1972). The effect of soil additions on uptake of molybdenum and selenium from different soil environments. *Plant Soil* 36:345–406.
Witt, H. H., and Jungk, A. (1977). Buerteilung der Molybdanversorgung von Pflanzen mit Hilfe der Mo-induzierbaren Nitratereduktase-Activitat. *Z. Pflanzenernaehr. Bodenkd.* 140:209–22.
Wu, D. H., Son, W. T., and Chen, J. R. (1986). A study of molybdenum adsorption characteristics of soils of Jilin province. *J. Soil Sci. (China)* 17:231–3.

9

Deficient, Sufficient, and Toxic Concentrations of Molybdenum in Crops

UMESH C. GUPTA

Introduction

Among the micronutrients essential for plant growth, molybdenum (Mo) is required in the smallest amounts. In most soils, the Mo requirements of plants can be met by liming the soil. Because of its low requirement, the deficiency and sufficiency concentrations of Mo in most plants are extremely small. Molybdenum toxicity to plants under field conditions seldom occurs, and usually it can be induced only under extreme experimental conditions (Johnson, 1966). Therefore, this chapter does not place great emphasis on the toxic concentrations of Mo in plants. However, plants can, under certain conditions, accumulate large concentrations of Mo and induce molybdenosis in ruminants that eat such material. That will be dealt with in Chapter 15.

The precision of modern analytical methods is such that even microquantities of Mo in plants can be detected accurately, and considerable data have accumulated over the past 40 years regarding Mo concentrations in a number of plant species. The purpose of this chapter is to report the sufficient, deficient, and toxic concentrations of Mo in a number of cultivated crop species as found by workers around the world.

Usually when one talks about the deficient, sufficient, and toxic concentrations of nutrients in crops, there is a range of values, rather than one definite number that can be considered as critical. Therefore, use of the term "critical concentration" in crops is somewhat misleading. A nutrient concentration considered critical by workers in some areas may not be critical under conditions in other areas. Likewise, use of the term "optimum concentration" of a nutrient, as used in the literature by some researchers to express a relationship to maximum crop yield, sometimes is not clear. Theoretically, such a concentration for a given nutrient should be sufficient to produce the best possible growth of a crop. Often a single value is published as the "optimum" concentration, when a range of concentrations would be equally appropriate; however, in practice, no single number or even a very narrow range of numbers can adequately describe this relationship. Thus, a range of values is more appropriate to describe the nutrient status of the crop. Dow and Roberts (1982) stated that a single point to describe a critical nutrient concentration is difficult to establish experimentally and can vary under different conditions. Those researchers likewise considered it desirable to deal with a critical range of nutrient concentrations rather than with a single concentration. They defined the critical nutrient range as that range of nutrient concentrations above which we can be reasonably confident that the crop is amply supplied with the nutrient, and below which we can be reasonably confident that the crop is deficient in the nutrient.

Therefore, in this chapter, wherever possible, the term "sufficient" or "sufficiency" will be used, rather than "critical" or "optimum."

Deficient, Sufficient, and Toxic Concentrations for Molybdenum in Crops

The deficient, sufficient, and toxic concentrations of Mo for specific crops, as reported by various researchers, are given in Table 9.1. The deficient plant Mo concentrations as reported were associated with plant disorders and/or reductions in the yields of crops. For some crops, the deficient and optimum Mo concentrations seem to differ markedly. Differences in the techniques used in various laboratories cannot be ruled out.

Toxic concentrations of plant Mo differ according to the crop species. Plants belonging to the dicotyledonous species generally are less tolerant of excess plant Mo than are the monocotyledonous species (Gramineae) (Kluge, 1983). Sometimes there may be slight toxicity symptoms on the foliage even though the yields may not be affected.

As is evident from Table 9.1, data on Mo deficiencies for many crops are missing, and the data on toxic Mo concentrations in crops are very scanty. Because of the very small requirements for Mo, there appear to be much greater variations in the deficiency and sufficiency concentrations of Mo, as shown by various workers, than for other micronutrients. Some of the differences likely are due to the differing procedures used to determine Mo. The percentage variations seem to be of much larger

Table 9.1. *Deficient, sufficient, and toxic concentrations of Mo in plants*

Crop	Part of plant tissue sampled	Mo in dry matter (ppm)			References
		Deficient	Sufficient	Toxic	
Alfalfa (*Medicago sativa* L.)	Leaves at 10% bloom	0.26–0.28	0.34		Reisenauer (1956)
	Whole plants at harvest	0.05			Bergmann (1983)
	Top 15.2 cm of plant prior to bloom	0.55–1.15			Evans and Purvis (1951)
		<0.4	1–5		Jones (1967)
	Upper-stem cuttings at early flowering stage		0.5[a]		Melsted et al. (1969)
	Shortly before flowering (top 1/3 of plants)	<0.2			Neubert et al. (1970)
	Whole tops at 10% bloom		0.5–5.0 0.12–1.29		Gupta (1970)
Barley (*Hordeum vulgare* L.), summer	Blades at 8 weeks old		0.03–0.07		Johnson et al. (1952)
	Whole tops at boot stage		0.09–0.18		Gupta (1971)
	Grain		0.26–0.32		Gupta (1971)
Beans (*Phaseolus vulgaris*)	Tops at 8 weeks old		0.4		Johnson et al. (1952)
	Upper fully developed leaves	1–5		500–900	Kluge (1983) Hall & Schwartz (1993)
Beets (*Beta vulgaris* L.)	Tops at 8 weeks old	0.05	0.62		Johnson et al. (1952)
Broccoli (*Brassica oleracea* L. Botrytis Group)	Tops at 8 weeks old	0.04			Johnson et al. (1952)

Crop	Plant part / growth stage			Reference
Brussels sprouts (*Brassica oleracea* L. Gemmifera Group)	Whole plants when sprouts began to form	<0.08	0.16	Gupta and Munro (1969)
	Whole plants when sprouts began to form	0.09	0.11–0.69	Gupta (1970)
	Leaves		0.61	Plant (1952)
Cabbage (*Brassica oleracea* var. *capitata* L.)	Leaves	0.09	0.42	Plant (1952)
Carrots (*Daucus carota* L.)	Whole tops at maturity		0.04–0.15	Gupta and LeBlanc (1990)
Cauliflower (*Brassica oleracea* var. *botrytis* L.)	Above-ground portion of plants at first appearance of a curd	<0.26	0.68–1.49	Chipman et al. (1970)
	Whole plants before appearance of curd	<0.11	0.56	640[b]	Gupta et al. (1978) Gupta (1969)
		0.01–0.05			Duval et al. (1991)
	Young leaves showing whiptail	0.07		390[b]	MacKay et al. (1966)
	Leaves	0.02–0.07	0.19–0.25	Peterson & Purvis (1961) Plant (1951b)
Corn (*Zea mays* L.)	Roots	0.023–0.3	2.8–11.9	Dios & Broyer (1965)
	Stems	0.013–0.11	1.4–7.0	Dios & Broyer (1965)
	At tassel middle of the first leaf opposite and below the lower ear	<0.1	>0.2	Neubert et al. (1970)
	Ear leaf at silk		0.2[a]	Melsted et al. (1969)
	Fully developed leaves when plants 40–60 cm high	0.6–1.0		Vos (1993)
			0.2–0.5	Bergmann (1992)
Cotton (*Gossypium hirsutum*)	65-day-old leaf blades	0.5[a]		Amin & Joham (1960)
	Fully matured leaves at bloom to boll development		0.6–2.0	Bergmann (1992)

Table 9.1. *(cont.)*

Crop	Part of plant tissue sampled	Mo in dry matter (ppm)			References
		Deficient	Sufficient	Toxic	
Cucurbits: cucumber (*Cucumis sativus* L.), muskmelon (*Cucumis melo* var. *reticulatus*)	Fully mature whole leaves		0.5–1.0	1,000	Locascio (1993)
Lettuce (*Lactuca sativa* L.)	Leaves	0.06	0.08–0.14		Plant (1951a)
	Leaves at head formation			277[b]	Gupta et al. (1978)
	Vegetative material			100–200	Kluge (1983)
Onions (*Allium cepa* L.)	Whole tops at maturity	<0.06	0.1		Gupta & LeBlanc (1990)
	Leaves			640[b]	Gupta et al. (1978)
Pasture grass (Gramineae)	First cut at first bloom		0.2–0.7		Neubert et al. (1970)
Peanuts (*Arachis hypogaea* L.)	Upper fully developed leaves	<1			Smith et al. (1993)
	Uppermost fully developed leaves during blossom		0.5–1.0		Bergmann (1992)
Peas (*Pisum sativum*)	Recent fully developed leaves at onset of blossom		0.4–1.0		Bergmann (1992)
Potatoes (*Solanum tuberosum*)	Above-ground vegetative material			100–200	Kluge (1983)
	Leaf blades	0.15			Ulrich (1993)
	Fully developed leaves at early bloom		0.2–0.5		Bergmann (1992)
Red clover (*Trifolium pratense* L.)	Total above-ground plants at bloom	<0.15	0.3–1.59		Neubert et al. (1970)

Crop	Plant part / growth stage				Reference
Rice (*Oryza sativa*)	First cut at flowering	0.03	0.26		Hawes et al. (1976)
	Plants at 10% bloom	0.1–0.2	0.27		Gupta & LeBlanc (1990)
	Whole plants at bud stage		0.45		Hagstrom & Berger (1965)
	Whole plants at bud stage		0.46–1.08		Gupta (1970)
	Whole plants at bud stage			192^b	Gupta et al. (1978)
	Uppermost fully developed leaves before flowering		0.4–1.0		Bergmann (1992)
Soybeans [*Glycine max* (L.) Merr.]	Plants when 26–28cm high	0.19			Peterson & Purvis (1961)
	Fully developed leaves at the top at the end of blossom		0.5–1.0		Bergmann (1992)
Spinach (*Spinacia oleracea* L.)	Leaves at 8 weeks old	0.1	1.61		Johnson et al. (1952)
	Whole tops at normal maturity		0.15–1.09		Gupta (1970)
Sugar beets (*Beta vulgaris* L.)	Leaf blades shortly after symptoms appear	0.01–0.15	0.2–20.0		Ulrich & Hills (1973)
		<0.1	0.2–2.0		Neubert et al. (1970)
	Fully developed leaf without stem (taken end June or early July)		>20		Kluge (1983)
		0.01–0.15	0.2–20.0	100–200	Ulrich et al. (1973)
Subterranean clover (*Trifolium subterraneum* L.)	Leaflets and petioles		0.1		Petrie & Jackson (1982)
Temperate-pasture legumes	Plant shoots		>0.1		Johansen (1978)
Timothy (*Phleum pratense* L.)	Whole tops at prebloom, head fully emerged from the panicle	0.11			Gupta & MacKay (1968)
	Whole aerial part of plants at onset of blossom		0.15–0.50		Bergmann (1992)

Table 9.1. (cont.)

Crop	Part of plant tissue sampled	Mo in dry matter (ppm)			References
		Deficient	Sufficient	Toxic	
Tobacco (*Nicotiana tabacum* L.)	Leaves at 8 weeks old		1.08		Johnson et al. (1952)
Burley tobacco	Cured leaves after harvest 50-day-old whole plants	0.42 0.38			Sims & Atkinson (1976) Sims et al. (1975)
Tomatoes (*Lycopersicon esculentum* Mill.)	Leaves at 8 weeks old	0.13	0.68		Johnson et al. (1952)
Tropical-pasture legumes in mixture with *Panicum maximum* cv. Gatton	Plant shoots		>0.02		Johansen (1978)
Wheat (*Triticum aestivum* L.)	Whole tops at boot stage Grain		0.09–0.18 0.16–0.20		Gupta (1971) Gupta (1971)
Winter wheat (*Triticum aestivum* L.)	Above-ground plants at ear emergence, when 40cm high Above-ground portion when 5-8cm high		>0.3 0.1–0.3	600–1,000	Neubert et al. (1970) Kluge (1983) Bergmann (1992)

[a] Considered critical.
[b] High but not toxic to crop yield.

magnitude, but the actual differences seem to be on the order of about 0.5 ppm or less.

Summary

This chapter provides a summary of the deficient, sufficient, and toxic concentrations of Mo in a variety of plant species, as reported by various researchers in different parts of the world. The Mo deficiency concentrations for several crops are still missing because Mo deficiencies have never been reported in those crops. Molybdenum toxicity to plants seldom occurs, and therefore data on the toxic concentrations of Mo in plant tissue are extremely limited. Some of the Mo concentrations reported have varied widely from one laboratory to another. Selection of the plant parts tested, the stage of plant growth at sampling, and the analytical technique used are some of the important reasons for such discrepancies. There is a need to establish specific guidelines regarding the stage of growth at the time of sampling, the method for sampling the plant parts, and the method for determining Mo, so that the reported concentrations of Mo can be compared universally.

References

Amin, J. V., and Joham, H. E. (1960). Growth of cotton as influenced by low substrate molybdenum. *Soil Sci.* 89:101–7.

Bergmann, W. (1983). Molybdenum. In *Farbatlas Ernährungsstörungen bei Kulturpflanzen für den Gebrauch im Feldbestand*, ed. W. Bergmann, pp. 146–56. Jena: VEB Gustav Fischer Verlag.

Bergmann, W. (ed.) (1992). *Nutritional Disorders of Plants. Visual and Analytical Diagnosis.* Jena: Gustav Fischer Verlag.

Chipman, E. W., MacKay, D. C., Gupta, U. C., and Cannon, H. B. (1970). Response of cauliflower cultivars to molybdenum deficiency. *Can. J. Plant Sci.* 50:163–7.

Dios, R. V., and Broyer, T. V. I. (1965). Deficiency symptoms and essentiality of molybdenum in crop hybrids. *Agrochimica* 9:273–84.

Dow, A. I., and Roberts, S. (1982). Proposal: critical nutrient ranges for crop diagnosis. *Agronomy J.* 74:401–3.

Duval, L., More, E., and Sicot, A. (1991). Observations on molybdenum deficiency in cauliflower in Brittany, *C. R. Acad. Agric. France* 78:27–34.

Evans, H. J., and Purvis, E. R. (1951). Molybdenum status of some New Jersey soils with respect to alfalfa production. *Agronomy J.* 43:70–1.

Gupta, U. C. (1969). Effect and interaction of molybdenum and limestone on growth and molybdenum content of cauliflower, alfalfa and bromegrass on acid soils. *Proc. Soil Sci. Soc. Am.* 33:929–32.

Gupta, U. C. (1970). Molybdenum requirement of crops grown on a sandy clay loam soil in the greenhouse. *Soil Sci.* 110:280–2.

Gupta, U. C. (1971). Boron and molybdenum nutrition of wheat, barley, and oats grown in Prince Edward Island soils. *Can. J. Soil Sci.* 51:415–22.

Gupta, U. C., Chipman, E. W., and MacKay, D. C. (1978). Effects of molybdenum and lime on the yield and molybdenum concentration of crops grown on acid sphagnum peat soil. *Can. J. Plant Sci.* 58:983–92.

Gupta, U. C., and LeBlanc, P. V. (1990). Effect of molybdenum applications on plant molybdenum concentration and crop yields on sphagnum peat soils. *Can. J. Plant Sci.* 70:717–21.

Gupta, U. C., and MacKay, D. C. (1968). Crop responses to applied molybdenum and copper on podzol soils. *Can. J. Soil Sci.* 48:235–42.

Gupta, U. C., and Munro, D. C. (1969). Influence of sulfur, molybdenum and phosphorus on chemical composition and yields of Brussels sprouts and of molybdenum on sulfur contents of several plant species grown in the greenhouse. *Soil Sci.* 107:114–18.

Hagstrom, G. R., and Berger, K. C. (1965). Molybdenum deficiencies of Wisconsin soils. *Soil Sci.* 100:52–6.

Hall, R., and Schwartz, H. F. (1993). Common bean. In *Nutrient Deficiencies and Toxicities in Crop Plants*, ed. W. F. Bennett, pp. 143–7. St. Paul, MN: APS Press.

Hawes, R. L., Sims, J. L., and Wells, K. L. (1976). Molybdenum concentration of certain crop species as influenced by previous applications of molybdenum fertilizer. *Agronomy J.* 68:217–18.

Johansen, C. (1978). Comparative molybdenum concentrations in some tropical pasture legumes. *Commun. Soil Sci. Plant Anal.* 9:1009–17.

Johnson, C. M. (1966). Molybdenum. In *Diagnostic Criteria for Plants and Soils*, ed. H. D. Chapman, pp. 286–301. Riverside, CA: University of California Press.

Johnson, C. M., Pearson, G. A., and Stout, P. R. (1952). Molybdenum nutrition of crop plants. II. Plant and soil factors concerned with molybdenum deficiencies in crop plants. *Plant Soil* 4:178–96.

Jones, J. B., Jr. (1967). Interpretation of plant analysis for several agronomic crops. In *Soil Testing and Plant Analyses*, ed. G. W. Hardy et al., pp. 49–58. Madison, WI: Soil Science Society of America.

Kluge, R. (1983). Molybdenum toxicity in plants. In *Proceedings Mengen- und Spurenelemente Arbeitstagung, Leipzig*, ed. M. Anke et al., pp. 10–17. Jena: Institut für Pflanzenernährung. Reprinted *Moly. Agric.* 7(1): 7, 1986 (Micronutrient Bureau, U.K.).

Locascio, S. J. (1993). Cucurbits: Cucumber, muskmelon, and watermelon. In *Nutrient Deficiencies and Toxicities in Crop Plants*, ed. W. F. Bennett, pp. 123–30. St. Paul, MN: APS Press.

MacKay, D. C., Chipman, E. W., and Gupta, U. C. (1966). Copper and molybdenum nutrition of crops grown on sphagnum peat soil. *Soil Sci. Soc. Am. Proc.* 30:755–9.

Melsted, S. W., Motto, H. L., and Peck, T. R. (1969). Critical plant nutrient composition values useful in interpreting plant analysis data. *Agronomy J.* 61:17–20.

Neubert, P., Wrazidlo, W., Vielemeyer, H. P., Hundt, I., Gollmick, F., and Bergmann, W. (1970). *Tabellen zur pflanzenanalyse – Erste orientierende "Übersicht,"* pp. 1–40. Jena: Institut für Pflanzenernährung.

Peterson, N. K., and Purvis, E. R. (1961). Development of molybdenum deficiency symptoms in certain crop plants. *Soil Sci. Soc. Am. Proc.* 25:111–17.

Petri, S. E., and Jackson, T. L. (1982). Effects of lime, P, and Mo application on Mo concentration in sub-clover. *Agronomy J.* 74:1077–81.

Plant, W. (1951a). The control of molybdenum deficiency in lettuce under field conditions. *Long Ashton Res. Sta. Ann. Rep.*, pp. 113–15.

Plant, W. (1951b). The control of "whiptail" in broccoli and cauliflower. *J. Hort. Sci.* 26:109–17.

Plant, W. (1952). The molybdenum content of some brassicae and lettuce crops in acid soils. *Trans. Int. Soc. Soil Sci. Comm. II–IV* 2:176–7.

Reisenauer, H. M. (1956). Molybdenum content of alfalfa in relation to deficiency symptoms and response to molybdenum fertilization. *Soil Sci.* 81:237–42.

Sims, J. L., and Atkinson, W. O. (1976). Lime, molybdenum and nitrogen source effects on yield and selected chemical components of burley tobacco. *Agronomy J.* 68:239–42.

Sims, J. L., Atkinson, W. O., and Smithbol, C. (1975). Mo and N effects on growth, yield, and Mo composition of burley tobacco. *Agronomy J.* 67:824–8.

Smith, D. H., Wells, M. A., Porter, D. M., and Cox, F. R. (1993). Peanuts. In *Nutrient Deficiencies and Toxicities in Crop Plants*, ed. W. F. Bennett, pp. 105–10. St. Paul, MN: APS Press.

Ulrich, A. (1993). Potato. In *Nutrient Deficiencies and Toxicities in Crop Plants*, ed. W. F. Bennett, pp. 149–56. St. Paul, MN: APS Press.

Ulrich, A., and Hills, F. J. (1973). Plant analysis as an aid in fertilizing sugar crops. Part I. Sugar beets. In *Soil Testing and Plant Analysis*, 2nd ed., ed. L. M. Walsh and J. D. Beaton, pp. 271–88. Madison, WI: Soil Science Society of America.

Ulrich, A., Monaghan, J. T., and Whitney, E. D. (1973). Sugarbeets. In *Nutrient Deficiencies and Toxicities in Crop Plants*, ed. W. F. Bennett, pp. 91–8. St. Paul, MN: APS Press.

Vos, R. D. (1993). Corn. In *Nutrient Deficiencies and Toxicities in Plants*, ed. W. F. Bennett, pp. 11–14. St. Paul, MN: APS Press.

10

Symptoms of Molybdenum Deficiency and Toxicity in Crops

UMESH C. GUPTA

Introduction

The symptoms of molybdenum (Mo) deficiency are common in certain crops under certain soil and climatic conditions. However, Mo toxicity is uncommon and is found only when unusually high concentrations of Mo are present.

Deficiency symptoms for most micronutrients appear on the young leaves at the top of the plant, because most micronutrients are not readily translocated. Molybdenum is an exception in that it is readily translocated, and its deficiency symptoms generally appear on the whole plant.

The symptoms associated with deficiency of Mo are closely related to nitrogen (N) metabolism. Because Mo is needed for nitrogenase activity, Mo deficiency prevents the fixation of N_2. This process involves higher plants and symbiotic organisms such as *Rhizobium*, and so Mo deficiency can produce symptoms associated with a deficiency of nitrogen. If the Mo is needed directly by the plant for nitrate reductase, then symptoms peculiar to Mo occur, although essentially the plant can be considered to be suffering from a shortage of protein due to failure of the initial processes of NO_3^- reduction.

The first type of symptom can be relieved by supplying fixed nitrogen. Even NO_3^- may serve the purpose, because the Mo requirement for nitrate reductase is lower than that for nitrogenase. The second type of symptom, however, indicates severe deficiency, and addition of NO_3^- in that case is likely to make things worse, because the plant is unable to process it, and excess accumulation of NO_3^- may occur as a symptom of Mo deficiency (Anderson, 1956; Freney and Lipsett, 1965). Accumulation of NO_3^- in plants can cause toxicity (Hanway and Englehorn, 1958; Moore and Hutchings, 1967); however, Hewitt and Jones (1948) earlier reported that the effects of true nitrogen deficiency differ from those of Mo deficiency in certain respects. Thus, Mo deficiency does not induce the tinting associated with nitrogen deficiency, and the latter does not cause the characteristic injury to leaves or growing points. The cause of this breakdown may perhaps be sought in the accumulation of nitrates or depletion of other nitrogenous metabolites.

The most striking symptom of Mo toxicity is a yellow or orange-yellow chlorosis, with some brownish tints that start in the youngest leaves (Bergmann, 1992). Further symptoms of damage include moribund buds, thick stems, and development of axillary buds and sometimes succulent older leaves.

This chapter presents the available information on Mo deficiencies for a variety of crops and Mo toxicities for a limited number of crops, as reported by researchers from various regions of the world.

Symptoms of Molybdenum Deficiency

Field and Horticultural Crops

Alfalfa (Medicago sativa L.)

Molybdenum deficiency is manifested as general yellowing of the whole plant (Bergmann, 1983).

Beans (Phaseolus vulgaris L.)

Molybdenum deficiency results in chlorosis, which is nearly always localized between the veins (Climax Molybdenum Company, 1956). In broad beans, the leaves are a paler, yellowish green and appear nitrogen-deficient (Hewitt and Jones, 1948).

Brussels Sprouts (Brassica oleracea L. Gemmifera Group)

Molybdenum deficiency results in cupped leaves, with reduced lamina (Bergmann, 1983). After high application of N fertilizer and a long dry-weather period, the margins of the light-green, youngest leaves will be "fired" because of accumulation of nitrates as a result of Mo deficiency (Bergmann, 1992).

Cauliflower (Brassica oleracea var. botrytis L.)

In the early stages of Mo deficiency there is marginal scorching, rolling or upward curling of leaves, and withering and crinkling of leaves (Vitosh et al., 1981). In later growth stages, the deficiency shows up as "whiptail," especially in younger leaves. Gupta (1969) observed cupping and drying of the edges of true leaves. During the later stages of growth those symptoms were associated with yellowing between the veins and leaves. There is distorted curding of cauliflower plants (Bergmann, 1983). Plant (1951b) stated that the chief symptom of whiptail is a reduction in the amount of lamina, usually with irregular and wavy margins. The lamina may be markedly thick and succulent or convoluted. Leaf color is generally not affected in whiptail, but some specimens are noticeably blue-green. Deficiency of Mo causes twisting of the youngest leaves (plants fail to heart) and chlorotic spots and blotches in the leaves, with the lamina of leaves restricted partly to the midrib (Bergmann, 1992).

Clover (Trifolium spp.)

Molybdenum deficiency in clover shows up as a general yellow or greenish yellow foliage color, stunting, and lack of vigor (Vitosh et al., 1981).

Corn (Zea mays L.)

Corn is rarely ever deficient in Mo. However, if deficiency occurs, older tips die at the ends, along the margins, and then between the veins (Vos, 1993). Molybdenum deficiency also has a striking effect on pollen formation. Tasseling, anthesis, and development of anthers in corn are inhibited by Mo deficiency (Römheld and Marschner, 1991).

With Mo deficiency, the internodes are short, the leaf area is reduced, and leaves appear chlorotic, which is more severe in younger leaves than in older leaves (Agarwala and Sharma, 1979). Agarwala and Sharma also reported a delay in cob emergence and reduction in cob size, and the cobs did not bear grains. Dios and Broyer (1965) reported symptoms of yellowish color in the young corn plants. These symptoms of Mo deficiency in corn differed from those in other plants in the lack of typical mottling and whitish appearance and softness of the edges. When there was severe Mo deficiency, young leaves failed to unroll, the older leaves shriveled, and the whole plant wilted and died (Weir, Noonan, and Boyle, 1966).

Cucumbers (Cucumis sativus L.)

Molybdenum deficiency resulted in upcurling and necrosis of the outer edges of cucumber leaves (Climax Molybdenum Company, 1956).

Lettuce (Lactuca sativa L.)

In the early stages of Mo deficiency, the plants were chlorotic and stunted (Plant, 1951a). The lamina failed to develop, and the leaves remained ovate rather than round as in healthy plants. Chlorosis of the leaves was followed by necrosis of the margins. With severe Mo deficiency, lettuce in the juvenile stage appears fired because of accumulation of nitrates (Bergmann, 1992).

Oats (Avena sativa L.)

In northern Tasmania, Mo deficiency in Algerian oats is known locally as "blue chaff disease" (Fricke, 1947). Affected plants show a bluish coloration of the outer glumes, and the grain is pinched. The Mo-deficient oat plants appear light green, as compared with the darker green of Mo-treated plots (Fitzgerald, 1954).

Onions (Allium cepa L.)

Molybdenum deficiency shows up as a dying of the leaf tips (Vitosh et al., 1981). Below the dead tips, the leaves show an inch or two of wilting. As the deficiency progresses, the wilting and dying advance down the leaves, and in severe cases the plant dies. Poor emergence and death of seedlings may occur on soils highly deficient in Mo (Bender, 1993).

Perennial Ryegrass (Lolium perenne L.)

In Mo-deficient ryegrass, growth is reduced, and the foliage turns pale green, suggestive of nitrogen deficiency, and there is considerable withering of older leaves (Hewitt and Jones, 1948).

Potatoes (Solanum tuberosum L.)

Molybdenum deficiency results in uniform yellowing of leaf blades, similar to that seen with nitrogen and sulfur deficiencies (Ulrich, 1993).

Red Clover (Trifolium pratense L.)

In Mo-deficient red clover, root-nodule weight was considerably lower in control plots than in those receiving Mo (Robinson, Lelacheur, and

Brossard, 1957). Studies by Hagstrom and Berger (1965) showed a yield response to Mo of 65% in red clover; however, the plants from check plots did not show any typical symptoms of nitrogen deficiency, and no differences in degree of nodulation were found. In severe cases, some symptoms of true Mo deficiency appear, such as incurled leaves and marginal scorch (Climax Molybdenum Company, 1956).

Rye (Secale cereale L.)

Molybdenum deficiency caused reduction in growth, and the foliage turned pale green. Leaves were white-tipped, with bleached irregular areas along lamina (Hewitt and Jones, 1948).

Spinach (Spinacea oleracea)

Severe symptoms of nitrate toxicity after high-nitrogen fertilization have been found to occur due to Mo deficiency (Bergmann, 1992).

Soybeans [Glycine max (L.) Merr.]

The symptoms of Mo deficiency resemble those of nitrogen deficiency and likely are caused by reduced nitrogen utilization. Deficiency reduces plant growth, number of pods, number of seeds per pod, seed size, nodulation, and total nitrogen and protein content of seeds (Sinclair, 1993). Leaves are pale green or yellow, necrotic and twisted, and the necrosis is confined largely to the margins, midribs, and interveinal areas. In Mo-deficient plots, the plants are stunted and quite yellow, as compared with plants receiving Mo or lime (Hagstrom and Berger, 1965). The symptoms resemble those of nitrogen starvation, which is quite typical of Mo-deficient legumes. The roots of Mo-deficient plants are very poorly nodulated (Hagstrom and Berger, 1965).

Sugar Beets (Beta vulgaris L.)

Symptoms of Mo deficiency first appear as a general yellowing, similar to that caused by sulfur deficiency or, to some extent, by nitrogen deficiency (Ulrich, Moraghan, and Whitney, 1993), and the center leaves are light green to yellow, as with sulfur deficiency, in marked contrast to the deep green of nitrogen deficiency. As the symptoms of Mo deficiency increase in severity, pitting develops along leaf veins. In Mo-deficient plants, the leaves turn golden yellow, with inward curling of margins, and leaves are arrow-shaped because of the curling of lamina (Agarwala and Sharma, 1979).

Sugarcane (Saccharum officinarum, S. spontaneum, and S. robustum)

In Mo-deficient sugarcane, older leaves die back, leaf beads are uniformly light green to yellow, stalks become slender, and vegetative growth rate is reduced (Gascho, Anderson, and Bowen, 1993). Short longitudinal chlorotic streaks begin on the apical third of the leaf, and older leaves dry prematurely from the middle toward the tip.

Tomatoes (Lycopersicon esculentum Mill.)

Molybdenum-deficient leaves show a characteristic cupping or curling of the margins, yellowing between the veins, and death of the growing tip (Wilcox, 1993). Marked yellow mottling between the veins and death of the growing tip have been reported (Climax Molybdenum Company, 1956).

Wheat (Triticum aestivum L.)

Molybdenum deficiency in wheat results in a golden yellow coloration of older leaves along the apex and the apical leaf margins (Agarwala and Sharma, 1979). Doyle et al. (1965) reported paler color and the presence of empty heads at maturity. Lipsett and Simpson (1971) described the symptoms of Mo deficiency in young wheat plants as white, necrotic areas extending back along the leaves from the tips. Normal tillering did not occur on affected plants, which were pale in color. Severely affected plants died. Mehrotra et al. (1982) reported marked decreases in tillering and leaf area, delay in flowering, and distortion of ears in wheat under conditions of Mo deficiency.

Other Crops

Cucurbits: Cucumber (Cucumis sativus L.), Muskmelon (Cucumis melo var. reticulatus L.), and Watermelon (Citrullus lanatus)

Molybdenum deficiency in the early stages in cucurbits leads to pale green leaves, with some interveinal mottling, and later the leaves become necrotic (Locascio, 1993). Fruit set is reduced, and younger leaves become cupped, twisted, and distorted.

Flax (Linum usitatissimum L.)

In Mo-deficient flax, growth is slightly reduced, shoots are pale, dull green in color, and upper leaves are pale and yellowish green at the tips, and subsequently the whole of the lamina becomes involved (Hewitt and Jones, 1948)

Kale (Brassica oleracea L. Acephala Group)

In Mo-deficient kale, growth is stunted, and the leaves are at first pale yellowish green, followed by diffuse bright yellow-green interveinal mottling and cupping of margins; scorching also occurs in extensive interveinal areas (Hewitt and Jones, 1948).

Melon (Cucumis melon L.)

Molybdenum deficiency results in poor and delayed flowering and reduced viability of the pollen grains, leading to reduction in fruit formation (Gubler, Grogan, and Osterli, 1982).

Tobacco (Nicotiana tabacum L.)

Molybdenum deficiency in burley tobacco results in chlorosis and spotting of lower leaves (Miner and Tucker, 1990). Older leaves show chlorosis and have reduced laminae (Bergmann, 1992).

Turfgrass

Symptoms of Mo deficiency typically are similar to those of nitrogen deficiency, although some interveinal chlorosis may occur (Turner, 1993).

Symptoms of Molybdenum Toxicity

Field and Horticultural Crops

Cauliflower (Brassica oleracea var. botrytis L.)

In conditions of Mo toxicity, young cauliflower seedlings turn purple (Vitosh et al., 1981). It was reported that application of 117g or more of Mo per kilogram of cauliflower seed resulted in phytotoxicity (Scheffer and Wilson, 1987). Such plants were smaller, and maturity was delayed.

Grapes (Vitis vinifera L.)

Molybdenum excess leads to a chlorosis that is difficult to differentiate visually from the symptoms of iron deficiency (Gärtel, 1993).

Kidney Beans (Phaseolus vulgaris var. nanus)

Molybdenum toxicity results in seeds of an intense yellow-orange color (Bergmann, 1992).

Lucerne (Medicago sativa L.)

Molybdenum toxicity produced light-colored leaves that turned golden yellow and then bronze (Falke, 1983). At very high rates of Mo, toxicity symptoms in alfalfa appeared as intense yellowing of leaves (Bergmann, 1992).

Sorghum [Sorghum bicolor (L.) Moench]

Symptoms of excess Mo in sorghum leaves appear as a dark violet coloration of the whole lamina and are distinguishable from the symptoms of phosphorus deficiency, which result in dark green leaves with overtones of dark red coloration (Clark, 1993).

Soybeans [Glycine max (L.) Merr.]

High rates of application of Mo can be detrimental to *Rhizobia* in the seed inoculum used in some pelleting procedures or can cause seedling injury under certain growing conditions (Sedberry, et al., 1973).

Tomatoes (Lycopersicon esculentum L.)

In a greenhouse experiment on Mo toxicity, tomato leaves turned golden yellow (Vitosh et al., 1981). In tomatoes receiving toxic amounts of Mo there are lots of anthocyanins in the cells, and purple tints also develop (Bergmann, 1992).

Other Crops

Flax (Linum usitatissimum L.)
High applications of Mo retarded the date of appearance and reduced the severity of lower-leaf necrosis, which is a characteristic symptom of the presence of excess Mn in the nutrient solution (Millikan, 1947). Excess Mo can also mitigate the deleterious effects of toxic levels of Mn, Zn, Cu, Ni, or Co.

Summary

This chapter describes Mo deficiency and toxicity symptoms in various crops as reported in the literature. Common symptoms of Mo deficiency in plants include a general yellowing and rolling, curling, and scorching of the leaves. In most cases, the symptoms are similar to those of nitrogen deficiency, because lack of Mo can interfere with nitrogen metabolism in plants. In severe cases of Mo deficiency, the plants may die. Plants can tolerate extraordinarily high amounts of Mo without exhibiting phytotoxicity, and Mo toxicity rarely occurs under field conditions. Large concentrations of Mo in feed crops can cause toxicity to animals, as discussed in Chapter 15. When Mo toxicity does occur, it is generally marked by chlorosis, yellowing, and other forms of leaf discoloration.

Acknowledgments

The author extends his gratitude and most sincere thanks to Barbara Burns of the Agriculture and Agri-Food Canada's Research Centre in Charlottetown for her tireless efforts in the typing of Chapters 1, 5, 9, 10, and 14 of this manuscript. Thanks are also due to the Charlottetown Research Centre librarian, Barrie Stanfield, for verifying and obtaining a number of scientific articles and books required in the preparation of the aforementioned chapters.

References

Agarwala, S. C., and Sharma, C. P. (1979). *Recognizing Micronutrient Disorders of Crop Plants on the Basis of Visible Symptoms and Plant Analysis*. Lucknow, India: Prem Printing Press.
Anderson, A. J. (1956). Molybdenum as a fertilizer. *Adv. Agron.* 8:163–202.

Bender, D. A. (1993). Onions. In *Nutrient Deficiencies and Toxicities in Crop Plants*, ed. W. F. Bennett, pp. 131–5. St. Paul, MN: APS Press.
Bergmann, W. (1983). Molybdenum. In *Farbatlas Ernährungsstörungen bei Kulturpflanzen für den Gebrauch im Feldbestand*, ed. W. Bergmann, pp. 146–56. Jena: VEB Gustav Fischer Verlag.
Bergmann, W. (ed.) (1992). *Nutritional Disorders of Plants. Visual and Analytical Diagnosis*. Jena: Gustav Fischer Verlag.
Clark, R. B. (1993). Sorghum. In *Nutrient Deficiencies and Toxicities in Crop Plants*, ed. W. F. Bennett, pp. 21–6. St. Paul, MN: APS Press.
Climax Molybdenum Company (1956). *Molybdenum Deficiency Symptoms in Crops*. New York: Climax.
Dios, R. V., and Broyer, T. V. I. (1965). Deficiency symptoms and essentiality of molybdenum in corn hybrids. *Agrochimica* 9:273–84.
Doyle, R. J., Parkin, R. J., Smith, J. A. C., and Gartrell, J. W. (1965). Molybdenum increases cereal yields on wheatbelt scrubplain. *J. Dept. Agric. West Aust.* 6:699–703.
Falke, H. (1983). Effect of increasing molybdenum applications on the molybdenum content of soils and plants In *Proceedings Mengen- und Spurenelemente Arbeitstagung*, ed. M. Anke et al., pp. 18–21. Jena: Institut für Pflanzenernährung.
Fitzgerald, J. N. (1954). Molybdenum on oats. *N.Z. J. Agric.* 89:619.
Freney, J. R., and Lipsett, J. (1965). Yield depression in wheat due to high nitrate applications, and its alleviation by molybdenum. *Nature (London)* 205:616–17.
Fricke, E. F. (1947). Molybdenum deficiency in oats. *J. Aust. Inst. Agric. Sci.* 13:75.
Gärtel, W. (1993). Grapes. In *Nutrient Deficiencies and Toxicities in Crop Plants*, ed. W. F. Bennett, pp. 177–83. St. Paul, MN: APS Press.
Gascho, G. J., Anderson, D. L., and Bowen, J. E. (1993). Sugarcane. In *Nutrient Deficiencies and Toxicities in Crop Plants*, ed. W. F. Bennett, pp. 37–42. St. Paul, MN: APS Press.
Gubler, W. D., Grogan, R. G., and Osterli, P. P. (1982). Yellows of melons caused by molybdenum deficiency in acid soil. *Plant Dis.* 66:449–51.
Gupta, U. C. (1969). Effect and interaction of molybdenum and limestone on growth and molybdenum content of cauliflower, alfalfa and bromegrass on acid soils. *Soil Sci. Soc. Am. Proc.* 33:929–32.
Hagstrom, G. R., and Berger, K. C. (1965). Molybdenum status of three Wisconsin soils and its effect on four legume crops. *Agronomy J.* 55:399–401.
Hanway, J. J., and Englehorn, A. J. (1958). Nitrate accumulation in some Iowa crop plants. *Agronomy J.* 50:331–4.
Hewitt, E. J., and Jones, E. W. (1948). Molybdenum as a plant nutrient. The effects of molybdenum deficiency on some vegetables, cereals and forage crops. *Long Ashton Res. Sta. Ann. Rep.*, pp. 81–90.
Lipsett, J., and Simpson, J. R. (1971). Wheat responses to molybdenum in southern New South Wales. *J. Aust. Inst. Agric. Sci.* 37:348–51.
Locascio, S. J. (1993). Cucurbits: cucumber, muskmelon, and watermelon. In *Nutrient Deficiencies and Toxicities in Crop Plants*, ed. W. F. Bennett, pp. 123–30. St. Paul, MN: APS Press.
Mehrotra, S. C., Chatterjee, C., Sharma, C. P., and Agarwala, S. C. (1982). Molybdenum deficiency effects in high yielding varieties of wheat and barley. *J. Indian Bot. Soc.* 61:317–19.

Millikan, C. R. (1947). Effect of molybdenum on the severity of toxicity symptoms in flax induced by an excess of either manganese, zinc, copper, nickel, or cobalt in the nutrient solution. *J. Aust. Inst. Agric. Sci.* 13:180–6.

Miner, G. S., and Tucker, M. R. (1990). Plant analysis as an aid in fertilizing tobacco. In *Soil Testing and Plant Analysis*, ed. R. L. Westermann, pp. 645–57. Madison, WI: Soil Science Society of America.

Moore, R. M., and Hutchings, R. J. (1967). Mortalities among sheep grazing *Phalaris tuberosa. Aust. J. Exp. Agric. Anim. Husb.* 7:17–21.

Plant, W. (1951a). The control of molybdenum deficiency in lettuce under field conditions. *Long Ashton Res. Sta. Ann. Rep.*, pp. 113–15.

Plant, W. (1951b). The control of "whiptail" in broccoli and cauliflower. *J. Hort. Sci.* 26:109–17.

Robinson, D. B., Lelacheur, K. E., and Brossard, G. A. (1957). Effect of molybdenum applications on leguminous hay crops in Prince Edward Island. *Can. J. Plant Sci.* 37:193–5.

Römheld, V., and Marschner, H. (1991). Function of micronutrients in plants. In *Micronutrients in Agriculture*, ed. J. J. Mortvedt, F. R. Cox, L. M. Shuman, and R. M. Welch, pp. 297–328. Madison, WI: Soil Science Society of America.

Scheffer, J. J. C., and Wilson, G. J. (1987). Cauliflower: molybdenum application using pelleted seed and foliar sprays. *N.Z. J. Exp. Agric.* 15: 485–90.

Sedberry, J. E., Jr., Sharmaputra, R. H., Brupbacher, S., Phillips, J. G., Marshall, J. G., Slvane, L. W., Melville, D. R., Ralb, J. I., and Davis, J. (1973). *Molybdenum Investigations with Soybeans in Louisiana.* Louisiana Agricultural Experiment Station bulletin 670.

Sinclair, J. B. (1993). Soybeans. In *Nutrient Deficiencies and Toxicities in Crop Plants*, ed. W. F. Bennett, pp. 99–103. St. Paul, MN: APS Press.

Turner, T. R. (1993). Turfgrass. In *Nutrient Deficiencies and Toxicities in Crop Plants*, ed. W. F. Bennett, pp. 187–96. St. Paul, MN: APS Press.

Ulrich, A. (1993). Potato. In *Nutrient Deficiencies and Toxicities in Crop Plants*, ed. W. F. Bennett, pp. 149–56. St. Paul, MN: APS Press.

Ulrich, A., Moraghan, J. T., and Whitney, E. D. (1993). Sugarbeet. In *Nutrient Deficiencies and Toxicities in Crop Plants*, ed. W. F. Bennett, pp. 91–8. St. Paul, MN: APS Press.

Vitosh, M. L., Warncke, D. D., Knezek, B. D., and Lucas, R. E. (1981). *Secondary and Micronutrients for Vegetable and Field Crops.* Bulletin E-486, Cooperative Extension Service, Michigan State University.

Vos, R. D. (1993). Corn. In *Nutrient Deficiencies and Toxicities in Crop Plants*, ed. W. F. Bennett, pp. 11–14. St. Paul, MN: APS Press.

Weir, R. G., Noonan, J. B., and Boyle, J. W. (1966). Molybdenum deficiency in maize. *Agric. Gaz. N.S.W.*, pp. 579–82.

Wilcox, G. E. (1993). Tomato. In *Nutrient Deficiencies and Toxicities in Crop Plants*, ed. W. F. Bennett, pp. 137–41. St. Paul, MN: APS Press.

11
Sources and Methods for Molybdenum Fertilization of Crops
JOHN J. MORTVEDT

Introduction

Molybdenum (Mo) deficiencies have been reported from many countries around the world, mainly in acidic soils. Sandy soils are Mo-deficient more often than are loam or clay soils. Most Mo deficiencies are associated with legume crops, because Mo is an essential constituent of enzymes necessary for fixation of nitrogen (N) by bacteria growing symbiotically with legumes. Molybdenum is also required in other enzyme systems in all plants.

The availability of Mo in soil increases with increasing soil pH. Therefore, liming a soil to the recommended pH range may increase the plant availability of soil Mo sufficiently that Mo fertilization may not be required. This chapter discusses the sources of Mo and the methods of applying Mo fertilizers to those crops that require additional Mo to produce optimum crop yields.

Molybdenum Sources

There are fewer sources of Mo for fertilizers than there are for the other micronutrients (Table 11.1). Ammonium and sodium molybdates and molybdic acid are soluble compounds. These sources of Mo are sometimes applied with other fertilizers or are used as foliar sprays. Both MoO_3 and Mo frits are insoluble in water, but are effective if applied as fine powders; MoO_3 is applied as a seed coating in many cases.

Municipal sewage sludges and fly-ash materials, both waste products, contain Mo that is available to crops. Because these two materials are applied to soils at relatively high rates ($5–40\,t\,ha^{-1}$), their Mo contents should be considered when determining their application rates. Although toxicity problems generally do not result from application

Table 11.1. *Mo sources*

Mo source	Chemical formula	Mo concentration (%)
Ammonium molybdate	$(NH_4)_6Mo_7O_{24} \cdot 4H_2O$	54
Molybdenum trioxide	MoO_3	66
Molybdenum frits	Fritted glass	20–30
Molybdic acid	$H_2MoO_4 \cdot H_2O$	53
Sodium molybdate	$Na_2MoO_4 \cdot 2H_2O$	39

Source: Date from Martens and Westermann (1991).

of sewage sludge, continued applications of fly ash as a liming material have resulted in Mo toxicities in some soils (Jarrell, Page, and Elseewi, 1980).

Ammonium molybdate and H_2MoO_4 were found to be equally effective for green beans (*Phaseolus vulgaris* L.) growing on an acidic soil (pH 5.1) in Brazil (Franco, 1980). However, a fritted-glass source of Mo was ineffective. Plants grew poorly without lime, even with applied Mo, but responded well to Mo at a low lime rate. There was no crop response to Mo when that soil was limed above pH 6.0.

Methods of Fertilizer Application

Methods of Mo application include soil application of Mo fertilizer alone or with phosphorus (P) or nitrogen-phosphorus-potassium (NPK) fertilizers (molybdenized fertilizers), foliar sprays, dusts, and seed treatment.

Soil Application

Molybdenum fertilizers can be applied broadcast or in bands incorporated into the soil. Because the amounts of Mo required are so low, molybdenized P or NPK fertilizers generally are used to add volume and thus provide a more uniform Mo application to the field. These products are made by spraying a soluble Mo source onto the fertilizer in the granulation process or by adding the Mo source to the acid used in the manufacture of the fertilizer. Because these Mo sources do not react with the components of the fertilizer, the availability of Mo to plants is not affected. There is also evidence that Mo uptake by plants is enhanced by the presence of adequate P. Molybdenized fertilizers generally are applied broadcast on pastures growing on Mo-deficient soils in Australia and New Zealand (Anderson, 1956). Soluble Mo sources also can be sprayed onto the soil surface before tillage to obtain a more uniform application. However, that method entails extra application costs; so it is not often used.

Sims and Wells (1990) reported that use of NPK fertilizers containing Mo resulted in higher leaf yields of burley tobacco (*Nicotiana tabacum* L.) with band application than with broadcast application. Including Mo in the transplant solution also increased tobacco-leaf yields in 1 of 3 years.

Seed Treatment

Seed treatment is the most common method of Mo application. Molybdenum sources are coated onto the seeds with some type of sticker and/ or conditioner. This method ensures a more uniform application in the field, and the amounts of Mo that can be coated onto seeds are sufficient to provide the required Mo. The data in Table 11.2 show the effectiveness of a very low rate of seed-applied Mo on soybean [*Glycine max* (L.) Merr.] yields, especially on acid soil. Soybean yields were similar at three soil pH values in the presence of seed-applied Mo.

Field experiments in New Zealand compared the effectiveness of seed-applied Mo and Mo applied by foliar spray on cauliflower (*Brassica oleracea* var. *botrytis* L.) growth (Scheffer and Wilson, 1987). Molybdenum (24 g Mo per 1 kg seed) incorporated into the seed pellet was as effective as foliar Mo spray (1,025 g ha^{-1}) for alleviating Mo deficiencies. However, Mo toxicity resulted when Mo was incorporated in the seed pellets at more than 115 g kg^{-1}.

Thompson and Hsieh (1972) compared Mo seed treatment of alfalfa (*Medicago sativa* L.) with soil-applied Mo on a pH-4.9 soil in greenhouse pots. Coating alfalfa seeds with $CaCO_3$ resulted in increased forage yields in the presence and in the absence of applied Mo (Table 11.3). There was no response to Mo application without $CaCO_3$ included in the seed coating. Alfalfa forage yields were similar with seed-applied Mo and soil-applied Mo when the seed coating included $CaCO_3$.

Most legume seeds are inoculated with the appropriate *Rhizobium* species to ensure that nodulation will occur. Seed treatment with soluble sources of Mo may decrease the effectiveness of the inoculum because of salt effects. Therefore, some suggest the use of

Table 11.2. *Soybean responses to Mo seed treatments as affected by soil pH values*

	Yield (kg ha⁻¹)	
Soil pH	No added Mo	Mo added (35 g ha⁻¹)
5.3	1,610	2,820
5.7	2,150	2,820
6.3	2,620	2,890

Source: Data from Segars (1981).

Table 11.3. *Responses of alfalfa to soil- and seed-applied Mo and CaCO₃*

Mo application (g ha⁻¹)	Method	Yield per pot (g)	
		No CaCO₃ added	CaCO₃ added
0	—	4.00	7.06
140	Seed	4.81	8.53
560	Soil	4.49	9.26

Source: Data from Thompson and Hsieh (1972).

MoO_3, an insoluble source that allows the use of higher Mo application rates.

Including Mo sources with bacterial inoculants as a seed treatment is a common practice. However, care must be taken when combining Mo sources with bacterial inoculants in seed treatments. The results of numerous experiments were summarized by Shivashankar and Hagstrom (1991). Application of 4–8 g of Mo per kilogram of seeds as Na_2MoO_4 was acceptable for most grain legumes with respect to seed germination, establishment, nodulation, and crop growth. Nodule weight, number of pods per plant, and yield of peanuts (*Arachis hypogaea* L.) and soybeans increased with increasing rates of Mo applied to the seeds in two experiments (Table 11.4); the findings showed that increasing the Mo application above 8 g of Mo per kilogram of seeds may reduce the rates of seed germination and nodulation.

Table 11.4. *Effects of Mo seed treatment on nodulation and yield parameters for peanuts and soybeans*

Seed Mo (g kg⁻¹)	Nodule dry weight per plant (g)	Pods per plant	Pod yield (kg ha⁻¹)	Stover (kg ha⁻¹)
Peanuts				
0	0.14	17.3	1,690	2,060
4	0.16	18.0	1,750	2,140
8	0.19	20.9	2,190	2,280
Soybeans				
0	0.12	62	2,330	1,940
4	0.15	80	2,830	2,050
8	0.16	82	2,940	2,170

Source: Data from Shivashankar and Hagstrom (1991).

Table 11.5. *Yields and Mo and N concentrations in red clover seeds as affected by Mo and lime applications*

Mo on seeds	Lime (kg ha⁻¹)	Yield (kg ha⁻¹)	Seed Mo (mg kg⁻¹)	Seed N (%)
−	0	3.0	0.45	2.11
−	560	5.4	0.30	2.15
+	0	50.6	0.57	2.02
+	560	61.5	0.65	1.96

Source: Data from Hagstrom and Berger (1965).

Yields of peas (*Pisum sativum* L.) were increased 98% by application of Mo at 55 g ha⁻¹ as Na_2MoO_4 applied to the seeds (Hagstrom and Berger, 1965). A foliar-spray application at the same Mo rate was equally effective on that pH-5.3 soil. However, a dry dust applied to the seeds was not very effective in correcting Mo deficiencies. Liming that soil at a rate of 2,240 kg ha⁻¹ prior to planting did not result in increased crop yields. A seed treatment of soaking red clover (*Trifolium pratense* L.) seeds in a 1% solution of Na_2MoO_4 for 30 minutes also resulted in increased seed yields on the same soil (Table 11.5). Red clover seed yields were higher if the soil was limed. Those authors also noted that Mo seed treatments had to be applied in a liquid or slurry form; dusting seed with Na_2MoO_4 and talc was ineffective for supplying Mo to seedlings. In another experiment, yields of sweet corn were increased 40% by soil application of Na_2MoO_4 at $900\,g\,ha^{-1}$.

Another method of Mo application is to include a soluble Mo source in the transplant solution. Sims, Suchy, and Cornelius (1983) reported that inclusion of Na_2MoO_4 in the transplant solution for burley tobacco resulted in higher Mo concentrations, nitrate reductase activity, and cured-leaf yields. Responses to Mo were greater with transplant-solution application than with broadcast application of the same Mo source.

Crop Responses to Molybdenum Sources

Many legume crops have responded dramatically to Mo applications on acid soils in various regions of the world. Liming some of these soils may provide sufficient plant-available Mo to the soil, so that Mo fertilization may not be required. Whereas legumes generally have been most responsive to Mo applications, responses have also been reported for broccoli (*Brassica oleracea* L. Botrytis Group), cabbage (*Brassica oleracea* var. *capitata* L.), corn (*Zea mays* L.), rutabaga (*Brassica napobrassica* L.), and wheat (*Triticum aestivum* L.) (Gupta and Lipsett, 1981).

Parker and Harris (1962) reported that soybean yields on some soils in Georgia were not affected by lime applications when Mo was also applied. In contrast, Giddens and Perkins (1972) showed that lime alone did not increase the available Mo to amounts that would be adequate for alfalfa production on highly oxidized soils of the Georgia piedmont (pH 5.8–6.0). Molybdenum applications ($110\,g\,ha^{-1}$) were needed to maintain the alfalfa stands.

Some soils are so acid that legume production is not possible without at least a low rate of lime application. Gupta (1969) reported that lime was needed for successful alfalfa production in eastern Canada. Whereas Mo applications did not increase forage yields on a limed, high-Mo soil, both Mo and lime were required to produce good yields on low-Mo soils.

Another consequence of the lime–Mo interaction in soils and plants is that applications of both lime and Mo fertilizers for forage legumes possibly could result in high Mo concentrations in plant tissues. Feeding high-Mo forage to ruminant animals that are on rations that are low in copper (Cu) could result in molybdenosis (Mo toxicity caused by Mo–Cu imbalance). This points to a need for monitoring of forage Mo concen-

Table 11.6. *Alfalfa responses to Mo applications as affected by soil pH values*

Lime (t ha⁻¹)	Soil pH	No Mo	Mo (50 g ha⁻¹)ᵃ	Mo (100 g ha⁻¹)ᵃ	Mo (400 g ha⁻¹)ᵃ
			Yield (kg ha⁻¹)		
0	5.3	770	1,520	1,290	1,580
2.2	5.5	2,240	2,620	2,320	3,240
9.0	6.5	3,210	3,370	3,140	3,300

ᵃ Applied annually.
Source: Data from Mortvedt and Anderson (1982).

trations when a program of Mo fertilization is being used. James, Jackson, and Harward (1968) found that alfalfa did not respond to Mo applications on limed soils in Oregon. Responses to Mo were obtained on some of those soils if they were not limed. However, application of Mo plus lime on two soils resulted in herbage Mo concentrations within the hazardous range for livestock (5–10 mg kg⁻¹).

Eleven field experiments were conducted in five southeastern states in the United States over a 3-year period to determine the responses of forage legumes to Mo and lime applications (Mortvedt and Anderson, 1982). Alfalfa and five clover (*Trifolium* spp.) species were included in those experiments. Liming resulted in increased forage yields from soils where the initial soil pH was less than 5.5. Alfalfa yields generally increased with increasing soil pH (Table 11.6), but clover yields were less influenced by soil pH (data not shown).

Yield responses to Na₂MoO₄ applications were infrequent, but alfalfa responses to Mo applications occurred on non-limed soils with pH < 5.5 (Mortvedt and Anderson, 1982). The findings suggested that the critical soil pH was 5.5 for obtaining forage responses to Mo applications. The maximum response to Na₂MoO₄, either broadcast or incorporated into the soil prior to planting, was at a rate of 400 g of Mo per hectare. However, annual Mo applications of 50 g ha⁻¹ sprayed on the soil surface resulted in significant forage responses. Concentrations of Mo in the forage generally increased with increasing Mo application rates or soil pH values. Maximum alfalfa yields were accompanied by plant Mo concentrations greater than 0.5 mg kg⁻¹. Plant N concentrations were better related to relative yields on Mo-deficient soils than were Mo concentrations. Mortvedt (1981) also found that plant responses to Mo applications were more closely related to uptake of N than to uptake of Mo and concluded that analyzing legume forages for N rather than Mo may be preferable for determining crop responses to Mo applications to acid soils.

Six field experiments were conducted in four southeastern states over a 3-year period to determine the responses of soybeans to Mo applications at several soil pH values (Anderson and Mortvedt, 1982). Soybean yields were increased by Mo application rates (as Na_2MoO_4) up to $800\,g\,ha^{-1}$ on relatively acid soils. Soybean yields also were increased by liming soils whose pH values were less than 5.7, with the highest yields occurring near pH 6.0. Yield responses were more closely related to soil pH than to the available soil Mo concentrations.

The concentrations of Mo in those experiments generally were higher in soybean seeds than in trifoliate leaves and ranged widely at various locations in that southeastern region. As with forage legumes, the relative yields were more closely related to leaf N concentrations than to leaf or seed Mo concentrations. Crop yields were not well related to the available soil Mo concentrations, suggesting that soil pH, rather than soil Mo concentration, is the best indicator of potential Mo problems with legumes.

Residual Effects of Applied Molybdenum

The residual effects of Mo fertilizers are related to the reactions of the applied Mo with the soil, Mo removal by cropping and by grazing animals, and the extent of Mo leaching from the root zone. One of the earliest reports of marked residual effects from applied Mo was by Anderson (1956). Growth effects due to Mo application at $70\,g\,ha^{-1}$ to subterranean clover (*Trifolum subterraneum* L.) pastures were still evident after 10 years.

Evidence of variable residual effects of Mo was reported by Riley (1987). Whereas an application of Mo at $110\,g\,ha^{-1}$ to an acidic soil in Western Australia provided sufficient Mo for wheat for 15 years, application of Mo to another soil at $140\,g\,ha^{-1}$ was effective for only 1 year. The different degrees of effectiveness of Mo applications in various soils have been found to be closely related to the degree of soil adsorption of Mo. Barrow et al. (1985) concluded that the effectiveness of applied Mo decreased about 50% annually; so residual responses may be significant for only a few years after application.

Crops with high Mo requirements may require higher rates of Mo application to obtain residual responses. For example, Mulder (1954) found that cauliflower required Mo application at 1,800 g ha⁻¹ for optimum yields over a 3-year period, but addition of Mo at 444 g ha⁻¹ was sufficient for cabbage over the same period.

Gupta (1979) reported that Mo applications of 400 g ha⁻¹ corrected Mo deficiencies in forage crops for 3 years in eastern Canada. In contrast, Jones and Ruckman (1973) noted a continuing response by subterranean clover for 8 years after Mo application at 280 g ha⁻¹ in California. Residual effects of applied Mo were evident in tobacco plants in the second year after application on only one of two acidic soils in Maryland (Khan, 1991).

More frequent Mo applications may be required on acidic soils, because Fe and Al oxides fix Mo in unavailable forms (Gupta and MacKay, 1968). Smith and Leeper (1969) reported that crops in the central highlands of eastern Australia continued to respond to annual applications of Mo, even after 20 years of Mo applications.

The residual effects of applied Mo may not always benefit the next crop. The findings in a long-term soil fertility experiment on acid soils (pH 5.2–6.0) in Alabama suggested that those soils were not Mo-deficient (Adams, Burmester, and Mitchell, 1990). Molybdenum had been applied to corn biennially in a corn–soybean rotation experiment that included a lime treatment. When the soybean foliage developed chlorosis, suggesting Mo deficiency, the treatment area was split, with half receiving Mo foliar sprays, and the other half receiving none. Soybean yields were dramatically increased by Mo applications. There were yield responses to lime only in the absence of applied Mo. Those findings demonstrate that Mo applications made at the wrong time in the crop rotation may not be adequate to supply the Mo needs of Mo-sensitive crops.

Summary

The main Mo sources are ammonium molybdate, sodium molybdate, and molybdenum trioxide. The three sources generally are equally effective, and the selection will depend on the method of application. The application methods are band or broadcast application (generally of Mo contained in P or NPK fertilizers) to soil, foliar sprays, and seed treatment.

Seed treatment is the primary method of application, because the recommended rates for Mo application are so low (50–400 g ha⁻¹). Molybdenum sources can be applied to seeds as a liquid or slurry, and some type of a sticking or conditioning agent may be included. Because legume seeds usually are treated with a bacterial inoculant, the Mo source must be compatible with the inoculant. Insoluble molybdenum trioxide is commonly used when legume seeds are also treated with inoculant, although low additions of the soluble molybdate sources also can be used.

Liming soils to reach the recommended pH values may provide sufficient plant-available Mo, because the availability of soil Mo increases with soil pH, and therefore Mo fertilization may not be needed. However, many acidic soils do not have sufficient available Mo for crop production, even when limed. The high cost or insufficient supplies of lime also may preclude its use in some regions. Surface applications of lime to permanent pastures generally are not as effective for increasing the soil pH. For all of these situations, application of Mo fertilizers at Mo rates of less than $1\,\mathrm{kg\,ha^{-1}}$ will provide sufficient Mo for optimum crop production. Caution should be used in Mo fertilization programs for forage legumes, because high Mo concentrations in forages can lead to animal toxicities.

References

Adams, J. F., Burmester, C. H., and Mitchell, C. C. (1990). Long-term fertility treatments and molybdenum availability. *Fert. Res.* 21:167–70.

Anderson, A. J. (1956) Molybdenum as a fertilizer. *Adv. Agron.* 8:163–202.

Anderson, O. E., and Mortvedt, J. J. (eds.) (1982). *Soybeans: Diagnosis and Correction of Manganese and Molybdenum Problems.* University of Georgia Southern Cooperative Series bulletin 281.

Barrow, N. J., Leahy, P. J., Southey, I. N., and Purser, D. B. (1985). Initial and residual effectiveness of molybdate fertilizer in two areas of southwestern Australia. *Aust. J. Agric. Res.* 36:579–87.

Franco, A. A. (1980). Effects of lime and molybdenum on nodulation and nitrogen fixation of *Phaseolus vulgaris* L. in acid soils of Brazil. *Turrialba* 30:99–105.

Giddens, J., and Perkins, H. F. (1972). Essentiality of molybdenum for alfalfa on highly oxidized Piedmont soils. *Agronomy J.* 64:819–20.

Gupta, U. C. (1969). Effect of the interaction of molybdenum and limestone on growth and molybdenum content of cauliflower, alfalfa and bromegrass on acid soils. *Soil Sci. Soc. Am. Proc.* 33:929–32.

Gupta, U. C. (1979). Effect of methods of application and residual effect of molybdenum on the molybdenum concentration and yield of forages on podzol soils. *Can. J. Soil Sci.* 59:183–9.

Gupta, U. C., and Lipsett, J. (1981). Molybdenum in soils, plants and animals. *Adv, Agron.* 34:73–115.

Gupta, U. C., and Mackay, D. C. (1968). Crop responses to applied molybdenum and copper on podzol soils. *Can. J. Soil Sci.* 48:235–42.

Hagstrom, G. R., and Berger, K. C. (1965). Molybdenum deficiencies of Wisconsin soils. *Soil Sci.* 100:52–6.

James, D. W., Jackson, T. L., and Harward, M. E. (1968). Effect of molybdenum and lime on the growth and molybdenum content of alfalfa grown on acid soils. *Soil Sci.* 105:397–402.

Jarrell, W. M., Page, A. L., and Elseewi, A. A. (1980). Molybdenum in the environment. *Residue Reviews* 74:1–43.

Jones, M. B., and Ruckman, J. E. (1973). Long-term effects of phosphorus, sulfur and molybdenum on a subterranean clover pasture. *Soil Sci.* 115:343–8.

Khan, M. A. (1991). Interactive effects of soil pH, boron and molybdenum on the growth and elemental composition of Maryland tobacco. University Microfilms, Inc., Ann Arbor, MI: Dissertation Abstracts 52:4, 1787B.

Martens, D. C., and Westermann, D. T. (1991). Fertilizer applications for correcting micronutrient deficiencies. In *Micronutrients in Agriculture*, 2nd ed., ed. J. J. Mortvedt et al., pp. 549–82. Madison, WI: Soil Science Society of America.

Mortvedt, J. J. (1981). Nitrogen and molybdenum uptake and dry matter relationships of soybeans and forage legumes in response to applied molybdenum on acid soils. *J. Plant Nutr.* 3:245–56.

Mortvedt, J. J., and Anderson, O. E. (eds.) (1982). *Forage Legumes: Diagnosis and Correction of Molybdenum and Manganese Problems*. University of Georgia Southern Cooperative Series bulletin 278.

Mulder, E. G. (1954). Molybdenum in relation to growth of higher plants and microorganisms. *Plant Soil* 5:368–415.

Parker, M. B., and Harris, H. B. (1962). Soybean response to molybdenum and lime, and the relationship between yield and chemical composition. *Agronomy J.* 54:480–3.

Riley, M. M. (1987). Molybdenum deficiency in wheat in Western Australia. *J. Plant Nutr.* 10:2117–23.

Scheffer, J. J. C., and Wilson, G. J. (1987). Cauliflower: molybdenum application using pelleted seed and foliar sprays. *N.Z. J. Exp. Agric.* 15:485–90.

Segars, W. I. (1981). Molybdenum – the unique element. *Agrichemical Age* 25:38–9.

Shivashankar, K., and Hagstrom, G. R. (1991). Molybdenum fertilizer sources and their use in crop production. In *Proceedings of the International Symposium on the Role of Sulphur, Magnesium and Micronutrients in Balanced Plant Nutrition*, ed. S. Portch, pp. 297–305. Washington, DC: Sulphur Institute.

Sims, J. L., Suchy, M. E., and Cornelius, P. L. (1983). Placement of molybdenum fertilizer in the transplant solution of burley tobacco. *Agronomy J.* 75:239–42.

Sims, J. L., and Wells, K. L. (1990). Response of burley and dark-cured tobacco to fertilizer placement methods and starter fertilizer. *Tobacco Sci.* 34:11–14.

Smith, B. H., and Leeper, G. W. (1969). The fate of applied molybdenum in acidic soils. *J. Soil Sci.* 20:246–54.

Thompson, L. F., and Hsieh, C. F. (1972). Coating legume seed with micronutrient fertilizer. *Arkansas Farm Res.* 21:11.

12

Yield Responses to Molybdenum by Field and Horticultural Crops

JAMES F. ADAMS

Introduction

The fact that higher plants need molybdenum (Mo) was recognized as early as 1930s, as described in many early review articles (Hewitt, 1956; Anderson, 1956; Stout and Johnson, 1956; Rubins, 1956; Evans, 1956; Davies, 1956; Purvis and Peterson, 1956; Reisenauer, 1956). A later review, by Gupta and Lipsett (1981), covered every aspect of soil fertility: Mo fertilizers and their application, the physiological roles of Mo, determinations of Mo in plants and soils, factors affecting plant uptake of Mo, and the problems of toxicity and deficiency of Mo. The importance of Mo was first noted in legumes, because clovers (*Trifolium* spp.) were extensively used in mixed pastures in Australia. Since that time, Mo deficiencies have been identified in many other legumes, such as soybeans, alfalfa, peas, and various beans. Molybdenum deficiencies have also been reported in nonlegume crops. The early experiments revealed that Mo not only is required for nitrogen (N) fixation by rhizobia for use by legumes but also is required for nitrate reductase utilization by legumes and nonlegumes. Field data on yield responses to applied Mo are more limited than for other essential elements; because of the low requirements for Mo, it has not been as thoroughly studied as other elements. Because Mo deficiencies are found in acid soils, other infertility factors have tended to mask Mo deficiencies, and thus we have fewer data on crop yields in response to Mo in field and horticultural crops. The plant part and age at testing are important in diagnosing Mo deficiencies. Table 12.1 shows deficient and sufficient Mo concentrations reported in the literature reviewed in this chapter. The literature reviewed here concerns field experiments, because greenhouse and growth-chamber data do not adequately predict yield responses to additions of essential nutrients under field conditions.

182

Table 12.1. *Molybdenum concentrations in various plant parts at various stages of growth for selected crops*

Plant	Part of plant or tissue sampled	Mo in dry matter (ppm)		References
		Deficient	Sufficient	
Barley	Grain, whole plant		0.28–0.66	Hawes et al. (1976)
	Joint stage		0.46–3.11	
	Leaves and straw		0.13–0.58	
	Whole plant (boot stage)		0.33–27.05	Gupta et al. (1978)
Cauliflower	Seedling	0.09–0.20	5.4–10.1	Scheffer and Wilson (1987)
	Leaf	0.06	0.07–0.11	
	Whole plant		0.07	Gupta et al. (1990)
Corn	Grain	0.03	0.05	Tanner (1978)
			0.08	Weir and Hudson (1966)
			0.07–0.73	Soon and Bates (1985)
	Leaves		1.4–2.2	Lutz and Lillard (1973)
Honeydew	Leaf & petiole	0.10	0.60–1.03	Gubler et al. (1982)
Oats	Leaves	0.40	0.60	Fitzgerald (1954)
		0.07	0.10	Riley (1987)
Peas	Whole plant	0.17	0.69	Hagstrom and Berger (1963)
Peanuts	Leaves	0.15	0.26–0.65	Boswell et al. (1967)
	Seeds	0.60	2.33–10.33	
	Leaves		0.06	Welch and Anderson (1962)
	Seeds		0.88	
	Leaves	0.20	0.30	Rebafka et al. (1993)
	Seeds	0.10	0.30	
		0.20	0.62	Hafner et al. (1992)
Rice	Leaves		0.22–7.20	Moore and Patrick (1991)
Soybeans	Leaves at 40 days after planting	0.32	1.13	Hashimoto and Yamasaki (1976)
	Whole plant	1.00	5.49	Hagstrom and Berger (1963)
Tobacco	Whole plants at 50 days after transplant	0.22	0.25–0.48	Sims et al. (1975)
	After transplant	0.20	0.27	Sims et al. (1979)

Table 12.1. *(cont.)*

| Plant | Part of plant or tissue sampled | Mo in dry matter (ppm) | | References |
		Deficient	Sufficient	
Tomatoes	Shoots at 30 days after planting		0.06–0.15	Ambak et al. (1991)
Wheat	Grain	0.29–0.44	0.41–1.02	Hawes et al. (1976)
	Whole grain, joint stage	0.16–0.40	0.64–1.32	
	Leaves and straw	0.27–0.32	0.33–0.93	
	Whole plant (boot stage)		0.30–21.04	Gupta et al. (1978)

Yield Responses of Field Crops to Added Molybdenum

Legumes

Soybeans

Molybdenum deficiency in legumes was first reported for subterranean clover (*Trifolium subterraneum* L.) in Australia (Anderson, 1942). It was not until much later that Mo deficiency in soybeans [*Glycine max* (L.) Merr.] was recognized. That may have been due to the fact that large seeded legumes often contain enough seed Mo to meet the requirements of the plant even when grown on soils that are Mo-deficient (Hewitt, 1956). Another possible explanation for such late recognition is that deficiencies occur mainly on acid soils, and Mo deficiency can be masked by other fertility problems, such as aluminum (Al) and manganese (Mn) toxicities and/or calcium (Ca) and magnesium (Mg) deficiencies.

Most soybean responses to Mo in the United States have been reported from the southeastern states on moderately to strongly acid soils. One of the earliest studies reported increases in yields due to Mo as high as 30% and 50% for two experiments in separate years (Parker and Harris, 1962). Those data showed Mo to be as effective as lime for increasing yields on soils that had initial soil pH values of 5.6 and 5.7. They also showed that a foliar spray was as effective as seed treatment for supplying Mo. On the other hand, Hagstrom and Berger (1963) reported yield increases as high as 80% following Mo application, an increase greater than that produced by lime alone, indicating that some soils are so low in Mo that liming alone may be ineffective. Burmester, Adams, and Odom (1988) conducted 15 field experiments with initial soil pH values ranging from 4.6 to 5.6 and observed similar yield increases with foliar-applied Mo, but they found that at soil pH values below 5.3, Al toxicity limited yields. Under those strongly acid conditions, Mo alone did not increase yields as much as did lime plus Mo. Forbes, Street, and Gammon (1986) adjusted soil pH values to 4.7, 5.2, and 6.1 and found that there were no responses to Mo except on plots also receiving lime treatment, indicating additional nutritional problems. Keogh et al. (1971) reported that in two of their three experiments there were higher yields with lime or lime plus Mo than with Mo alone, but in another experiment the Mo treatment alone was equal to the lime and lime-plus-Mo treatments. Rhoades and Nangju (1979) found that at a soil pH of 4.5, soybeans did not respond to Mo, whereas cowpeas [*Vigna unguiculata* (L.) Walp] in the same experiment had increased yields. This illustrates not only that other fertility factors may limit yields at low soil pH values but also that crops have differing Mo requirements.

Often Mo deficiencies in legumes are manifested as N deficiencies. This element is essential for the enzyme nitrate reductase and is required by rhizobia for N fixation. Parker and Harris (1977) compared nodulating and non-nodulating soybeans at four rates of N application and two rates of Mo application. Nitrogen and Mo studies had shown that the responses of legumes to Mo are primarily due to the Mo requirements of the rhizobia, rather than the direct requirements of the legumes. They found that non-nodulating soybeans did not respond to Mo, but soybean yields were increased by application of N. In the same experiment with nodulating soybeans, yields were increased by addition of Mo without addition of N, but its effect disappeared with increasing applications of N. A similar study by Hashimoto and Yamasaki (1976) showed that non-nodulating soybeans did not respond to applied Mo.

The residual amounts of Mo following Mo applications at traditional rates appear to be ineffective for correcting Mo deficiencies. Adams, Burmester, and Mitchell (1990) found that soybeans would respond to Mo even when the Mo had been applied the previous year. Their findings indicated that Mo application at $280 \, g \, ha^{-1}$ was an insufficient rate when applied to the preceding crop in the rotation [corn (*Zea mays* L.)], but that the following crop of soybeans continued to respond to the residual Mo from that application on soils of pH 6.1. Hawes, Sims, and Wells (1976) reported that Mo must be applied at a rate of 880 g ha⁻¹ to have a residual effect on yields. Their rates of 220 and 440 g ha⁻¹ were ineffective.

Peanuts

A recent review by Davis and Rhoades (1994) and an earlier review by Cox, Adams, and Tucker (1982) noted that there have been very few reports of yield increases due to Mo application to peanuts (*Arachis hypogaea* L.). The earliest report of an increased yield due to Mo application was by Parker (1964), who conducted 15 experiments from 1956 to 1963 in which only one experiment showed a yield increase (31%) due to Mo, which had been applied at a rate of 280 g ha⁻¹ on a Ruston loamy sand of pH 6.3. Boswell, Anderson, and Welch (1967) conducted 30 Mo experiments on soils whose pH values ranged from 5.0 to 6.8, and they found no increased yields following application of Mo. Sixteen of the experiments were continued for a second year, with no significant yield increases. They had only one experiment in which the method and timing of application had an effect on yield. They found that the combination of seed treatment and foliar application of Mo at 3 weeks after early bloom was superior to foliar application at 6 weeks after early bloom.

A more recent study on a soil in western Africa showed increased yields of peanuts following Mo application (Hafner et al., 1992). In two experiments they found that addition of Mo increased the yields by 37% and 86% in the first and second years, respectively, with the soil's initial pH at 4.9. A micronutrient fertilizer was used at a rate of 100 kg ha⁻¹ containing 4% manganese (Mn), 4.0% iron (Fe), 1.5% copper (Cu), 1.5% zinc (Zn), 0.5% boron (B), 1% molybdenum (Mo), and 0.005% cobalt (Co) in the first experiment. In the second year, Mo pelleted with seeds was the only micronutrient applied. The second-year data confirm that the yield response in the first year had been due to Mo. In an adjacent location, Rebafka, Ndunguru, and Marschner (1993) found that some types of phosphate fertilizers can induce Mo deficiencies. They found that when using normal superphosphate there was no increase in yield, and the plants were lighter in color, but when using triple super-phosphate there was a yield increase and improved color. Their findings clearly demonstrate the effects of sulfur (S) and phosphorus (P) on Mo uptake. Such antagonistic and synergistic effects had been documented earlier by Stout et al. (1951).

The absence of any Mo effect on peanut crops reported by the workers in Georgia and the limited reports of peanut responses to Mo in other growing areas are the reasons that currently there is no recommendation for Mo application to peanut crops in the southeastern states. A contributing factor is the use of lime as the Ca source, thus keeping soil pH values up and increasing the availability of Mo.

Beans

Molybdenum deficiencies in large seeded legumes were first reported by Wilson (1949) as scald disease. Stout and Johnson (1956) reviewed the earlier literature concerning many horticultural and field crops and reported that large seeded legumes contain sufficient Mo for adequate plant growth even when grown on Mo-deficient soils. That was supported by Hagstrom and Berger (1963), who found no yield increase for peas (*Pisum sativum* L.) following application of Mo, whereas in the same experiment soybeans and clover showed large yield increases. Later work by Rhoades and Nangju (1979) showed that seeds pelletized with Mo increased the yield of cowpeas (*Vigna sinensis* Endl.) by 25% on a soil of pH 4.5. Rhoades and Kpaka (1982) found that pelleting seeds with nitramolybdenum at 0.4 g per 100 g of seeds was sufficient for maximum yield and was more effective than foliar application of Mo. They reported a 21% increase over the yield on control plots, with a soil pH of 5.0.

In a study comparing selected legumes for their Mo requirements, Gladstone, Loneragan, and Goodchild (1977) found that on soil of pH 5.0, narrow-leaf lupin (*Lupinus angustifolius*), yellow lupin (*Lupinus luteus*), and sand-plain lupin (*Lupinus cosentinii*) did not show increased yields with addition of Mo, but did respond to Co. For the 2-year study, only purple vetch (*Vicia benghalensis*) showed a significant yield increase, and that only in the first year. In the second year, the response of narrow-leaf lupin to addition of Mo was only an increase in leaf N. Hagstrom and Berger (1963) also noted an increase in lupin yield, from 3,910 to 6,800 kg ha^{-1}, with application of "Moly-Gro" to seeds.

Other field studies by Garg, Sharma, and Tucker (1971) and Sharma and Garg (1973) with cluster beans and cowpeas as fodder showed no responses to added Mo. That was expected, because the soil pH ranged from 8.0 to 8.2. Dalal and Quilt (1977) found no response of pigeon peas (*Cajanus cajan* L.) to added Mo even on a soil that had a pH of 5.0.

Nonlegumes

Field experiments in the early 1950s found Mo deficiencies for small grains such as wheat (*Triticum aestivum* L.) and oats (*Avena sativa* L.) (Fitzgerald, 1954; Mulder, 1954). Even corn was found in field experiments to respond to added Mo in the early 1950s (Noonan, 1953).

Wheat

There have been fewer field experiments in which wheat has responded to applied Mo than there have been for soybeans. The acidic soils in the Australian wheat belt were where the first wheat responses to added Mo were reported. Mulder (1954), knowing of the Mo deficiencies reported for wheat in Australia, conducted field experiments with Mo additions on similar soils in The Netherlands. He found that Mo applied at a rate of 390 g ha^{-1} increased wheat yields, but higher rates of Mo application decreased yields from the maximum, though yields were still higher than those on untreated plots. Freney and Lipsett (1965) found that added Mo, with high rates of N application, prevented yield decreases and maintained high yields. Later, Lipsett and Simpson (1973), in a study of N and Mo additions on those same soils, found that even Mo alone could increase yields. They also confirmed the findings from earlier work in which Mo was applied with either urea or ammonium nitrate and yields increased. That demonstrated the requirement for both Mo and nitrate reductase in the production of protein, as described by Price, Clark, and Funkhouser (1972).

Coventry et al. (1987) reported increases in wheat yields due to added Mo in 2 of the 5 years in their study, but increases in grain weight and N content in 3 of the 5 years. Leaf N was increased during one of those years because of addition of lime, which they assumed to have caused increased mineralization of N and increased availability of Mo, which would have facilitated assimilation of N. That site was also reported to have insufficient Mo for growing subterranean clover (Coventry et al., 1985).

The residual effects of applied Mo will vary for different soil types, and Mo may have to be applied annually. Riley (1987) summarized the data from a 3-year study and reported that Mo applied at a rate of 75 g ha^{-1} would give, during the second year, only 86% of the yield seen during the first year. But the residual effects of Mo application at a rate of 140 g ha^{-1} gave maximum wheat yields for 15 years. The decline in the residual effectiveness of Mo when applied at low and moderate rates was also seen in the work of Adams et al. (1990). Hawes et al. (1976) found that Mo applied at a rate of 880 g ha^{-1} to a preceding crop was necessary to maintain maximum wheat yields.

Barley and Oats

Few yield responses to application of Mo have been reported for either barley (*Hordeum vulgare* L.) or oats. Hawes et al. (1976) applied Mo at several rates to barley and wheat. The wheat showed increased yields, but the barley did not. Gupta and MacLeod (1978) found that barley would not respond to added Mo, but plant Mo concentrations increased with increased liming. One of the few responses to Mo was reported by Fitzgerald (1954). Yields increased from 39.5 to 58.5 bushels per acre with seed-applied Mo at 137 g ha^{-1}. Another early study (Gartrel, 1966) reported that oats and wheat would respond to Mo application.

Corn

One of the earliest reported responses of corn (*Zea mays* L.) to Mo fertilization was by Noonan (1953). Since that time there have been few field experiments that have recorded yield increases due to Mo application. Tanner and Grant (1977) found that corn responded to seed-applied Mo in 5 of their 13 field experiments. In one experiment, with application of either Mo alone or a low rate of lime alone, yields were not higher than those for control plots, but when Mo and lime were combined, there was a significant yield increase. That would indicate that in the presence of strongly acidic conditions, there is more than one factor limiting growth. Later work by Tanner (1978) showed that Mo-deficient corn suffered from premature sprouting on the cob. Seed-applied Mo reduced the early sprouting, but an additional foliar spray was needed to completely eliminate the problem.

A prime reason for Mo deficiency being rare in corn crops is the very low Mo requirement of the corn plant. Weir and Hudson (1966) found that a second generation of corn grown in a nutrient solution without Mo was required to reduce the Mo concentration in the grains sufficiently that the seedlings would respond to addition of Mo on Mo-deficient soil. They also found that the Mo concentration in the grains had to be less than 0.08 mg kg^{-1} for the grains to show responses to added Mo when planted on Mo-deficient soils. Because of the lack of yield responses to added Mo, most of the literature has focused mainly on Mo uptake in field experiments, mainly the effect of soil pH on the uptake of Mo. For example, Ambak, Bakar, and Tadano (1991) found increasing Mo concentrations in grain as liming rates were increased. Soon and Bates (1985) reported the same effect, but they used a Ca sludge that raised pH and supplied Mo. It has also been reported that phosphorus fertilization enhances Mo concentrations in corn (Lutz and Lillard, 1973). There is a generally accepted interaction between Mo and P, as reviewed by Adams (1980).

Rice

There have been few reports of rice (*Oryza sativa* L.) responding to Mo application. Because most rice is grown in flooded paddies, the chemical interactions involving Mo are not the same as in dryland agriculture. When acid soils are flooded, there are increases in soil pH and changes in redox potential, thus changing the solubilities of Fe minerals (Ponnamperuma, 1972). Moore and Patrick (1991) reported increases in the amounts of MoO_4^{2-} in soil solution with increasing pH in flooded soils, and those increases resulted in increased Mo uptake. The conditions in flooded soils may explain why there have been few reports of increased yields due to addition of Mo.

Panda, Reddy, and Sharma (1991), in a 4-year study of how N affected production by nursery rice seedlings, found that Mo application increased yields in the resultant transplanted crop in only 1 of the 4 years. Others have reported increased leaf N concentrations and increased protein content following Mo application (Gupta and Basuchaudhuri, 1974).

Tobacco

Molybdenum deficiencies in tobacco (*Nicotiana tabacum* L.) were reported in the mid-1950s, as reviewed by Stout and Johnson (1956). Later research by Sims, Atkinson, and Smitobol (1975) showed Mo deficiencies on a soil with a pH value of 6.0 that was induced by the use of S fertilizers. Their highest yields were recorded when K_2SO_4 was omitted and Mo was applied at 880 g ha^{-1}. Later work by Sims, Leggett, and Pal (1979) showed that the same soil was also deficient in S and that a combination of Mo at 440 g ha^{-1} and 224 g of SO_4 gave the highest yield. The interactions involving Mo and S had been well documented earlier.

Eivazi et al. (1983), using several liquid fertilizers with different K sources, found that KH_2PO_4 and KCl increased Mo uptake by plants, but K_2SO_4 decreased Mo uptake. Whereas the synergistic effect of P on Mo uptake is well known, there are few data that might help explain the synergistic effect of Cl.

Yield Responses of Horticultural Crops to Added Molybdenum

Cauliflower

Molybdenum deficiency in cauliflower (*Brassica oleracea* var. *botrytis* L.) leads to the abnormality known as "whiptail." In an early review by Stout and Johnson (1956) it was noted that Mo deficiency in cauliflower was first reported in the mid-1920s and was studied for the next 17 years. There is limited documentation of Mo deficiencies in field experiments on cauliflower, but more work has been done in greenhouses. Because this chapter deals mainly with reported yield increases, greenhouse data will be discussed only in terms of plant uptake, rather than yields.

Scheffer and Wilson (1987) treated cauliflower seeds with Mo and seedlings with Mo seed treatment and foliar spray prior to transplanting them into field plots and found that curd yields increased by as much as 35% and 50%, respectively. They found that seed application of Mo at more than $117 \, g \, ha^{-1}$ caused toxicities, as demonstrated by reduced establishment of plant stands. The Mo concentrations in deficient seedling plants ranged from 0.09 to $0.20 \, mg \, kg^{-1}$. Earlier, Mulder (1954) had reported the responses of cauliflower to Mo addition in terms of failure to develop a heart or the extent of heart development. In the control groups in seven experiments, 50–100% of cauliflower plants developed no heart. In every experiment, application of Mo decreased the number of plants that developed no heart and also increased the sizes of the hearts.

Gupta, Chapman, and MacKay (1978) found an interaction between additions of lime and Mo in a greenhouse study and reported that at high lime rates the highest rate of soil-applied Mo ($144 \, mg \, kg^{-1}$) decreased yields. In another study, Gupta, LeBlanc, and Chapman (1990) found that a Mo concentration of $0.07 \, mg \, kg^{-1}$ in whole-plant tissue before development of curd was sufficient for optimum yield. In surveys to determine Mo status in cauliflower grown on Prince Edward Island, Canada, Gupta and Cutcliffe (1976) found that Mo concentrations in leaf tissue ranged from 0.1 to $9.8 \, mg \, kg^{-1}$. The mean value found by Gupta (1991) also fell within that range.

Brussels Sprouts, Cabbage, Rutabaga, Broccoli, and Turnips

Field data for the other crops in the *Brassica* genus are very limited. Most of the data have come from surveys of plant concentrations or uptake experiments in greenhouses or growth chambers. Mulder (1954) presented yield data for cabbage (*Brassica oleracea* var. *capitata* L.) and turnips (*Brassica rapa* L.) showing that those crops responded to Mo as much as did cauliflower. Their rate of Mo application of 780 g ha^{-1} for turnips gave lower yields than did a lower rate of applied Mo, indicating the possibility of toxicity problems at high rates of application. In a survey of field-grown Brussels sprouts (*Brassica oleracea* L. Gemmifera Group), rutabaga (*Brassica campestris* var. *napobrassica* DC.), broccoli (*Brassica oleracea* L. Botrytis Group), and cauliflower grown on Prince Edward Island, Canada, Mo concentrations were found to range from 0.1 to 11.4 mg kg^{-1}, 0.06 to 1.1 mg kg^{-1}, 0.1 to 16.5 mg kg^{-1}, and 0.1 to 9.8 mg kg^{-1}, respectively (Gupta and Cutcliffe, 1976). They found that only for rutabaga in that survey was the tissue concentration of Mo positively correlated with soil pH. The absence of correlations for the other crops was attributed to the farmers' practice of using Mo foliar sprays.

Melons

Most of the recently reported Mo deficiencies for melons (*Cucumis melon* L.) have involved California honeydews (*Cucumis melo* L.) (Gubler, Grogan, and Osterli, 1982). Gubler et al. (1982) reported a yield of 254 melons from one Mo-treated row in a plot versus 19 from an untreated row. Molybdenum deficiencies had been reported earlier, as reviewed by Stout and Johnson (1956), who stated that 0.01–0.1 g of $Na_2MoO_4 \cdot 2H_2O$ per hill would correct the Mo deficiency. There are very few more recent data on Mo deficiencies in melons.

Molybdenum Fertilizers and Application

Molybdenum fertilizers are discussed elsewhere in this volume and will be discussed only briefly here. In the majority of the studies cited in this chapter, $Na_2MoO_4 \cdot 2H_2O$ was the Mo source most commonly used for experiments. Murphy and Walsh (1972) reviewed Mo fertilizers and their rates and application methods. The methods, sources, and rates of application for all cited studies that showed positive yield responses are given in Table 12.2. In many of the studies in which different methods of application were investigated, there were no differences in yield responses that could be attributed to method of application. There were exceptions for some crops. For example, when foliar applications of Mo to peanuts were delayed until 6 weeks after early bloom, the yields were lower than when Mo was applied to the seed or was applied as foliar treatment 3 weeks after early bloom (Boswell et al., 1967). Rhoades and Kpaka (1982) found that seed treatment was more effective than foliar application for cowpeas. Most Mo application rates were between 100 and 300 g ha^{-1}, but some of the earlier studies had used much higher rates, which for some crops proved to be toxic. For example, Mulder (1954) found that cauliflower responded well to Mo at 1,560 g ha^{-1}, but that same rate decreased cabbage yields. He also found that turnip, sugarbeet, and wheat yields decreased at high Mo rates.

Soil Molybdenum Requirements for Maximum Yield

Soil pH is the parameter most frequently used to predict Mo deficiency in soils, but that method has not always been successful. In an early review of methods for determining Mo availability, Davies (1956) cited the use of a bioassay method in which the growth of *Aspergillus niger* reflected Mo availability. Donald, Passey, and Swaby (1952) pointed out several disadvantages of that method, which resulted in its limited acceptance.

Resins have also been used to determine Mo availability, and Bhella and Dawson (1972) reported a strong correlation between the Mo concentrations in resins and the concentrations of N and Mo in plant tissues. Karimian and Cox (1979) compared the Grigg method (ammonium oxalate extraction) to the resin method and found that with the Grigg method there was no correlation between the amount of Mo that could be extracted from the soil and the magnitude of the plant response or the Mo concentration in the plant, although they were able to derive a positive correlation between yield and Mo concentration in the plant on the basis of the ratio of amorphous Fe to free Fe and the soil pH in a 1-N KCl suspension. Burmester et al. (1988) found a much stronger correlation ($R^2 = 0.88$) to relative yield by using the ratio between oxalate-extractable Fe and Mo (Figure 12.1). Their assumption was that if the concentration of Mo increased on the outer surface of the Fe oxide, then its solubility would also increase, thereby increasing Mo availability. Their work showed the necessity to exclude data obtained for plants

Table 12.2. *Application rates, methods, and sources of Mo for selected crops that responded to Mo*

Plant	Method of application	Source of Mo	Rate of application	References
Cauliflower	Seed	$Na_2MoO_4 \cdot 2H_2O$	$65\,g\,ha^{-1}$	Scheffer and Wilson (1987)
Cauliflower, cabbage, turnips, lettuce	Soil	$Na_2MoO_4 \cdot 2H_2O$	$390–1,560\,g\,ha^{-1}$	Mulder (1954)
Corn	Seed	$Na_2MoO_4 \cdot 2H_2O$	25 g per 50kg of seed	Tanner and Grant (1977)
Cowpeas	Seed	"Nitramolybdenum" (4.85% Mo)	10–20mg per 100 seed	Rhoades and Nangju (1979)
	Foliar spray	$(NH_4)_6Mo_7O_{24} \cdot 4H_2O$	$60\,g\,ha^{-1}$	Rhoades and Kpaka (1982)
	Seed	"Nitramolybdenum" (4.85% Mo)	20mg per 100 seed	
Honeydew	Foliar spray	"Nitramolybdenum" (4.85% Mo)	$20\,g\,ha^{-1}$	Gubler et al. (1982)
Lupin	?	MoO_3	$200–1,200\,g\,ha^{-1}$	Gladstone et al. (1977)
	Seed	"Moly-Gro" (38% Mo)	$53\,g\,ha^{-1}$	Hagstrom and Berger (1963)
Rice	Foliar spray	$(NH_4)_6Mo_7O_{24} \cdot 4H_2O$	$110\,g\,ha^{-1}$	Panda et al. (1991)
Peas	Seed	$Na_2MoO_4 \cdot 2H_2O$	$55\,g\,ha^{-1}$	Hagstrom and Berger (1963)
Peanuts	?	$Na_2MoO_4 \cdot 2H_2O$	$100\,g\,ha^{-1}$	Parker (1964)
	Seed	"Moly-Gro" (38% Mo)	?	Boswell et al. (1967)
	Foliar spray 3wk after early bloom	$Na_2MoO_4 \cdot 2H_2O$	$218\,g\,ha^{-1}$	
	Seed	MoO_3	$100\,g\,ha^{-1}$	Hafner et al. (1992)
	Soil	"Fetrilon Combi 1" (0.1% Mo)		

Crop	Application	Compound	Rate	Reference
Pigeon peas	Soil	?	250 g ha⁻¹	Dalal and Quilt (1977)
Soybeans	Seed and foliar spray	$Na_2MoO_4 \cdot 2H_2O$	224 g ha⁻¹	Parker and Harris (1962)
	Foliar spray	$Na_2MoO_4 \cdot 2H_2O$	100 g ha⁻¹	Burmester et al. (1988)
	Soil	$Na_2MoO_4 \cdot 2H_2O$	200–800 g ha⁻¹	Forbes et al. (1986)
	Seed and foliar spray	$Na_2MoO_4 \cdot 2H_2O$	100 g ha⁻¹	Adams et al. (1990)
	Soil	$Na_2MoO_4 \cdot 2H_2O$	220–2,460 g ha⁻¹	Hawes et al. (1976)
	Soil	$Na_2MoO_4 \cdot 2H_2O$	874 g ha⁻¹	Hagstrom and Berger (1963)
Tobacco	Soil	$Na_2MoO_4 \cdot 2H_2O$	220–440 g ha⁻¹	Sims et al. (1975)
	Soil	$Na_2MoO_4 \cdot 2H_2O$	220–880 g ha⁻¹	Sims et al. (1979)
Wheat	Seed	$Na_2MoO_4 \cdot 2H_2O$	137 g ha⁻¹	Fitzgerald (1954)
	Foliar spray	$Na_2MoO_4 \cdot 2H_2O$	137 g ha⁻¹	
	?	$Na_2MoO_4 \cdot 2H_2O$	195–1,560 g ha⁻¹	Mulder (1954)
	Soil	$Na_2MoO_4 \cdot 2H_2O$	1,310 g ha⁻¹	Freney and Lipsett (1965)
	Foliar spray	$Na_2MoO_4 \cdot 2H_2O$	117 g ha⁻¹	Lipsett and Simpson (1973)
	Soil	?	62 g ha⁻¹	Coventry et al. (1987)

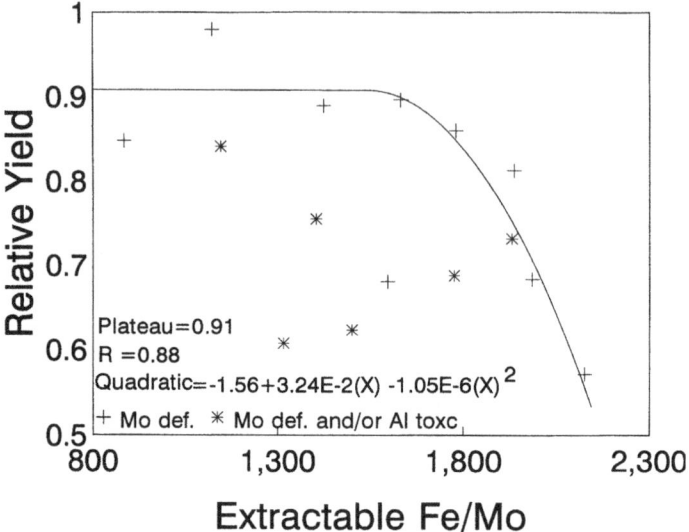

Figure 12.1. Yields for untreated soybeans relative to yields with lime and Mo applied as a function of the extractable Fe/Mo ratio. The relationship shown is for only Mo-deficient conditions.

grown on strongly acidic soils, for such data can be reflecting other nutrition problems.

Cox and Kamprath (1972) reviewed several other methods for testing soils. They cited several researchers who had investigated the method of water-extractable Mo in soil. Water-extractable Mo was found to be proportional to plant Mo concentration, but only within a certain pH range (4.7–7.5). A method for hot-water-extractable Mo was found to be better correlated with plant-tissue Mo concentration than was the traditional method of oxalate-extractable Mo (Grigg's method). Other researchers have used neutral 1-N ammonium acetate, 0.1-N NaOH, 0.03-N NH_4F, and 0.1-N HCl + 0.03-N NH_4F, with mixed results. A good soil-test method to adequately predict crop yields still eludes us today.

Summary

Molybdenum deficiencies in field and horticultural crops have been re- ported and described for more than 60 years, but yield responses to Mo application have been more limited than the reported yield responses to most other essential nutrients. Most of the early yield responses were reported for clover; later, Mo deficiency was recognized as a problem in the larger seeded legumes. Early experiments found that legumes were more susceptible to Mo deficiencies because of the greater requirements for Mo by their rhizobia, but there were also differences in Mo requirements among legumes. Seed size may play a role in determining Mo requirements, because plants that grow from seeds that already contain high Mo concentrations may not exhibit deficiency symptoms even when planted on Mo-deficient soils.

Nonlegumes respond to added Mo because of the requirements for Mo by their nitrate reductase enzymes. Molybdenum additions in the presence of high amounts of N in the soil have prevented depressed crop yields due to excess N. Grains such as corn, wheat, and rice have shown improved utilization of N following Mo application, but again, not all cultivars have the same Mo requirements. Many crops in the *Brassica* family respond to added Mo. Some of the early studies describing whiptail in cauliflower also found that different *Brassica* crops required different amounts of Mo.

Most Mo deficiencies are corrected with sodium molybdate, but there are several other sources that can also supply sufficient amounts of Mo. The application methods also vary. Because Mo is required in very small amounts ($100–300\,\mathrm{g\,ha^{-1}}$) it is broadcast only when it has been added to a bulk-blended fertilizer. More often, Mo is applied to the seeds prior to planting, or is applied as a foliar spray, or is applied to the soil by spraying. When differences in yields following the various application methods have been reported, seed application of Mo has been superior.

Several soil-test methods to determine Mo availability and to predict yield responses on a variety of soils have met with little success. The Grigg method and the method of anion-exchange resin can predict Mo uptake to some extent, but they are less accurate at predicting yield. Extraction tests must take into account Fe as well as Mo, because it has been found that the amount of Fe and the degree of crystallinity of Fe can influence Mo availability. Currently, there is no method available that would be a practical soil-test method and could be easily applied by soil-test laboratories. Because of these problems, the recommendations for addition of Mo must be based on soil PH, soil type, and the crop to be grown.

References

Adams, F. (1980). Interaction of phosphorus with other elements in soil and plants. In *The Role of Phosphorus in Agriculture*, ed. F. E. Khasawneh,

E. C. Sample, and E. J. Kamprath, pp. 665–80. Madison, WI: Soil Science Society of America.

Adams, J. F., Burmester, C. H., and Mitchell, C. C. (1990). Long-term fertility treatments and molybdenum availability. *Fert. Res.* 21:167–70.

Ambak, K., Bakar, Z. A., and Tadano, T. (1991). Effect of liming and micronutrient application on the growth and occurrence of sterility in maize and tomato plants in a Malaysian deep peat soil. *Soil Sci. Plant Nutr.* 37:689–98.

Anderson, A. J. (1942). Molybdenum deficiency on a South Australian ironstone soil. *J. Aust. Inst. Agric. Sci.* 8:73–75.

Anderson, A. J. (1956). Molybdenum deficiencies in legumes in Australia. *Soil Sci.* 81:173–82.

Bhella, H. S., and Dawson, M. D. (1972). The use of anion exchange resin for determining available soil molybdenum. *Soil Sci. Soc. Am. Proc.* 36:177–9.

Boswell, F. C., Anderson, O. E., and Welch, L. F. (1967). *Molybdenum Studies with Peanuts in Georgia.* University of Georgia Agricultural Experiment Station research bulletin 9.

Burmester, C. H., Adams, J. F., and Odom, J. W. (1988). Response of soybean to lime and molybdenum on Ultisols in northern Alabama. *Soil Sci. Soc. Am. J.* 52:1391–4.

Coventry, D. R., Hirth, J. R., Reeves, T. G., and Burnett, V. F. (1985). Growth and nitrogen fixation by subterranean clover in response to inoculation, molybdenum application and soil amendment with lime. *Soil Biol. Biochem.* 17:791–6.

Coventry, D. R., Morrison, G. R., Reeves, T. G., Hirth, J. R., and Fung, K. K. H. (1987). Mineral composition and responses to fertilizer of wheat grown on a limed and deep ripped soil in north-eastern Victoria. *Aust. J. Agric.* 27:687–94.

Cox, F. R., Adams, F., and Tucker, B. B. (1982). Liming, fertilization and mineral nutrition. In *Peanut Science and Technology*, ed. H. E. Pattee and C. T. Young, pp. 139–64. Yoakum, TX: American Research and Education Society.

Cox, F. R., and Kamprath, E. J. (1972). Micronutrient soil tests. In *Micronutrients in Agriculture*, ed. J. J. Mortvedt, P. M. Giordano, and W. L. Lindsay, pp. 289–317. Madison, WI: Soil Science Society of America.

Dalal, R. C., and Quilt, P. (1977). Effects of N, P, liming, and Mo nutrition and grain yield of pigeon pea. *Agronomy J.* 69:854–7.

Davies, E. B. (1956). Factors affecting molybdenum availability in soils. *Soil Sci.* 81:209–21.

Davis, J. G., and Rhoades, F. M. (1994). Micronutrients. In *Research-based Soil Testing Interpretation and Fertilizer Recommendations for Peanuts on Coastal Plain Soils*, ed. C. C. Mitchell, pp. 26–33. University of Georgia Southern Cooperative Series bulletin No. 380.

Donald, C. B., Passey, B. I., and Swaby, R. J. (1952). Bioassay of available trace metals from Australian soils. *Aust. J. Agric. Res.* 3:305–25.

Eivazi, F., Sims, J. L., Casey, M., Johnson, G. D., and Leggett, J. E. (1983). Growth and molybdenum concentration of burley tobacco as influenced by potassium, molybdenum, and chloride in transplant fertilizer solutions. *Can. J. Plant Sci.* 63:531–8.

Evans, H. J. (1956). Role of molybdenum in plant nutrition. *Soil Sci.* 81:199–208.

Fitzgerald, J. N. (1954). Molybdenum on oats. *N.Z. J. Agric.* 89:619.

Forbes, R. B., Street, J. J., and Gammon, N., Jr. (1986). Response of soybean to molybdenum, lime, and sulfur on flatwoods soils. *Soil Crop Sci. Soc. Fla. Proc.* 45:33–6.

Freney, J. R., and Lipsett, J. (1965). Yield depression in wheat due to high nitrate applications, and its alleviation by molybdenum. *Nature* 205:616–17.

Garg, K. P., Sharma, A. K., and Tucker, B. S. (1971). Studies on the effect of different rates of phosphorus and molybdenum on the growth and yield of cowpea fodder and residual effect on wheat. *Indian J. Agron.* 16:185–8.

Gartrel, J. W. (1966). Field responses of cereals to molybdenum. *Nature* 209:1050.

Gladstone, J. S., Loneragan, J. F., and Goodchild, N. A. (1977). Field responses to cobalt and molybdenum by different legume species, with inferences on the role of cobalt in legume growth. *Aust. J. Agric. Res.* 28:619–28.

Gubler, W. D., Grogan, R. G., and Osterli, P. P. (1982). Yellows of melons caused by molybdenum deficiency in acid soil. *Plant Disease* 66:449–51.

Gupta, D. K. D., and Basuchaudhuri, P. (1974). Effect of molybdenum on the nitrogen metabolism of rice. *Exp. Agric.* 10:251–5.

Gupta, U. C. (1991). Boron, molybdenum and selenium status in different plant parts in forage legumes and vegetable crops. *J. Plant Nutr.* 14:613–21.

Gupta, U. C., Chapman, E. W., and MacKay, D. C. (1978). Effects of molybdenum and lime on the yield and molybdenum concentration of crops grown on acid sphagnum peat soil. *Can. J. Plant Sci.* 58:983–92.

Gupta, U. C., and Cutcliffe, J. A. (1976). Micronutrient status of rutabaga, broccoli, brussels sprouts, and cauliflower and its relationship with soil pH in Prince Edward Island. *Can. J. Plant Sci.* 56:759–61.

Gupta, U. C., LeBlanc, P. V., and Chapman, E. W. (1990). Effect of molybdenum applications on plant molybdenum concentration and crop yields on sphagnum peat soils. *Can. J. Plant Sci.* 70:717–21.

Gupta, U. C., and Lipsett, J. (1981). Molybdenum in soils, plants, and animals. *Adv. Agron.* 34:73–115.

Gupta, U. C., and MacLeod, J. A. (1978). Response to molybdenum and limestone on wheat and barley. *Commun. Soil Sci. Plant Anal.* 9:897–904.

Hafner, H., Ndunguru, B. J., Bationo, A., and Marschner, H. (1992). Effect of nitrogen, phosphorus and molybdenum application on growth and symbiotic N_2-fixation of groundnut in an acid sandy soil in Niger. *Fert. Res.* 31:69–77.

Hagstrom, G. R., and Berger, K. C. (1963). Molybdenum status of three Wisconsin soils and its effect on four legume crops. *Agronomy J.* 55:399–401.

Hashimoto, K., and Yamasaki, S. (1976). Effects of molybdenum application on the yield, nitrogen nutrition and nodule development of soybeans. *Soil Sci. Plant Nutr.* 22:435–43.

Hawes, R. L., Sims, J. L., and Wells, K. L. (1976). Molybdenum concentration of certain crop species as influenced by previous applications of molybdenum fertilizer. *Agronomy J.* 68:217–18.

Hewitt, E. J. (1956). Symptoms of molybdenum deficiency in plants. *Soil Sci.* 81:159–71.

Karimian, N., and Cox, F. R. (1979). Molybdenum availability as predicted from selected soil chemical properties. *Agronomy J.* 71:63–5.

Keogh, J. L., Maples, R. M., Hardy, G. W., and Huey, B. (1971). *Limestone and Molybdenum Studies on Soybeans.* University of Arkansas Agricultural Experiment Station.

Lipsett, J., and Simpson, J. R. (1973). Analysis of the wheat response by wheat to application of molybdenum in relation to nitrogen status. *Aust. J. Inst. Agric. Anim. Husb.* 13:563–6.

Lutz, J. A., Jr., and Lillard, J. H. (1973). Effect of fertility treatments on the growth, chemical composition and yield of no-tillage corn on orchardgrass sod. *Agronomy J.* 65:733–6.

Moore, P. A., and Patrick, W. H., Jr. (1991). Aluminum, boron and molybdenum availability and uptake by rice in acid sulfate soils. *Plant Soil* 136:171–81.

Mulder, E. G. (1954). Molybdenum in relation to growth of higher plants and micro-organisms. *Plant Soil* 5:368–415.

Murphy, L. S., and Walsh, L. M. (1972). Corrections of micronutrient deficiencies with fertilizers. In *Micronutrients in Agriculture*, ed. J. J. Mortvedt, P. M. Giordano, and W. L. Lindsay, pp. 347–87. Madison, WI: Soil Science Society of America.

Noonan, H. D. A. (1953). Molybdenum deficiency in maize and other crops in the Taree district. *Agric. Gaz. New South Wales* 64:422–4.

Panda, M. M., Reddy, M. D., and Sharma, A. R. (1991). Yield performance of rainfed lowland rice as affected by nursery fertilization under conditions of intermediate deepwater (15–50 cm) and flash floods. *Plant Soil* 132:65–71.

Parker, M. B. (1964). Molybdenum. In *Micronutrients and Crop Production in Georgia*, ed. R. L. Carter, pp. 42–52. University of Georgia Agricultural Experiment Station bulletin 126.

Parker, M. B., and Harris, H. B. (1962). Soybean response to molybdenum and lime and the relationship between yield and chemical composition. *Agronomy J.* 54:480–3.

Parker, M. B., and Harris, H. B. (1977). Yield and leaf nitrogen of nodulating and nonnodulating soybeans as affected by nitrogen and molybdenum. *Agronomy J.* 69:551–4.

Ponnamperuma, F. N. (1972). The chemistry of submerged soils. *Adv. Agron.* 24:29–96.

Price, C. A., Clark, H. E., and Funkhouser, E. A. (1972). Functions of micronutrients in plants. In *Micronutrients in Agriculture*, ed. J. J. Mortvedt, P. M. Giordano, and W. L. Lindsay, pp. 231–42. Madison, WI: Soil Science Society of America.

Purvis, E. R., and Peterson, N. K. (1956). Methods of soil and plant analysis for molybdenum. *Soil Sci.* 81:223–8.

Rebafka, F. P., Ndunguru, B. J., and Marschner, H. (1993). Single superphosphate depresses molybdenum uptake and limits yield response to phosphorus in groundnut (*Arachis hypogaea* L.) grown on an acid sandy soil in Niger, West Africa. *Fert. Res.* 34:233–42.

Reisenauer, H. M. (1956). Molybdenum content of alfalfa in relation to deficiency symptoms and response to molybdenum fertilization. *Soil Sci.* 81:237–42.

Rhoades, E. R., and Kpaka, M. (1982). Effects of nitrogen, molybdenum and cultivar on cowpea growth and yield on an oxisol. *Commun. Soil Sci. Plant Anal.* 13:279–83.

Rhoades, E. R., and Nangju, D. (1979). Effects of pelleting cowpea and soyabean seed with fertilizer dusts. *Exp. Agric.* 15:27–32.

Riley, M. M. (1987). Molybdenum deficiency in wheat in Western Australia. *J. Plant Nutr.* 10:2117–23.

Rubins, E. J. (1956). Molybdenum deficiencies in the United States. *Soil Sci.* 81:191–7.

Scheffer, J. J. C., and Wilson, G. J. (1987). Cauliflower: molybdenum application using pelleted seed foliar sprays. *N.Z. J. Exp. Agric.* 15:485–90.

Sharma, A. K., and Garg, K. P. (1973). Effect of phosphorus and molybdenum on growth and fodder yield of legumes. *Indian J. Agron.* 18:1–5.

Sims, J. L. Atkinson, W. O., and Smitobol, C. (1975). Mo and N effects on growth, yield, and Mo composition of burley tobacco. *Agronomy J.* 67:824–8.

Sims, J. L., Leggett, J. E., and Pal, U. R. (1979). Molybdenum and sulfur interaction effects on growth, yield, and selected chemical constituents of burley tobacco. *Agronomy J.* 71:75–8.

Soon, Y. K., and Bates, T. E. (1985). Molybdenum, cobalt and boron uptake from sewage-sludge-amended soils. *Can. J. Soil Sci.* 65:507–17.

Stout, P. R., and Johnson, C. M. (1956). Molybdenum deficiency in horticultural and field crops. *Soil Sci.* 81:183–90.

Stout, P. R., Meagher, W. R., Pearson, G. A., and Johnson, C. M. (1951). Molybdenum nutrition of crop plants. I. The influence of phosphate and sulfate on the absorption of molybdenum from soils and solution cultures. *Plant Soil* 3:51–87.

Tanner, P. D. (1978). A relationship between premature sprouting on the cob and the molybdenum and nitrogen status of maize grain. *Plant Soil.* 49:427–32.

Tanner, P. D., and Grant, P. M. (1977). Response of maize (*Zea mays* L.) to lime and molybdenum on acid red and yellow-brown clays and clay loams. *Rhod. J. Agric. Res.* 15:143–9.

Weir, R. G., and Hudson, A. (1966). Molybdenum deficiency in maize in relation to seed reserves. *Aust. J. Exp. Agric. Husb.* 6:35–41.

Welch, L. F., and Anderson, O. E. (1962). Molybdenum content of peanut leaves and kernels as affected by soil pH and added molybdenum. *Agronomy J.* 54:215–17.

Wilson, R. D. (1949). A field response of rock melons to molybdenum. *J. Aust. Inst. Agric. Sci.* 1949:155–6.

13

Responses of Forage Legumes and Grasses to Molybdenum

CHRIS JOHANSEN, PETER C. KERRIDGE,
and ADIB SULTANA

Introduction

In this chapter we consider the role of molybdenum (Mo) in the nutrition of legume and grass species cultivated to feed livestock, whether in the form of direct grazing in pastures or as plants cut and removed for use as fodder. It is recognized that crop residues and plant species other than legumes and grasses (e.g., brassicas) are important sources of livestock fodder worldwide and that Mo nutrition can markedly affect their production and quality, but this review is restricted to forage legumes and grasses. We concentrate mainly on legume responses to Mo, because a prime role for this element in plant growth is its involvement in the symbiotic process of nitrogen fixation (Chatt et al., 1969), although Mo imbalances in grasses have also been reported (e.g., Lipsett, 1975).

The probability of Mo deficiency generally increases with increasing soil acidity (Gupta and Lipsett, 1981), and thus we further focus toward legume responses to the Mo-deficiency component of the acid-soil syndrome. It is difficult to diagnose Mo deficiencies, largely because of the relatively small quantities of Mo required for normal plant functions, and thus there is a higher probability than for other elements that Mo deficiencies in legumes will remain undetected. Consequently, legume biomass production and additions of fixed nitrogen (N) to the soil often remain suboptimal. Thus, in this chapter we emphasize the importance of accurate diagnosis of Mo deficiency and suggest appropriate corrective measures for sustaining pastures and forage production. We mention briefly how Mo imbalances in legume and grass forage can affect the livestock that consume the fodder; further details can be found in Chapter 15.

202

Responses to Molybdenum and Their Global Distribution

Deficiencies

The likelihood of Mo deficiency in legumes and grasses increases with increasing soil acidity (e.g., Gupta and Lipsett, 1981) because of a decrease in the availability of molybdate ions with decreasing pH (Lindsay, 1972). Soil factors other than soil pH are of lesser importance in predicting the likelihood of a Mo-responsive situation. It is difficult to identify a clear-cut relationship between the parent material of a soil and its Mo responsiveness, but some correlations have been suggested. Barrow and Spencer (1971) found that Mo contents were higher in soils derived from acid magmas than in soils derived from granodiorite or sediments, but that the negative relationship between the response of *Trifolium subterraneum* to Mo in pots and the Mo content in the parent material of the derived soil was modified by the ability of the soil to adsorb Mo. Aubert and Pinta (1977) noted that the low total Mo content of weathered sandstones would tend to increase the likelihood of a response to Mo application on soils derived from them. The nature and amount of organic matter in a soil can have a range of effects on Mo availability (Stevenson and Ardakani, 1972), making it difficult to frame generalizations in this regard.

The first field response of a pasture species to Mo application was reported in 1942 for subterranean clover (*Trifolium subterraneum* L.) grown on acid soil in southern Australia (Anderson, 1956). Thereafter, Mo deficiency was commonly found to be limiting the legume component (usually subterranean clover) of the temperate-legume-based ley pasture systems on the acid soils prevalent across southern Australia (e.g., Anderson, 1956; Gladstones, Loneragan, and Goodchild, 1977). The problem in that region has become more acute in recent years because of further gradual acidification of those legume-based pasture soils (Coventry et al., 1985). Outside of Australia, there have been numerous reports of field-grown pasture and forage temperate legumes responding to Mo application at soil pH values less than 6.0 (e.g., Kubota and Allaway, 1972), though an exhaustive listing of those reports will not be attempted.

Because acid soils are common in the humid and subhumid tropics, the pasture and forage legumes grown in tropical areas are particularly prone to Mo deficiency. For example, there have been many reports on the responses to Mo by legumes grown on the acid soils of tropical and subtropical Australia (Johansen et al., 1977; Ross and Calder, 1990).

Elsewhere in the tropics, there are increasing reports of responses by legumes to Mo, including tropical legumes in Brazil (Werner, Monteiro, and Meirelles, 1983; Mattos and Colozza, 1986), black gram (*Vigna mungo* L. Hepper) and groundnut (*Arachis hypogaea* L.) in Thailand (Bell et al., 1990), and groundnut in Niger (Hafner et al., 1992). In many instances in South America it is believed that the responses to fritted trace elements are primarily due to Mo (De-Polli et al., 1979). Some tropical acidic soils cannot support the growth of tropical legumes that are particularly sensitive to Mo deficiency, such as *Neonotonia* spp., without addition of Mo fertilizer (Johansen et al., 1977). Reductions in biomass yield due to Mo deficiency can range up to 100% in the first year of testing, and for perennial tropical legumes, Mo responsiveness increases in subsequent years after establishment (Johansen et al., 1977).

As trace amounts of Mo are required for nitrate reductase activity, Mo responses can also be expected in grasses. Lipsett (1975) found that the temperate pasture grass *Phalaris aquatica* L. responded to Mo application. In Brazil, *Brachiaria decumbens* cv. Basilisk showed a 19% increase in dry matter and a 29% increase in total N production Following application of Mo at $160\,g\,ha^{-1}$ under field conditions. Larger populations of *Azospirillum* spp. appeared when Mo was applied (Miranda, Seiffert, and Dobereiner, 1985). Johansen (1978a) tested the responses to Mo for three tropical pasture grasses grown in pots. A marginal response was recorded for *Panicum maximum* var. *trichoglume* cv. Gatton (panic), but there was no response in *Cenchrus ciliaris* cv. Biloela (buffel) or *Setaria sphacelata* cv. Nandi (setaria). However, Mo application markedly reduced the high concentrations of nitrate accumulated in buffel. Excessive nitrate accumulation in forage can be toxic to ruminants (Wright and Davison, 1964); more than 0.5% was recorded in buffel grass deficient in Mo (Johansen, 1978a), in comparison with a critical level of 0.4% (Wright and Davison, 1964). The purpose of a study by Johansen (1978a) was to ascertain whether the responses of the grasses (as well as the legume components in mixtures) in field plots to Mo application were direct effects of Mo on the growth of the grasses or were due to the N fixed and transferred by the legume (with more N fixed at higher rates of Mo application). The marginal-to-negligible Mo responses of grasses in pots (Johansen, 1978a) indicated that the latter was the case.

Toxicities

Although excessive accumulation of Mo in plants is not known to adversely affect plant growth, high concentrations of Mo in plant tissue

(10–20 mg kg⁻¹) can induce copper (Cu) deficiency in ruminants that consume the material – the problem of molybdenosis (Scott, 1972). The occurrence of molybdenosis has been documented in the United States in some western states and in Florida (Kubota and Allaway, 1972). It occurs in areas with poorly drained neutral or alkaline soils formed in the granitic alluvium of wet, narrow floodplains and alluvial fans of small streams. In recent years, molybdenosis resulting from pollution of farmlands, such as from processing of Mo ores, is being increasingly reported (e.g., Schalscha, Morales, and Pratt, 1987; Smith, Brown, and Deuel, 1987). Excessive accumulation of Mo in berseem clover (*Trifolium alexandrinum* L.) has also been reported to be widespread in the wet, poorly drained, calcareous, alkaline soils of the Indo-Gangetic Plain (Nayyar, Randhawa, and Pasricha, 1977).

Species and Genotype Differences in Responses

Legumes

Although there have been many studies comparing the responses of different legumes to lime (e.g., Munns and Fox, 1977; Barnard and Fölscher, 1988), few have partitioned out the specific effect of Mo. Among temperate legumes, Gladstones et al. (1977) found that *Lupinus cosentinii*, *Vicia atropurpurea*, and subterranean clover responded to Mo in a sandy lateritic soil, but *Lupinus angustifolius*, *Lupinus luteus*, and *Trifolium hirtum* did not. Among tropical legumes, on the basis of multilocation field studies, *Neonotonia wightii* cv. Tinaroo and *Desmodium intortum* cv. Greenleaf were most responsive to Mo application, followed by *Macroptilium atropurpureum* cv. Siratro and *Medicago sativa* cv. Hunter River; *Lotononis bainesii* cv. Miles and *Stylosanthes guianenis* cv. Cook were least responsive (Johansen et al., 1977). In field studies in Malaysia, *Centrosema pubescens* was more responsive than *Pueraria phaseoloides*, with *Stylosanthes guianensis* showing no response (Tham and Kerridge, 1982). Similar differences in species responses to Mo have been observed in Brazil. For example, in one pot experiment, *Neonotonia wightii* was shown to be more responsive than *Centrosema pubescens* and *Macrotyloma axillare* (Colozza and Werner, 1984).

Among legume species, there have been some reports of genotype differences in their responses to Mo. For example, Franco and Munns (1981) reported differences in the abilities of common bean (*Phaseolus vulgaris* L.) genotypes to accumulate Mo. Among forage species,

Younge and Takahashi (1953) noted genotype differences in the responses of alfalfa (*Medicago sativa* L.) to Mo application.

Grasses

In a pot study, Lipsett (1975) found that *Phalaris tuberosa* and wheat (*Triticum aestivum* cv. Robin) responded markedly to Mo application in soil known to support subterranean clover without additional Mo in the field. No responses to Mo could be found for oats (*Avena sativa* cv. Avon), barley (*Hordeum vulgare* cv. Abyssinian), ryegrass (*Lolium perenne*), kikuyu grass (*Pennisetum clandestinum*), rape (*Brassica napus*), or subterranean clover. Among tropical grasses, there is an indication that panic is more responsive to Mo application than is buffel or setaria (Johansen, 1978a).

Diagnosis of Molybdenum Disorders

There are several possible methods for diagnosing imbalances of Mo in forage and pasture legumes and grasses, involving the use of soil maps, plant symptoms, soil analysis, plant analysis, and plant growth tests, listed here in their order of increasing precision regarding the information that they can provide about the nature and extent of the problem. It is suggested that these methods be used in a stepwise and complementary manner and that the eventual diagnosis and prognosis take account of information from as many sources as possible. Efficient design of experiments in plant growth depends on information from the first four methods listed. However, in the case of Mo, because of the low amounts involved and the limited information on critical values, particular caution is needed in interpreting analyses of Mo in soils and plants.

Soil Maps

A first step in identifying areas more likely to be prone to either Mo deficiency or Mo toxicity is to examine soil maps of the target region. Because the availability of Mo to plants is determined primarily by soil pH, knowledge of the distribution of acidic (say, pH < 6.0) and alkaline (say, pH > 7.0) soils will give a first indication of where Mo deficiencies and toxicities, respectively, are likely to be found. Care must be taken to account for the pH changes with depth in the soil profile. Some acid-soil maps of general nature are available (e.g., Wambeke, 1976), but most soil maps, and the classifications they use, do not specifically pinpoint soils with acidity problems. Sites where excessive Mo is likely to be accumulated in plants, posing a threat of molybdenosis, are easier to discern, as they comprise alkaline soils of high organic-matter content and where water is likely to accumulate (Kubota and Allaway, 1972).

Symptoms

Because the prime requirement of legumes for Mo is for symbiotic N fixation, symptoms of Mo deficiency in legumes appear as symptoms of N deficiency, generally as chlorosis, beginning with the older leaves. Of course, N deficiency in legumes can be caused by any one of the many factors that combine to determine optimum N fixation. If N-deficient legumes are poorly nodulated, then inadequate infection by rhizobia can be surmised. However, if Mo deficiency is the primary factor limiting N fixation, then the plants should be adequately nodulated, although deficiencies of other nutrients such as sulfur (S) may limit N fixation in plants that apparently are adequately nodulated.

In solution culture, at least, it is possible to produce symptoms of Mo deficiency in tropical pasture legumes that are fed on mineral N but are non-nodulated. Analogous situations can occur in the field when soil mineral N levels are high and nodulation is suppressed. As an example, in *Desmodium intortum*, symptoms of Mo deficiency appear on the leaves, ranging from those positioned at the middle to the fully expanded leaves at the top (Andrew and Pieters, 1972a). The leaflets become pale, develop concave curvature, and then become interveinally chlorotic. A broad, almost white, interveinal necrosis can develop suddenly, without any prior chlorosis. Similar such symptoms have been described in detail for the tropical pasture legumes *Macroptilium atropurpureum* (Andrew and Pieters, 1972b), *Lotononis bainesii* (Andrew and Pieters, 1976a), and *Neonotonia wightii* (Andrew and Pieters, 1976b).

Molybdenum deficiency in *Phalaris tuberosa* is manifested as pale and stunted plants with scorched and necrotic leaves, similar to those seen after frost damage (Lipsett, 1975). The symptoms appear within 3–4 weeks of seedling emergence, and subsequent growth is severely retarded. No such symptoms have been noted in conjunction with the slightly reduced growth of panic plants without Mo (Johansen, 1978a).

In summary, the symptoms of Mo deficiencies in pasture legumes and grasses are rather ambiguous and thus are of little value in diagnosing Mo deficiencies.

Soil Analysis

Analysis of soils for plant-available Mo is inherently more difficult than analysis for other essential elements because of the small quantities of Mo involved and the strong interactions between the available Mo in soil and other factors such as soil pH, organic matter, and anions. The traditional, most commonly used method of estimating plant-available Mo in soils is extraction with acid ammonium oxalate. Summarizing the prior studies, Reisenauer (1965), using the oxalate method, concluded that the critical Mo concentrations for pasture soils were around $0.2 \, mg \, kg^{-1}$ at a soil pH of 5 and $0.05 \, mg \, kg^{-1}$ at pH 6. However, there have been many reports of poor correlation between oxalate-extractable Mo and yield responses or Mo uptake by pasture and crop species (e.g., Cox and Kamprath, 1972; Little and Kerridge, 1978).

Reasonable correlations have been reported between the ability of a soil to adsorb Mo, as determined by equilibration of soil in 0.01-M $CaCl_2$ solution, and the yield responses of pasture legumes. Such correlations have been reported for subterranean clover grown on a range of soils in New South Wales, Australia (Barrow and Spencer, 1971), and for tropical pasture legumes grown on nine different soils in Queensland, Australia (Little and Kerridge, 1978). In the latter study, it was found that a soil-solution Mo concentration of around $4 \, \mu g \, L^{-1}$ was critical with respect to the responses of those legumes to Mo. The soil solution for soil equilibrated at field capacity was extracted at 15 bar, and the Mo was analyzed by atomic-absorption spectrophotometry using a graphite-furnace analyzer. That method effectively distinguished the degrees of response to Mo by plants grown on soils varying in their absolute requirements for Mo to overcome deficiency.

Several other extractants have been used to estimate plant-available Mo in soils (e.g., Cox and Kamprath, 1972; Gupta and Lipsett, 1981), but none has achieved widespread acceptance, and the ammonium oxalate extraction remains as frequently used as any other method. Biological assays have also been used to detect Mo deficiencies in soil. Franco, Peres, and Nery (1978) found that the nitrogenase activity of *Azotobacter paspali* was correlated with yield responses to Mo by the tropical forage legume *Centrosema pubescens*.

When determinations are made by ammonium oxalate extraction, excessive concentrations of Mo in herbage (leading to molybdenosis) can occur when the plant-available Mo concentrations in a soil are around $0.8 \, mg \, kg^{-1}$ at a soil pH of 5 and around $0.3 \, mg \, kg^{-1}$ in slightly acid to alkaline soils (Reisenauer, 1965). Thus the difference between deficient and toxic concentrations of Mo in a soil is relatively small.

Soil pH appears to be the main factor interacting with the estimates of Mo made using the various extractants tested, but soil iron (Fe) and other elements (e.g., S, P) can also interact (Cox and Kamprath, 1972). This limits our ability to extrapolate from the critical Mo concentrations found in one study to a range of agroenvironmental conditions; any wide extrapolation would require a considerably improved understanding of the interactions involved.

Plant Analysis

The inherent limitations and the precautions that must be taken in interpreting chemical analyses of plants to diagnose nutrient imbalances (e.g., Bates, 1971; Smith, 1986a) apply particularly to Mo. One reason is that fewer calibration studies have been conducted for this element than for the other essential elements. For legumes, it has been argued that because the major Mo requirement is for N fixation, and because nodules accumulate Mo in higher amounts than other plant parts, the Mo status of the nodules ideally would be the appropriate indicator of Mo sufficiency (Johansen, 1978b). However, the difficulties in sampling nodules and separating them from the soil limit the practicability of establishing and using critical values for Mo in nodules.

In deciding the appropriate above-ground plant parts to be used for plant analysis, the phloem mobility of the element of concern needs to be considered. In general, younger plant parts are favored for phloem-immobile elements (e.g., Ca), and older parts for mobile elements (e.g., N). The phloem mobility of Mo in legumes was originally considered as "intermediate" (Bukovac and Wittwer, 1957), but more recent reports indicate considerable phloem mobility (e.g., Kannan and Ramani, 1978; Brodrick and Giller, 1991). In *Vigna mungo*, Jongruaysup, Dell, and Bell (1994) found that Mo was variably mobile, being phloem-immobile at a low Mo supply, but phloem-mobile at an adequate Mo supply. Given these considerations and this uncertainty, the whole shoot may be the safest choice for a sample from which to interpret Mo status.

The foregoing reservations notwithstanding, a summary of the critical values to describe Mo deficiencies, determined for various forage and pasture legumes and grasses, is given in Table 13.1. Most of those determinations were made for lucerne, where a wide range of critical values has been found (0.1–$0.5\,\mathrm{mg\,kg^{-1}}$). A similar range has been reported for red clover. However, the critical values for white and subterranean clovers are more often in the vicinity of 0.10 mg kg^{-1}, and the values for tropical pasture legumes appear to be an order of magnitude lower than those for temperate legumes (Table 13.1). There is a similar scale of differences in critical Mo concentrations between temperate and tropical pasture grasses (Table 13.1). At best, the data in Table 13.1 should be considered as offering only a guideline as to whether or not the growth of a particular species is likely to be limited by a deficiency of Mo. For example, a value of 0.3 mg Mo per kilogram of dry matter in subterranean clover would indicate a high probability of Mo sufficiency, but a value of 0.05 mg kg^{-1} would suggest that growth tests would be needed to determine whether or not the plants would respond to Mo application in that situation. Because of the low concentrations of Mo in plant tissue, it is essential that precautions be taken to minimize contamination during sampling, drying, and grinding of samples. Some reports of high Mo concentrations may have been due to sample contamination.

Molybdenum can accumulate to high concentrations (e.g., several hundred milligrams per kilogram of tissue) without obvious damage to plants (Johnson, 1966; Jones, 1972), but concentrations exceeding 20 mg kg^{-1} can result in molybdenosis in ruminants grazing the herbage (Gupta and Lipsett, 1981). Thus, because of the calibrations that have already been carried out, and because the higher Mo concentrations that are toxic to ruminants can be measured relatively easily, plant analyses are more useful for diagnosing potentially toxic (to ruminants) concentrations of Mo than for detecting Mo deficiencies that can limit plant growth.

Plant Growth Tests

In view of the imprecision involved in using the aforementioned methods, tests of plant growth ultimately are necessary to determine if any forage/pasture legume or grass is limited by Mo deficiency. The technique of pot culture, conducted in environmental conditions as nonlimiting as possible, can provide reasonable estimates of any limitations due to Mo deficiency (e.g., Foy and Barber, 1959; Johansen et al., 1977; Standley, Bruce, and Webb, 1990). Deficiencies of Mo can be secondary to other nutrient deficiencies, such as P deficiency in acid soils, and care must be taken to ensure optimum supplies of other essential elements so as to permit maximum expression of any Mo limitation. Half-factorial designs, with presence or absence of elements most likely to be limiting in a particular soil, are recommended for this purpose (e.g., Standley et al., 1990). Such designs also permit detection of nutrient interactions; for example, Mo strongly interacts with applications of lime (Kerridge, Andrew, and Murtha, 1972), P, and S (Gupta and Lipsett, 1981). A factorial design involving Mo and lime will allow the extent of Mo deficiency to be determined; a strong Mo response in the presence of lime will indicate gross Mo deficiency (Kerridge et al., 1972). For legumes, dependence on N fixation should be ensured by inoculating with appropriate rhizobial strains and by not adding mineral N.

Prior soil and plant analyses for Mo will assist in determining whether or not pot studies are justified. The responses of plants in pots to Mo will give some indication of the likelihood of a field response and the appropriate rates of Mo application to be used in field trials. The responses in pots normally are greater than field responses, though generally there is good correspondence between the extent of the response to Mo in pots and the extent of field responsiveness with respect to soils that vary in their degrees of Mo deficiency or plants with varying sensitivities to Mo deficiency (e.g., Johansen et al., 1977). Field trials are necessary to determine if there is a need for fertilization with Mo and to decide on the appropriate corrective measures for different Mo-deficient situations. Application of excessive amounts of Mo as a corrective measure can lead to Mo concentrations in the herbage high enough to cause molybdenosis, and thus one should determine or estimate the nature of the Mo response curve under field conditions.

Field trials can be carried out on freshly sown pastures or on established pastures. On freshly sown pastures, Mo deficiency can be masked at first by high soil concentrations of N mineralized during cultivation. Thus the severity of the Mo deficiency can increase over time, not only because of adsorption of applied Mo but also because of reduced availability of N in the soil. Legume responses to Mo are stronger when legumes are grown in association with grasses, which generally are more competitive for the available N in the soil. Thus, in Malaysia, where *Centrosema pubescens* responded to Mo in pot trials (Watson, 1960), there was no field response when it was grown as a component in a leguminous cover-crop mixture (Watson, Wong, and Narayanan, 1963), but there was a strong response when it was grown in association with a grass (Tham and Kerridge, 1982).

It may be difficult to demonstrate a response on established grass-legume pastures because of a variable legume composition. Under such conditions, it has been demonstrated that the Mo requirement can be

Table 13.1. *Critical concentrations or ranges for Mo deficiency in some pasture and forage legumes and grasses*

Plant	Critical Mo concentration or range in dry matter (mg kg^{-1})	Plant part	Growth stage	Reference
Legumes				
Lucerne (*Medicago sativa* L.)	0.26–0.28	Leaves	10% bloom	Reisenauer (1956)
	<0.4	Top 15 cm	Prior to bloom	Jones (1967)
	<0.5	Upper stem	Early flowering	Melsted et al. (1969)
	<0.2	Top third of shoot	Pre-flowering	Neubert et al. (1970)
	<0.5	Top 15 cm	Vegetative	Cornforth and Sinclair (1982), quoted by Smith (1986b)
	0.3–0.5	Top half of shoot	First cut	James et al. (1968)
	0.10–0.14	Whole shoot	Hay	Cornforth and Sinclair (1982), quoted by Smith (1986b)
	<0.2	Whole shoot	Early flowering	Peverill (1984), quoted by Smith (1986b)
	<0.12	Whole shoot	10% bloom	Gupta (1970)
	0.55–1.15	Whole shoot	Harvest	Evans and Purvis (1951)
Red clover (*Trifolium pratense* L.)	<0.46	Whole shoot	Bud stage	Gupta (1970)
	0.1–0.2	Whole shoot	Bud stage	Hagstrom and Berger (1965)
	<0.22	Whole shoot	10% bloom	Gupta and MacLeod (1975)
	<0.26	Whole shoot	Flowering	Hawes et al. (1976)
	<0.15	Whole shoot	Flowering	Neubert et al. (1970)
White clover (*Trifolium repens* L.)	0.10–0.14	Leaf blades + petioles	Vegetative	Cornforth and Sinclair (1982), quoted by Smith (1986b)
Subterranean clover (*Trifolium subterraneum* L.)	0.05–0.10	Youngest open leaf	Not known	Robson (1984), quoted by Smith (1986b)
	0.15	Whole shoot	Mid-flowering	Gartrell (1980), quoted by Smith (1986b)

Species	Value	Plant part	Stage	Reference
Siratro (*Macroptilium atropurpureum*)	<0.02	Whole shoot	Harvest	Johansen (1978b)
Glycine (*Neonotonia wightii*)	<0.02	Whole shoot	Harvest	Johansen (1978b)
Lotononis (*Lotononis bainesii*)	<0.02	Whole shoot	Harvest	Johansen (1978b)
Desmodium (*Desmodium intortum*)	0.03	Whole shoot	70 days	Johansen (1978a)
Grasses				
Perennial ryegrass (*Lolium perenne*)	<0.2 <0.15	Leaf Whole shoot	Flowering Vegetative	Peverill (1984), quoted by Smith (1986b) Cornforth and Sinclair (1982), quoted by Smith (1986b)
Timothy (*Phleum pratense* L.)	<0.11	Whole shoot	Pre-bloom	Gupta and MacKay (1968)
Green panic (*Panicum maximum* var. *trichoglume*)	<0.02	Whole shoot	5 weeks	Johansen (1978a)
Buffel grass (*Cenchrus ciliaris*)	<0.02	Whole shoot	5 weeks	Johansen (1978a)
Setaria (*Setaria sphacelata*)	<0.02	Whole shoot	5 weeks	Johansen (1978a)

determined with greater precision by measuring the legume N concentrations before and after application of Mo fertilizer (Kerridge, 1981). Lack of a field response to Mo can also be due to deficiencies of other nutrients or to Mo contamination in commercial fertilizers. Another factor that can mask a soil Mo deficiency is the use of test plants that are not highly sensitive to Mo deficiency. Thus, although *Stylosanthes* spp. have shown responses to Mo in pot experiments, Mo deficiency has not been verified in the field using *Stylosanthes* under the same conditions in which strong responses have been demonstrated by other legumes (Johansen et al., 1977; Tham and Kerridge, 1982). Conversely in plant-evaluation studies on acid soils where Mo has not been applied, it is likely that some species that otherwise might be will adapted to that environment have been discarded.

Corrective Options

Once Mo deficiencies have been detected, it is relatively easy and inexpensive to correct them, primarily because of the low amounts of Mo required by plants and the many sources available. However, the major impediment to alleviation of Mo deficiency in forage species is its correct diagnosis in the first instance.

Species and Genotype Selection

Unlike the situation for crop plants, for a given agroenvironment there usually are several choices among adapted legume and grass species to use in a pasture or for forage purposes. Thus, for a region prone to Mo deficiency, there is scope for avoiding species more susceptible to Mo deficiency. For example, among tropical pasture legumes, *Macroptilium atropurpureum* could be substituted for *Desmodium intortum* or *Neonotonia wightii* (Johansen et al., 1977), and *Stylosanthes* spp. for *Centrosema* spp. (Tham and Kerridge, 1982). Similarly, within a species (e.g., lucerne) it is possible to select and use cultivars less susceptible to Mo deficiency (Younge and Takahashi, 1953). However, currently our lack of knowledge of species and genotype differences in response to Mo limits this approach.

Liming

Because Mo availability in a soil is largely determined by soil pH (Lindsay, 1972), liming is a means of alleviating Mo deficiency. Figure

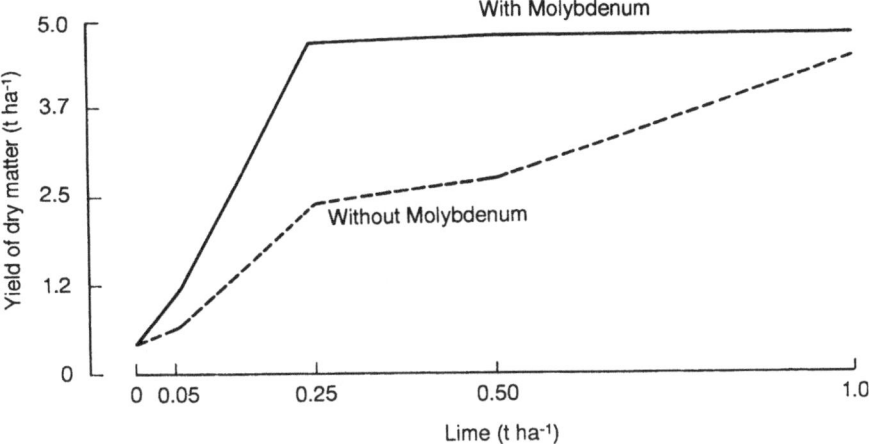

Figure 13.1. Effects of applying molybdenum trioxide at 140 g ha^{-1} in the presence of different rates of lime on the dry-matter yield of subterranean clover (Grabben Gullen, New South Wales, Australia, 1948–49). (Reprinted from Anderson and Moye, 1952, with permission from the Commonwealth Scientific and Industrial Research Organization, Melbourne.)

13.1 shows an example of the classic negative interaction between lime and Mo. There have been many reports of lime being used to correct known Mo deficiencies (e.g., Anderson, 1956; Adams, 1978). However, if Mo deficiency is the main limitation imposed by an acidic soil condition, direct application of Mo fertilizer may be more economical than trying to lime the soil to a pH value where sufficient Mo will be available in the soil solution (Doerge, Bottomley, and Gardner, 1985). For many acid soils, quantities of lime on the order of several tons per hectare would be needed to shift the soil pH above 6.0 (Adams, 1978), which would be difficult to justify and execute in low-input pasture situations. Optimum rates of lime application to correct Mo deficiencies need to be established on a site-specific basis, because of the multiple interactions that can occur when a soil is limed (Anderson, 1956). In some situations, plant responses to added Mo can be seen even after liming, suggesting an absolute deficiency of Mo (Foy and Barber, 1959), and that can be overcome only by direct application of Mo (Coventry, Hirth, and Fung, 1987; Burmester, Adams, and Odom, 1988).

Molybdenum Fertilizers

Application of only small amounts of Mo fertilizer, in soluble form as molybdate (Mo at less than $100 g ha^{-1}$), can correct Mo deficiencies in pasture and forage legumes, at least in the first year following application (Anderson, 1956; Murphy and Walsh, 1972; Reuter, 1975; Johansen et al., 1977). That allows many options for effective application of Mo. In formulating fertilizers, the soluble molybdate salts (e.g., ammonium or sodium molybdate) are generally preferred to the less soluble molybdenum trioxide (MoO_3) or molybdenite (MoS_2) (Murphy and Walsh, 1972), though in practice the fertilizer companies will use the cheaper sources of Mo. Because of the small amounts of Mo required, Mo is usually mixed with a macronutrient fertilizer carrier, the most common form of which is molybdenized superphosphate (Reuter, 1975). However, careful mixing and regular checking are required to ensure uniform distribution of the Mo throughout the carrier (Lipsett and David, 1977). It is preferable that the Mo be incorporated into the product, rather than added and mixed at the conclusion of the manufacturing process.

A more effective method for applying Mo to forage legumes is by seed application (Reisenauer, 1963; Murphy and Walsh, 1972). Molybdenum is normally applied to the seed in a liquid or slurry form, as part of the lime pelleting, rock phosphate pelleting, *Rhizobium* inoculation, and/or fungicidal treatment. However, when inoculating legume seeds with rhizobia, the more soluble molybdate forms of Mo can inhibit nodulation, and thus MoO_3 is preferred (Date and Hillier, 1968; Gartrell, 1969; Tripathi and Edward, 1978; Gault and Brockwell, 1980). Nevertheless, the adverse effects of molybdate on rhizobia can be minimized if the rhizobial inoculation is applied to the seed first and is separated from the Mo by a layer of lime (Sims, Sigafus, and Tiaranan, 1974). Kerridge, Cook, and Everett (1973), Johansen et al. (1977), and Kerridge (1978) found that seed pelleting with MoO_3 and soil application at the same rate of Mo had similar effects in alleviating Mo deficiency in a range of tropical pasture legumes.

Foliar sprays can effectively deliver Mo provided that appropriate precautions (e.g., use of an effective wetting agent) are taken (Murphy and Walsh, 1972; Gupta and Lipsett, 1981), although this method is perhaps more feasible for high-value crops with high Mo requirements, such as brassicas (Gupta and Lipsett, 1981). Use of foliar sprays containing Mo may be considered for forage crops if the Mo deficiency is detected only during the course of growth, as well as for perennial stands.

For black gram (*Vigna mungo* L. Wilczek) and green gram (*V. radiata* L. Wilczek), foliar sprays were found to be only slightly less effective than seed treatment for correcting Mo deficiencies (Anwarulla and Shivashankar, 1987). Application in a spray is appropriate in extensive pastures where legumes are introduced by band seeding accompanied by herbicide application (Cook, MacLeod, and Walsh, 1992).

Another way of ensuring adequate Mo for a crop or sward is to sow with seeds that naturally have a high Mo content, as Mo is preferentially accumulated in the seeds. For example, Gurley and Giddens (1969) found that the Mo requirement of soybeans grown on Mo-deficient soil could be met by using seeds with a high natural concentration of Mo. They concluded that foliar application of Mo to a seed crop, or else growing the seed crop at a high soil pH, might be a practical means of producing seeds for Mo-deficient target areas. However, that would be feasible only for species that have large seeds and could accumulate a large quantity of Mo per seed; most pasture and forage legumes have small seeds, as compared with grain crops.

Residual Effects

In addition to determining the optimum Mo requirements for establishment of a forage crop or pasture, it is important to understand the residual effects of applied Mo in order to determine the appropriate practices for repeated applications. In general, small initial applications of Mo will have long-lasting residual effects. Anderson (1956) reported that an application of Mo at 140 g ha^{-1} was fully effective for correcting severe deficiencies in subterranean clover pastures in southern Australia 10 years after application. For subterranean clover pastures in Western Australia, Barrow et al. (1985) determined that the effectiveness of Mo, applied as sodium molybdate, declined over time in a sigmoidal fashion, with the residual Mo in the second year being about half as effective as the original application. To the contrary, Swain (1959), using pot trials to evaluate responses, could find no residual effect 3 years after Mo application at 80 g ha^{-1} on subterranean clover growing on a krasnozem soil. For white clover (*Paspalum dilatatum*) pastures on a krasnozem soil developed on basalt, an initial application of Mo at 75 g ha^{-1} was fully effective for 2 years and gave 70% of maximum yield after 4 years (Kerridge and White, 1977). For rotations of rice and berseem clover (*Trifolium alexandrinum*) in West Bengal, India, application of Mo to one crop has been shown to have residual effects on the following

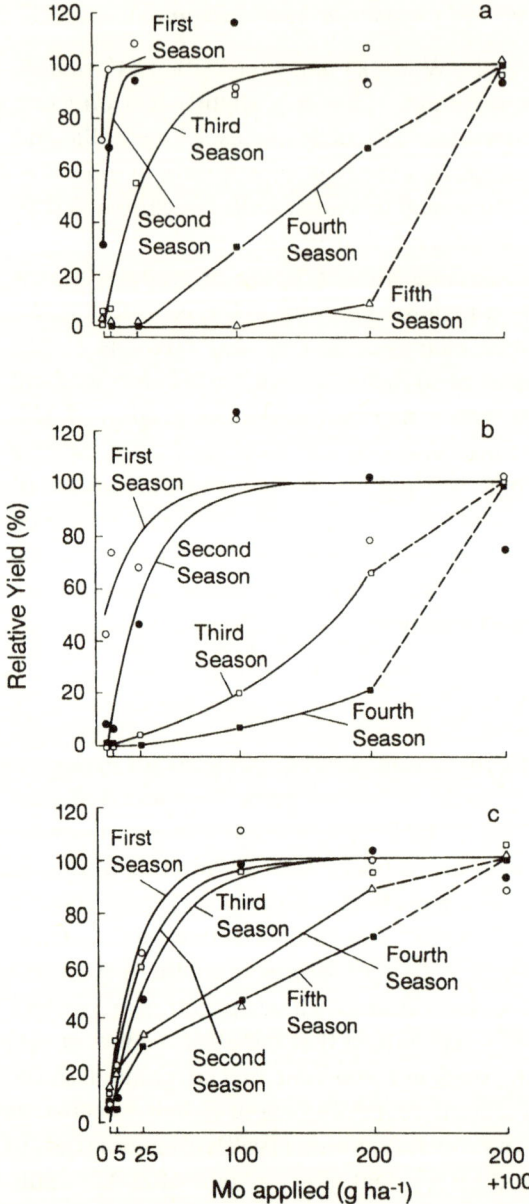

Figure 13.2. Responses to initial and maintenance (200 + 100; 200 g Mo ha^{-1} in first year and 100 g Mo ha^{-1} in subsequent years) applications of Mo as molybdenum trioxide by (a) desmodium (*Desmodium intortum*) at Cooroy, (b) glycine (*Neonotonia wightii*) at Cooroy, and (c) glycine at Glastonbury in southeastern Queensland, 1971–76. (Reprinted from Johansen et al., 1978, with permission from the Commonwealth Scientific and Industrial Research Organization, Melbourne.)

crop of the other species (Samui and Bhattacharya, 1982; Samui and Dasgupta, 1982).

Johansen et al. (1977) calculated the residual effects of added Mo over a 5-year period for several tropical legumes at a range of sites, from highly responsive to marginally responsive, in subtropical Queensland. Figure 13.2 shows some examples of the changes over time in the response functions for the responsive species *Desmodium intortum* and *Neonotonia wightii*. By using fitted Mitscherlich equations, it was possible to calculate the initial Mo applications required to support maximum growth of the legume-grass swards over 5 years (Johansen et al., 1977). For example, at the most responsive site, Mo at $200\,g\,ha^{-1}$ supported maximum growth of *Macroptilium atropurpureum* over 5 years, *D. intortum* over 3 years, and *N. wightii* only for 2 years. There was no response of *M. atropurpureum* to Mo at another site over the entire 5-year period. In general, the residual effectiveness of Mo was inversely proportional to the initial response during the first 2 years. The response to Mo by the companion grass (*Panicum maximum*) followed that of the legume, because of the residual effects of N fixed by the legume (Johansen et al., 1977; Johansen, 1978a; Johansen and Kerridge, 1979). In those studies, therefore, both the initial response and the decline in residual effectiveness were inversely related to soil pH in the effective rooting-soil profile and were directly related to the ability of the soil to adsorb Mo (Johansen et al., 1977; Little and Kerridge, 1978). As for the initial responsiveness to Mo, there appear to be large site and species differences in long-term requirements for Mo. These cannot be readily predicted on the basis of our current knowledge, and thus a continuation of site and species-specific field experimentation is required. We need studies of the relationships between field and pot responses to soil types and species sensitivities to Mo deficiencies. In the future, effort should also be directed at relating the requirements for Mo and the residual effects of Mo to those for phosphorus, as the processes of adsorption to soil are similar for the two elements (Barrow and Shaw, 1974).

Forage Quality

Although a low dietary intake of Mo can result in Mo deficiency in some animals (Scott, 1972), the main problem faced by animals grazing pastures and forage is consumption of excessive amounts of Mo (Kubota and Allaway, 1972; Suttle, 1975). Excessive intake of Mo will induce Cu deficiency in ruminants (Suttle, 1975). Although herbage Mo concentrations above $20\,\text{mg}\,\text{kg}^{-1}$ are generally considered as toxic, toxicity is actually determined by the ratio between Cu and Mo (Gupta and Lipsett, 1981). Ratios of Cu:Mo in the range of $2:1$ to $7:1$ have been reported as critical, but Cu deficiency generally occurs if feed intake of Cu is less than $5\,\text{mg}\,\text{kg}^{-1}$ (Gupta and Lipsett, 1981). Molybdenosis can be alleviated by supplementing the animal's intake with Cu or with sulfate, which immobilizes Mo (Suttle, 1975). Similarly, application of a sulfur compound to the forage crop, as when gypsum is added to alleviate alkaline soil conditions, can substantially reduce the Mo concentration in the foliage from a toxic to a safe amount (e.g., Sisodia, Sawarkar, and Rai, 1975; Gupta and MacLeod, 1975; Pasricha et al., 1977; Gupta and Mehla, 1980). Application of N fertilizer can similarly reduce foliar Mo from toxic to safe concentrations (Reith et al., 1984).

As the converse of molybdenosis, Cu toxicity can develop in ruminants grazing on pastures containing Cu at $10-15\,\text{mg}\,\text{kg}^{-1}$ and very low amounts of Mo (e.g., $<0.2\,\text{mg}\,\text{kg}^{-1}$) (Underwood, 1966).

Molybdenum concentrations in grasses can affect animal nutrition indirectly, through effects on nitrate content. Because Mo is required for nitrate reductase activity, low concentrations of Mo can cause accumulation of nitrate to concentrations toxic to ruminants ($>0.4\%$) (Wright and Davison, 1964; Johansen, 1978a). This is likely to be a more serious problem in young plants.

Research Needs

The major emphasis in our research to optimize Mo nutrition for pasture and forage legumes and grasses should be more widespread use of uniform, systematic methods of diagnosis for Mo deficiencies. This particularly applies to the humid, tropical acid-soil regions of Asia, Africa, and South America, where relatively few such diagnoses have been attempted. As discussed earlier, particular care must be taken in the diagnostic process because of the minute levels of Mo involved. The reported critical concentrations of Mo in soils and plants can provide only broad indications of the likelihood of deficiency, because of the limited number of species tested and the unique characteristics of the sites and experimental procedures used to determine such concentrations. It is suggested that pot screening techniques combined with some form of field validation should receive the main emphasis in our efforts to diagnose Mo deficiencies. In South America, though there have been many reports of responses to Mo in pot experiments, there is a general feeling that Mo is not required, because of lack of field validation of the pot experiments. As it is likely that half of the 200 million hectares of savannas in South America will be developed for pasture or cropland in the next few decades (Vera et al., 1992), there is some sense of urgency that we should undertake more extensive diagnoses. There is also a need for further refinement of our techniques for soil and plant analysis, principally in the area of a better understanding of the interactions involved, which ultimately determine the available Mo in the soil solution and its concentration in plant tissue. This will allow wider extrapolation from the critical values than is now possible.

In conducting studies on Mo applications and other treatments (e.g., lime) designed to alleviate Mo deficiencies, we must pay due attention in our experimental designs to calculating the residual effects of applied Mo. These are usually substantial, and thus a major limitation to legume production in acid soils can be overcome by judicious application of only small quantities of Mo. Because of the threat of molybdenosis, it is also important to avoid excessive applications aimed at prolonging the residual effects.

There are indications of large differences between species, and between genotypes within a species, in the ability to take up Mo and accumulate it in plant tissue. Greater knowledge of such differences will allow exploitation of the genetic option to address Mo deficiency and toxicity problems. This is particularly so for pasture legumes and grasses, for which there usually are many possible choices of species and genotypes that can be used in a particular agroenvironment.

The increasing uses of industrial by-products as soil additives for pastures and forage crops, and their increasing cultivation on land exposed to actual and potential pollution, highlight the need for continual monitoring for Mo toxicity effects. Fortunately, the soil and plant concentrations of Mo that are diagnostic for toxicity are more precisely known than are those for Mo deficiency.

Summary

Molybdenum deficiencies are widespread in areas with acid soils, and they are increasingly being detected in pasture and forage legumes. They have been detected in some grass species as well. In alkaline soils, where molybdate is much more readily available in the soil solution, legumes and grasses are prone to accumulate toxic amounts of Mo, leading to molybdenosis, or induced Cu deficiency, in ruminants. The limited evidence thus far available indicates large species and genotype differences in the responses of plants to Mo application and in the ability to accumulate Mo in plant tissue.

This review emphasizes the importance of accurate diagnosis of Mo deficiency, in view of the effects that even small quantities of Mo (compared with other essential elements) can have on plant responses. A systematic and conservative approach is recommended when using and interpreting soil information, plant symptoms, soil analysis, plant analysis, pot trials, and field trials. Although there have been several localized analytical determinations of the critical Mo concentrations that define deficiency for soils and plants, such values cannot be extrapolated widely because of the many interactions affecting Mo concentrations in soils and plants. Pot techniques are emphasized as effective diagnostic tools.

Once Mo deficiencies are detected, there are many possible corrective measures, including species/genotype selection and liming. Serious Mo deficiencies in legumes can be corrected with small quantities of Mo by applying it directly to the soil in a carrier fertilizer, by applying it to the seeds, by applying it as a foliar spray, or by producing seeds of high Mo content. On most soils, the applied Mo will have a good residual effect. In summary, there are many effective means to alleviate both Mo deficiencies and toxicities. The emphasis in future research should be on accurate, systematic diagnoses and on better delineation of those areas where Mo deficiencies and toxicities are most often encountered.

References

Adams, F. (1978). Liming and fertilization of ultisols and oxisols. In *Mineral Nutrition of Legumes in Tropical and Subtropical Soils*, ed. C. S. Andrew and E. J. Kamprath, pp. 377–94. Melbourne: Commonwealth Scientific and Industrial Research Organization.

Anderson, A. J. (1956). Molybdenum as a fertilizer. *Adv. Agron.* 8:163–202.

Anderson, A. J., and Moye, D. V. (1952). Lime and molybdenum in clover development on acid soils. *Aust. J. Agric. Res.* 3:95–110.

Andrew, C. S., and Pieters, W. H. J. (1972a). *Foliar Symptoms of Mineral Disorders in Desmodium intortum*. Melbourne: CSIRO Division of Tropical Pastures technical paper 10.

Andrew, C. S., and Pieters, W. H. J. (1972b). *Foliar Symptoms of Mineral Disorders in Phaseolus atropurpureus*. Melbourne: CSIRO Division of Tropical Pastures technical paper 11.

Andrew, C. S., and Pieters, W. H. J. (1976a). *Foliar Symptoms of Mineral Disorders in Lotononis bainesii*. Melbourne: CSIRO Division of Tropical Agronomy technical paper 17.

Andrew, C. S., and Pieters, W. H. J. (1976b). *Foliar Symptoms of Mineral Disorders in Glycine wightii.* Melbourne: CSIRO Division of Tropical Agronomy technical paper 18.

Anwarulla, M. S., and Shivashankar, K. (1987). Influence of seed treatment and foliar nutrition of molybdenum on green gram and black gram. *J. Agric. Sci.* 108:627–34.

Aubert, H., and Pinta, M. (1977). Molybdenum. In *Trace Elements in Soils*, ed. H. Aubert and M. Pinta, pp. 55–62. Amsterdam: Elsevier.

Barnard, R. O., and Fölscher, W. J. (1988). Growth of legumes at different levels of liming. *Trop. Agric.* 65:113–16.

Barrow, N. J., Leahy, P. J., Southey, I. N., and Purser, D. B. (1985). Initial and residual effectiveness of molybdate fertilizer in two areas of south Western Australia. *Aust. J. Agric. Res.* 36:579–87.

Barrow, N. J., and Shaw, T. C. (1974). Factors affecting the long-term effectiveness of phosphate and molybdate fertilizers. *Commun. Soil Sci. Plant Anal.* 5:355–64.

Barrow, N. J., and Spencer, K. (1971). Factors in the molybdenum and phosphorus status of soils on the Dorrigo Plateau of New South Wales. *Aust. J. Exp. Agric. Anim. Husb.* 11:670–76.

Bates, T. E. (1971). Factors affecting critical nutrient concentrations in plants and their evaluation: a review. *Soil Sci.* 112:116–30.

Bell, R. W., Rerkasem, B., Keerati-Kasikorn, P., Phetchawee, S., Hiranburana, N., Ratanarat, S., Pongsakul, P., and Loneragan, J. F. (1990). *Mineral Nutrition of Food Legumes in Thailand with Particular Reference to Micronutrients.* ACIAR technical report 16. Canberra: Australian Centre for International Agricultural Research.

Brodrick, S. J., and Giller, K. E. (1991). Root nodules of *phaseolus*: efficient scavengers of molybdenum for N_2 fixation. *J. Exp. Bot.* 42:679–86.

Bukovac, M. J., and Wittwer, S. H. (1957). Absorption and mobility of foliar applied nutrients. *Plant Physiol.* 32:428–35.

Burmester, C. H., Adams, J. F., and Odom, J. W. (1988). Response of soybean to lime and molybdenum on ultisols in northern Alabama. *Soil Sci. Soc. Am. J.* 52:1391–4.

Chatt, J., Dilworth, J. R., Richards, R. L., and Sanders, J. R. (1969). Chemical evidence concerning the function of molybdenum in nitrogenase. *Nature* 224:1201–2.

Colozza, M. T., and Werner, J. C. (1984). Aplicacao de nutrientes em tres leguminosas forrageiras cultivadas num solo da regiao do Vale do Ribeira. *Zootecnia (Brasil)* 22:327–53.

Cook, S. J., MacLeod, N. D., and Walsh, P. A. (1992). Reliable and cost-effective legume establishment in black speargrass grazing lands. In *Proceedings of the 6th Australian Society of Agronomy Conference*, *"Looking Back–Planning Ahead,"* University of New England, Armidale 10–14 February, Australian Society of Agronomy, Parkville, Victoria 3052, pp. 406–409.

Coventry, D. R., Hirth, J. R., and Fung, K. K. H. (1987). Nutritional restraints on subterranean clover grown on acid soils used for crop-pasture rotation. *Aust. J. Agric. Res.* 38:163–76.

Coventry. D. R., Hirth, J. R., Reeves, T. G., and Burnett, V. F. (1985). Growth and nitrogen fixation by subterranean clover in response to inoculation, molybdenum application and soil amendment with lime. *Soil Biol. Biochem.* 17:791–6.

Cox, F. R., and Kamprath, E. J. (1972). Micronutrient soil tests. In *Micronutrients in Agriculture*, ed. J. J. Mortvedt, P. M. Giordano, and W. L. Lindsay, pp. 289–317. Madison, WI: Soil Science Society of America.

Date, R. A., and Hillier, G. R. (1968). Molybdenum application in the lime of lime-pelleted subterranean clover seed. *J. Aust. Inst. Agric. Sci.* 34:171–2.

De-Polli, H., de Carvalho, S. R., Lemos, P. F., and Franco, A. A. (1979). Efeito de micronutrientes no estabelecimento e persistencia de leguminosas em pastagens de morro em solo Podzolico Vermelho-Amarelo. *Rev. Bras. Cien. Solo* 3:154–7.

Doerge, T. A., Bottomley, P. J., and Gardner, E. H. (1985). Molybdenum limitations to alfalfa growth and nitrogen content on a moderately acid, high-phosphorus soil. *Agronomy J.* 77:895–901.

Evans, H. J., and Purvis, E. R. (1951). Molybdenum status of some New Jersey soils with respect to alfalfa production. *Agronomy J.* 43:70–1.

Foy, C. D., and Barber, S. A. (1959). Molybdenum response of alfalfa on Indiana soils in the greenhouse. *Soil Sci. Soc. Am. J.* 23:36–9.

Franco, A. A., and Munns, D. N. (1981). Response of *Phaseolus vulgaris* L. to molybdenum under acid conditions. *Soil Sci. Soc. Am. J.* 45:1144–8.

Franco, A. A., Peres, J. R. R., and Nery, M. (1978). The use of *Azotobacter paspali* N_2-ase (C_2H_2-reduction activity) to measure molybdenum deficiency in soils. *Plant Soil* 50:1–11.

Gartrell, J. W. (1969). The effect of sodium molybdate mixed in the lime seed pellet on nodulation, nitrogen content and growth of subterranean clover. *Aust. J. Exp. Agric. Anim. Husb.* 9:432–6.

Gault, R. R., and Brockwell, J. (1980). Studies on seed pelleting as an aid to legume inoculation. 5. Effects of incorporation of molybdenum compounds in the seed pellet on inoculant survival, seedling nodulation and plant growth of leucerne and subterranean clover. *Aust. J. Exp. Agric. Anim. Husb.* 20:63–71.

Gladstones, J. S., Loneragan, J. F., and Goodchild, N. A. (1977). Field responses to cobalt and molybdenum by different legume species, with inferences on the role of cobalt in legume growth. *Aust. J. Agric. Res.* 28:619–28.

Gupta, U. C. (1970). Molybdenum requirement of crops grown on a sandy clay loam soil in the greenhouse. *Soil Sci.* 110:280–2.

Gupta, U. C., and Lipsett, J. (1981). Molybdenum in soils, plants and animals. *Adv. Agron.* 34:73–115.

Gupta, U. C., and MacKay, D. C. (1968). Crop responses to applied molybdenum and copper on podzol soils. *Can. J. Soil Sci.* 48:235–42.

Gupta, U. C., and MacLeod, L. B. (1975). Effects of sulfur and molybdenum on the molybdenum, copper and sulfur concentrations of forage crops. *Soil Sci.* 119:441–7.

Gupta, V. K., and Mehla, D. S. (1980). Influence of sulphur on the yield and concentration of copper, manganese, iron and molybdenum in berseem (*Trifolium alexandrinum*) grown on two different soils. *Plant Soil* 56:229–34.

Gurley, W. H., and Giddens, J. (1969). Factors affecting uptake, yield response and carryover of molybdenum in soybean seeds. *Agronomy J.* 61:7–9.

Hafner, H., Ndunguru, B. J., Bationo, A., and Marschner, H. (1992). Effect of nitrogen, phosphorus and molybdenum application on growth and symbiotic N₂-fixation of groundnut in an acid sandy soil in Niger. *Fert. Res.* 31:69–77.

Hagstrom, G. R., and Berger, K. C. (1965). Molybdenum deficiencies of Wisconsin soils. *Soil Sci.* 100:52–6.

Hawes, R. L., Sims, J. L., and Wells, K. L. (1976). Molybdenum concentration of certain crop species as influenced by previous applications of molybdenum fertilizer. *Agronomy J.* 68:217–18.

James, D. W., Jackson, T. L., and Harward, M. E. (1968). Effect of molybdenum and lime on the growth and molybdenum content of alfalfa grown on acid soils. *Soil Sci.* 105:397–402.

Johansen, C. (1978a). Response of some tropical grasses to molybdenum application. *Aust. J. Exp. Agric. Anim. Husb.* 18:732–6.

Johansen, C. (1978b). Comparative molybdenum concentrations in some tropical pasture legumes. *Commun. Soil Sci. Plant Anal.* 9:1009–17.

Johansen, C., and Kerridge, P. C. (1979). Nitrogen fixation and transfer in tropical legume-grass swards in south-eastern Queensland. *Trop. Grassl.* 13:165–70.

Johansen, C., Kerridge, P. C., Luck, P. E., Cook, B. G., Lowe, K. F., and Ostrowski, H. (1977). The residual effect of molybdenum fertilizer on growth of tropical pasture legumes in a sub-tropical environment. *Aust. J. Exp. Agric. Anim. Husb.* 17:961–8.

Johansen, C., Kerridge, P. C., Merkley, K. E., Luck, P. E., Cook, B. G., Lowe, K. F., and Ostrowski, H. (1978). *Growth and Molybdenum Response of Tropical Legume/Grass Swards at Six Sites in South-eastern Queensland over a Five Year Period.* Tropical agronomy technical memorandum 10, Division of Tropical Crops and Pastures, CSIRO, Queensland.

Johnson, C. M. (1966). Molybdenum. In *Diagnostic Criteria for Plants and Soils*, ed. H. D. Chapman, pp. 286–301. University of California Press.

Jones, J. B., Jr. (1967). Interpretation of plant analysis for several agronomic crops. In *Soil Testing and Plant Analysis*, part 2, ed. G. W. Hardy, pp. 49–58. Madison, WI: Soil Science Society of America.

Jones, J. B., Jr. (1972). Plant tissue analysis for micronutrients. In *Micronutrients in Agriculture*, ed. J. J. Mortvedt, P. M. Giordano, and W. L. Lindsay, pp. 319–46. Madison, WI: Soil Science Society of America.

Jongruaysup, S., Dell, B., and Bell, R. W. (1994). Distribution and redistribution of molybdenum in black gram (*Vigna mungo* L. Hepper) in relation to molybdenum supply. *Ann. Bot.* 73:161–7.

Kannan, S., and Ramani, S. (1978). Studies on molybdenum absorption and transport in bean and rice. *Plant Physiol.* 62:179–81.

Kerridge, P. C. (1978). Fertilization of acid tropical soils in relation to pasture legumes. In *Mineral Nutrition of Legumes in Tropical and Subtropical Soils*, ed. C. S. Andrew and E. J. Kamprath, pp. 395–415. Melbourne: Commonwealth Scientific and Industrial Research Organization.

Kerridge, P. C. (1981). Determination of molybdenum requirement on established grass-legume pasture, Malaysia. *MARDI Res. Bull.* 9:197–201.

Kerridge, P. C., Andrew, C. S., and Murtha, G. G. (1972). Plant nutrient status of soils of the Atherton Tableland, North Queensland. *Aust. J. Exp. Agric. Anim. Husb.* 12:618–27.

Kerridge, P. C., Cook, B. G., and Everett, M. L. (1973). Application of molybdenum trioxide in the seed pellet for sub-tropical pasture legumes. *Trop. Grassl.* 7:229–32.

Kerridge, P. C., and White, R. E. (1977). The residual effect of molybdenum fertilizer on a krasnozem on basalt at Maleny, south-east Queensland. *Aust. J. Exp. Agric. Anim. Husb.* 17:669–73.

Kubota, J., and Allaway, W. H. (1972). Geographic distribution of trace element problems. In *Micronutrients in Agriculture*, ed. J. J. Mortvedt, P. M. Giordano, and W. L. Lindsay, pp. 525–54. Madison, WI: Soil Science Society of America.

Lindsay, W. L. (1972). Inorganic phase equilibria of micronutrients in soil. In *Micronutrients in Agriculture*, ed. J. J. Mortvedt, P. M. Giordano, and W. L. Lindsay, pp. 41–57. Madison, WI: Soil Science Society of America.

Lipsett, J. (1975). A comparison of the responses by six grasses, rape and subterranean clover to application of molybdenum. *Aust. J. Exp. Agric. Anim. Husb.* 15:227–30.

Lipsett, J., and David, D. J. (1977). Amount and distribution of molybdenum in a bag of molybdenized superphosphate. *J. Aust. Inst. Agric. Sci.* 43:149–51.

Little, I. P., and Kerridge, P. C. (1978). A laboratory assessment of the molybdenum status of nine Queensland soils. *Soil Sci.* 125:102–6.

Mattos, G. B., and Colozza, M. T. (1986). Micronutrientes em pastagens. In *Simposio sobre Calagem e Adubacao de Pastagens, 1. Nova Odessa-SP*, ed. H. B. Mattos, J. C. Werner, T. Yamada, and E. Malavolta, pp. 233–256. Anais Piracicaba-SP, Brasil: Asociacao Brasileira para Pesquisa da Potassa e do Fosfato.

Melsted, S. W., Motto, H. L., and Peck, T. R. (1969). Critical plant nutrient composition values useful in interpreting plant analysis data. *Agronomy J.* 61:17–20.

Miranda, C. H. B., Seiffert, N. F., and Dobereiner, J. (1985). Efeito de aplicacao de molibdenio no numero de Azospirillum e na producao de *Brachiaria decumbens*. *Pesq. Agropec. Bras.* 20:509–13.

Munns, D. N., and Fox, R. L. (1977). Comparative lime requirements of tropical and temperate legumes. *Plant Soil* 46:533–48.

Murphy, L. S., and Walsh, L. M. (1972). Correction of micronutrient deficiencies with fertilizers. In *Micronutrients in Agriculture*, ed. J. J. Mortvedt, P. M. Giordano, and W. L. Lindsay, pp. 347–87. Madison, WI: Soil Science Society of America.

Nayyar, V. K., Randhawa, N. S., and Pasricha, N. S. (1977). Molybdenum accumulation in forage crops. I. Distribution of molybdenum, copper, sulphur and nitrogen in berseem (*Trifolium alexandrinum* L.) grown on calcareous flood plains. *J. Res. Punjab Agric. Univ.* 14:245–51.

Neubert, P., Wrazidlo, W., Vielemeyer, H. P., Hundt, I., Gollmick, Fr., and Bergmann, W. (1970). *Tabellen zur Pflanzenanalyse – Erste orientierende Uebersicht*. Jena: Institut für Pflanzenerähnrung, Jena, der Deutschen Akademie der Landwirtschaftswissenschaften zu Berlin.

Pasricha, N. S., Nayyar, V. K., Randhawa, N. S., and Sinha, M. K. (1977). Influence of sulphur fertilization on suppression of molybdenum uptake by berseem (*Trifolium alexandrinum* L.) and oats (*Avena sativa* L.) grown on a molybdenum-toxic soil. *Plant Soil* 46:245–50.

Reisenauer, H. M. (1956). Molybdenum content of alfalfa in relation to deficiency symptoms and response to molybdenum fertilization. *Soil Sci.* 81:237–42.

Reisenauer, H. M. (1963). Relative efficiency of seed-sand-soil applied molybdenum fertilizer. *Agronomy J.* 55: 459–60.

Reisenauer, H. M. (1965). Molybdenum. In *Methods of Soil Analysis. Part 2. Chemical and Microbiological Properties*, ed. C. A. Black, D. D. Evans, L. E. Ensminger, J. L. White, F. E. Clark, and R. C. Dinauer, pp. 1050–8. Agronomy series 9. Madison, WI: American Society of Agronomy.

Reith, J. W. S., Burridge, J. C., Berrow, M. L., and Caldwell, K. S. (1984). Effects of fertilizers on the contents of copper and molybdenum in herbage cut for conservation. *J. Sci. Food Agric.* 35:245–56.

Reuter, D. J. (1975). The recognition and correction of trace element deficiencies. In *Trace Elements in Soil-Plant-Animal Systems*, ed. D. J. D. Nicholas and A. R. Egan, pp. 291–324. New York: Academic Press.

Ross, B. J., and Calder, G. J. (1990). Nutrient studies on sandy red earths in the Douglas-Daly area, Northern Territory. *Trop. Grassl.* 24:121–3.

Samui, R. C., and Bhattacharya, A. K. (1982). Effect of Zn, B and Mo on rice and their residual effects on berseem and rice. *Indian Agric.* 26:101–6.

Samui, R. C., and Dasgupta, S. K. (1982). Effect of soil and foliar application of zinc, boron and molybdenum on direct seeded rice, transplanted rice and berseem grown in sequence. *Indian J. Agron.* 27:35–40.

Schalscha, E. B., Morales, M., and Pratt, P. F. (1987). Lead and molybdenum in soils and forage near an atmospheric source. *J. Environ. Qual.* 16:313–15.

Scott, M. L. (1972). Trace elements in animal nutrition. In *Micronutrients in Agriculture*, ed. J. J. Mortvedt, P. M. Giordano, and W. L. Lindsay, pp. 555–91. Madison, WI: Soil Science Society of America.

Sims, J. L., Sigafus, R. E., and Tiaranan, N. (1974). Effect of lime, inoculant and molybdenum pelleting of seed on growth and nitrogen content of crownvetch. *Agronomy J.* 66:446–9.

Sisodia, A. K., Sawarkar, N. J., and Rai, M. M. (1975). Effect of sulphur and molybdenum on yield and nutrient uptake in berseem (*Trifolium alexandrinum*). *J. Indian Soc. Soil Sci.* 23:96–102.

Smith, C., Brown, K. W., and Deuel, L. E., Jr. (1987). Plant availability and uptake of molybdenum as influenced by soil type and competing ions. *J. Environ. Qual.* 16:377–82.

Smith, F. W. (1986a). Interpretation of plant analysis: concepts and principles. In *Plant Analysis. An Interpretation Manual*, ed. D. J. Reuter and J. B. Robinson, pp. 1–12. Melbourne: Inkata Press.

Smith, F. W. (1986b). Pasture species. In *Plant Analysis. An Interpretation Manual*, ed. D. J. Reuter and J. B. Robinson, pp. 101–19. Melbourne: Inkata Press.

Standley, J., Bruce, R. C., and Webb, A. A. (1990). A nutrient survey of legume evaluation sites on red earths, solodic soils and black earths in central Queensland. *Trop. Grassl.* 24:15–23.

Stevenson, F. J., and Ardakani, M. S. (1972). Organic matter reactions involving micronutrients in soils. In *Micronutrients in Agriculture*, ed. J. J. Mortvedt, P. M. Giordano, and W. L. Lindsay, pp. 79–114. Madison, WI: Soil Science Society of America.

Suttle, N. F. (1975). Trace element interactions in animals. In *Trace Elements in Soil-Plant-Animal Systems*, ed. D. J. D. Nicholas and A. R. Egan, pp. 271–89. New York: Academic Press.

Swain, F. G. (1959). Responses to molybdenum three years after previous application on red basaltic soils on the far north coast of New South Wales. *J. Aust. Inst. Agric. Sci.* 25:51.

Tham Kah Cheng and Kerridge, P. C. (1982). Responses to lime, K, Mo and Cu by grass-legume pasture on some Ultisols and Oxisols of Peninsular Malaysia. *MARDI Res. Bull.* 10:350–69.

Tripathi, S. K., and Edward, J. C. (1978). Response of rhizobium culture inoculation, zinc and molybdenum application on rhizosphere and phyllosphere microbial population of soyabean (*Glycine max* Merill). *Curr. Sci.* 47:503–4.

Underwood, E. J. (1966). *The Mineral Nutrition of Livestock.* Aberdeen: Central Press.

Vera, R. R., Thomas, R., Sanint, L., and Sanz, J. I. (1992). Development of sustainable ley-farming systems for the acid-soil savannas of tropical America. *Annals Academie Brasileira de Ciencias* 64:105–25.

Wambeke, A. (1976). Formation, distribution and consequences of acid soils in agricultural development. In *Plant Adaptation to Mineral Stress in Problem Soils*, ed. M. J. Wright, pp. 15–24. Ithaca, NY: Cornell University Press.

Watson, G. A. (1960). Interaction of lime and molybdenum on the nutrition of *Centrosema pubescens* and *Pueraria phaseoloides. J. Rubber Res. Inst. Malaya* 16:126–38.

Watson, G. A., Wong Phui Weng, and Narayanan, R. (1963). Effect of cover plants and soil nutrient status and on growth of Heavea. II. The influence of application of rock phosphate, basic slag and magnesium limestone on the nutrient content of leguminous cover plants. *J. Rubber Res. Inst. Malaya* 18:28–37.

Werner, J. C., Monteiro, F. A., and Meirelles, N. M. F. (1983). Efeito das adubacoes com fosforo, potassio e molibdenio mais cobre na consorciacao de capimgordura com *Centrosema. Zootecnia (Brasil)* 21:109–34.

Wright, M. J., and Davison, K. L. (1964). Nitrate accumulation in crops and nitrate poisoning in animals. *Adv. Agron.* 16:197–247.

Younge, O. R., and Takahashi, M. (1953). Response of alfalfa to molybdenum in Hawaii. *Agronomy J.* 45:420–8.

14

Molybdenum and Sulfur Relationships in Plants

J. A. MACLEOD, UMESH C. GUPTA,
and BARRIE STANFIELD

Introduction

The observation that molybdenum (Mo) uptake by plants decreases with increasing concentrations of sulfate was first reported for tomato plants (*Lycopersicon esculentum* Mill.) in solution culture (Stout and Meagher, 1948). Stout et al. (1951) confirmed that observation with tomato plants grown in tissue culture (Table 14.1) and with tomatoes and peas (*Pisum sativum* L.) in soil (Tables 14.2 and 14.3). They attributed the action of sulfate ions in suppressing Mo uptake to direct competition between two divalent anions of similar sizes.

Since the report by Stout and co-workers, the effects of sulfur (S) to decrease Mo uptake have been reported in many species grown under a wide range of conditions, including vegetable crops such as beans (*Phaseolus vulgaris* L.) (Widdowson, 1966), Brussels sprouts (*Brassica oleracea* L. Gemmifera Group) (Gupta and Cutcliffe, 1968; Gupta, 1969; Gupta and Munro, 1969), cauliflower (*Brassica oleracea* L. Botrytis Group) (Mulder, 1954), cauliflower and lettuce (*Lactuca sativa* L.) (Plant, 1956), peas (Reisenauer, 1963; Gupta and Gupta, 1972), and peas and tomatoes (Stout et al. 1951). Similar relations between Mo and S have been found in forages such as berseem (*Trifolium alexandrinum* L.) (Pasricha and Randhawa, 1972; Shukla and Pathak, 1973; Sisodia, Sawarkar, and Rai, 1975; Pasricha et al., 1977; Gupta and Mehla, 1979, 1980), alfalfa (*Medicago sativa* L.) (Phillips and Meyer, 1993), timothy (*Phleum pratense* L.) and red clover (*Trifolium pratense* L.) (Gupta and MacLeod, 1975), subterranean clover (*Trifolium subterraneum* L.) (Jones and Ruckman, 1973), tall fescue (*Festuca arundinacea* Schreb.) (Bush et al., 1981), mixed-species pastures (Neenan and Walsh, 1956), and forage weeds such as dandelion (*Taraxacum officinale* Weber)

Table 14.1. *Effects of solution composition on uptake of radioactive Mo by tomatoes*

| Solution anion | Mo uptake (cpm per gram of fresh weight)[a] | | | |
| | Solution cation: NH_4^+ | | Solution cation: K^+ | |
	Blade	Stem	Blade	Stem
NO_3^-	140	15	89	21
Cl^-	110	22	38	7
SO_4^{2-}	21	1	5	3

[a] $1\,\mu g$ of Mo is equivalent to $5,170\,cpm$.

Table 14.2. *Effects of sulfate in absorption solution on concentrations of Mo in various parts of tomato plants*

| Sulfate (SO_4^-) concentration in absorbing solution ($\mu g\,mL^{-1}$) | Mo content (μg per kilogram fresh weight) | | |
	Blade	Stem	Root
0	11.0	10.7	20.1
200	5.5	2.2	16.8

Table 14.3. *Effects of $CaSO_4$ application to soil on concentrations of Mo in tomatoes and peas*

| $CaSO_4$ added to soil ($\mu g\,g^{-1}$) | Mo concentration in tissue ($\mu g\,g^{-1}$) | |
	Tomato	Pea
0	5.25	12.80
100	3.52	8.05
400	2.45	5.70

(Feely, 1990). Seed legumes, including mung beans (*Phaseolus aureus* Roxb.) (Chaphale, Naphade, and Kene, 1991), groundnut (*Arachis hypogaea* L.) (Singh, Prasad, and Prasad, 1977), and soybeans [*Glycine max* (L.) Merr.] (Kumar and Singh, 1980; MacLeod and Gupta, 1994), and other field and some non-food crops, including raya (*Brassica juncea* L.) (Pasricha and Randhawa, 1972; Dhankar et al., 1993), sorghum [*Sorghum bicolor* (L.) Moench] (Olsen and Watanabe, 1979), and tobacco (*Nicotiana tabacum* L.) (Sims and Atkinson, 1974; Pal et al., 1976; Sims, Leggett, and Pal, 1979), have shown decreased plant uptake of Mo in the presence of increased S concentrations.

This relationship between Mo and S has practical significance for crop production, because it permits the manipulation of Mo concentrations in plant tissue through S application. This mechanism can also lead to the development of Mo deficiency in crops when sulfate-containing materials are added to the soil to correct S deficiency. Olsen (1972) reviewed the literature on Mo–S relations and concluded that in some situations fertilization with both Mo and S may be required. Thus it is useful to consider the conditions that influence the relationship between Mo and S in plant and soil systems.

Factors Affecting the Relationship between Molybdenum and Sulfur

Soil and Climate Factors

Stout and Meagher (1948) showed that addition of sulfate to the culture medium reduced uptake of radioactive Mo by tomatoes (Table 14.1). Stout et al. (1951) confirmed the action of S to decrease Mo uptake from solution culture and showed that the effect was more pronounced in shoots and leaves than in roots (Table 14.2). They also demonstrated that S decreased Mo uptake by tomatoes and peas in soil culture (Table 14.3). They attributed the effect of S in decreasing Mo uptake to direct competition for plant uptake between divalent anions of the same size. Bush et al. (1981) found that addition of 0.05 mEq of sulfate per liter of solution decreased Mo uptake by tall fescue, but that further increases in S concentration had little additional effect (Table 14.4).

Barshad (1951) found that the Mo content in ladino clover grown in pots with high-pH soil increased with addition of gypsum (Table 14.5). He attributed that to decreases in bicarbonate concentrations and decreased competition between bicarbonate and Mo. Barshad (1951) also found that application of gypsum decreased the solubility of Mo, but that on the high-pH soil the Mo content of birdsfoot trefoil (*Lotus corniculatus* L.) was not influenced by gypsum application (Table 14.6).

Plant (1956) found that on acid soils, calcium carbonate increased Mo concentrations in marrow-stem kale, but calcium sulfate had no signifi-

Table 14.4. *Effects of addition of SO_4^{2-} on Mo concentration in tall fescue*

SO_4^{2-} concentration (mEq L^{-1})	Mo accumulation (dpm × 10^3 per 100mg)	
	At 6 hours	At 24 hours
0.00	60	130
0.05	32	60
0.10	30	35
1.00	20	30

Table 14.5. *Effects of CaSO$_4$ application on Mo content of ladino clover and soil pH*

Treatment	Mo content of dry clover leaves (µg g^{-1})		Soil pH after cut 1
	Cut 1	Cut 2	
Check	12.7	6.5	7.2
CaSO$_4$[a]	21.5	10.6	6.8

[a] Treatment equivalent to 8.4 tonnes of CaSO$_4$·2H$_2$O per hectare.

Table 14.6. *Effects of gypsum on water-soluble Mo, soil pH, and Mo content of birdsfoot trefoil*

Soil treatment	Water-soluble Mo in dry soil (µg g^{-1})	Soil pH	Mo content of dry leaves (µg g^{-1})
Check	0.21	8.8	27
CaSO$_4$[a]	0.08	8.2	28

[a] CaSO$_4$·2H$_2$O added at 11 tonnes per hectare.

cant effect (Table 14.7). Plant (1956) found that on the Bromham site, gypsum decreased the soil pH and decreased Mo concentrations in cauliflower, but on the Grower site the gypsum increased the soil pH and increased cauliflower Mo concentrations (Table 14.8). On those acid soils, the symptoms of Mo deficiency and the responses to gypsum were also confounded by toxic amounts of manganese (Mn). Mulder (1954)

Table 14.7. *Effects of calcium sources on Mo content of marrow-stem kale on two soil types*

Treatment	Mo content ($\mu g\,g^{-1}$)	
	Old Red Sandstone (pH 4.5)	Granite (pH 5.0)
Control	0.10	0.15
$CaCO_3$	0.81	0.71
$CaSO_4$	0.10	0.13

Table 14.8. *Effects of gypsum on soil pH and Mo concentration in cauliflower at two locations*

Location and treatment	Soil pH	Mo content of dry cauliflower ($\mu g\,g^{-1}$)
Bromham		
Control	5.2	0.10
Gypsum	4.9	0.06
Grower		
Control	4.8	0.06
Gypsum	5.2	0.10

attributed the decreases in Mo uptake by cauliflower following application of manganese sulfate partially to manganese ions and partially to sulfate ions.

Gupta and Cutcliffe (1968) found that the Mo content of Brussels sprouts decreased with increased rates of application of ordinary superphosphate. Under similar conditions, Gupta and Munro (1969) found that the Mo content of Brussels sprouts decreased with S application, but increased with increased rates of phosphorus (P) application (Table 14.9).

Olsen and Watanabe (1979) found that the Mo content of sorghum growing on a series of soils ranging in pH from 7.09 to 7.85 decreased with application of gypsum (Table 14.10). Such decreases in Mo concentrations were associated with decreases in soil pH.

Widdowson (1966) found that superphosphate decreased the Mo content of French beans grown on fine sandy loam soil, but that the magni-

Table 14.9.	*Effects of P and S on Mo content of Brussels sprouts with and without added Mo*

	Mo in dry matter (μg g^{-1})	
Treatment	No Mo	Mo added to soil (2.5 μg g^{-1})
P added to soil (μg g^{-1})		
6	0.11	9.66
33	0.14	13.35
66	0.18	15.52
S added to soil (μg g^{-1})		
0	0.19	20.52
50	0.10	3.16

Table 14.10.	*Variations in soil and plant Mo concentrations in relation to soil pH and soil type*

	pH		Untreated-soil Mo (μg g^{-1})	Concentration of Mo in sorghum (μg g^{-1})	
Soil type	Check	+SO$_4^{2-}$		Check	+S
Stoneham	7.27	7.12	0.20	1.6	0.9
Platner	7.85	7.81	1.35	1.5	0.9
Otero	7.27	7.09	0.09	1.7	1.0
Anselmo	7.65	7.37	0.10	2.0	0.9
Keith	7.67	7.59	0.45	2.4	1.2
Bridgeport	7.62	7.60	0.18	4.6	2.3

tude of the effect was less on limed soil than on unlimed soil (Table 14.11).

Gupta and Mehla (1980) found that Mo concentrations in berseem decreased with increasing rates of application of S on normal and re-claimed saline-sodic soils (Table 14.12). In studies of 30 samples of berseem grown on alkaline soils of high Mo content there was a strong negative correlation between S uptake by berseem and the concentrations of oxalate-extractable Mo in the 0–15- and 15–30-cm layers (Pasricha and Randhawa, 1971).

Table 14.11. *Effects of superphosphate and Mo application on soil pH and Mo content of French beans*

Soil type	Treatment	pH	Mo content ($\mu g\,g^{-1}$)	
			Tops	Beans
Ahuriri f.s.l.	Control	7.7	3.2	7.9
	Superphosphate	7.4	1.2	0.9
	Super + Mo	7.6	11.0	8.9
Twyford f.s.l.	Control	5.6	3.3	3.6
(unlimed)	Superphosphate	6.0	0.6	0.9
	Super + Mo	5.7	6.9	6.9
Twyford f.s.l.	Control	7.5	4.3	4.5
(limed)	Superphosphate	7.4	3.9	3.5
	Super + Mo	7.3	13.0	9.9

Table 14.12. *Effects of S application on Mo uptake by berseem on normal and reclaimed soils*

S application to soil ($\mu g\,g^{-1}$)	Mo content ($\mu g\,g^{-1}$)	
	Normal soil (initial pH 8.7)	Reclaimed soil (initial pH 7.9)
0	17	20
25	14	13
50	12	10
75	8	8

Plant Type and Plant Part

Sulfur has been shown to decrease Mo uptake in a wide range of species and in various plant parts. MacLeod and Gupta (1994) showed that application of S decreased the Mo concentrations in the leaves and seeds of a standard variety and a high-protein variety of soybeans (Table 14.13). The decreases in Mo content in leaves and seeds were evident with foliar-applied Mo as well as with seed-applied Mo, indicating that the interaction was not limited to uptake through root systems. Singh and Kumar (1979) found that Mo concentrations in the leaves, stems, pods, and seeds of soybeans decreased with increasing rates of S applica-

Table 14.13. *Effects of S and method of Mo application on Mo contents of two varieties of soybean seeds and leaves*

| | Mo content in dry matter ($\mu g\,g^{-1}$) | | | |
| | Seeds | | Leaves | |
Mo application	Without S	With S ($50\,kg\,ha^{-1}$)	Without S	With S ($59\,kg\,ha^{-1}$)
Standard variety				
Check	2.64	2.54	1.34	1.17
$50\,g\,ha^{-1}$ (with seed)	5.43	4.60	2.32	1.88
$50\,g\,ha^{-1}$ (foliar)	17.09	13.88	11.34	8.50
High-protein variety				
Check	3.39	2.58	2.25	1.67
$50\,g\,ha^{-1}$ (with seed)	6.03	4.63	3.55	2.36
$50\,g\,ha^{-1}$ (foliar)	17.20	15.06	11.25	8.70

Table 14.14. *Effects of S application on Mo content of soybeans*

| S application to soil ($\mu g\,g^{-1}$) | Mo in dry matter ($\mu g\,g^{-1}$) | | | |
	Leaves	Stems	Pods	Seeds
0	6.53	2.92	4.50	13.51
40	5.98	2.37	3.84	13.63
80	5.43	2.18	3.45	12.17
120	5.64	2.43	3.13	12.35

Table 14.15. *Effects of P and Mo on uptake of S by soybeans*

| Mo applied to soil ($mg\,kg^{-1}$) | S uptake (mg per pot) | | |
	No P added	P added to soil ($40\,mg\,kg^{-1}$)	P added to soil ($80\,mg\,kg^{-1}$)
0	25.4	34.0	37.7
1	23.7	34.2	38.7

Table 14.16. *Effects of S application on Mo contents of mung bean grain and straw*

	Mo content of dry matter (μg g^{-1})	
S applied (kg ha^{-1})	Grain	Straw
0	11.2	4.3
25	10.9	4.1
50	10.7	4.0

Table 14.17. *Effects of S and Mo on recovery of Mo from raya*

	Mo recovered (mg) per gram of Mo applied			
S applied to soil (μg g^{-1})	Mo applied to soil: 0.5 μg g^{-1}	Mo applied to soil: 1.0 μg g^{-1}	Mo applied to soil: 1.5 μg g^{-1}	Mo applied to soil: 3.0 μg g^{-1}
0	38.2	36.8	29.7	24.9
12.5	35.4	32.0	25.0	23.7
25.0	31.0	21.6	15.9	16.4
37.5	20.1	17.7	13.8	15.1
50.0	9.1	13.4	10.9	11.8

tion (Table 14.14). Kumar and Singh (1980) found that application of P reduced the ability of Mo to decrease the S content (Table 14.15).

Chaphale et al. (1991) found that S application decreased the Mo content in seeds and straw of mung beans (Table 14.16). Pasricha and Randhawa (1972) reported decreased Mo concentrations in the grain and tops of raya with increasing rates of S application. They reported that the amount of applied Mo (0.5 g ha^{-1}) recovered from the grain and tops dropped from 38.2 mg Mo recovered per gram of Mo applied to 9.1 mg Mo recovered per gram of Mo applied when the rate of S application was increased from zero to 50 g ha^{-1} (Table 14.17). Similar trends were observed when Mo was applied at rates up to 3 g ha^{-1}.

Sims et al. (1979) found that S fertilization decreased Mo concentrations in tobacco plants sampled at various times from 25 to 110 days after planting (Table 14.18). They found that at low Mo concentrations, the decrease in Mo concentration in tobacco plants due to S application also decreased nitrate reductase activity.

Table 14.18. *Effects of S fertilization on Mo concentration in tobacco leaves*

S added to soil (kg ha^{-1})	Mo content in dry matter (μg g^{-1})			
	25 days[a]	50 days	75 days	110 days
0	0.39	0.31	0.23	0.51
112	0.41	0.22	0.28	0.47
224	0.26	0.19	0.19	0.34
448	0.33	0.14	0.15	0.34

[a] Days after transplanting.

Table 14.19. *Effects of P, S, and Mo on Mo content of subclover*

Treatment	Mo in dry matter (μg g^{-1})
Check	0.6
P	0.5
S	0.1
Mo	6.7
P-S	0.5
S-Mo	5.4
P-Mo	4.8
P-S-Mo	6.6

Other Nutrients

Application of P can influence the effect of S application on Mo. Jones and Ruckman (1973) showed that application of P in combination with S could overcome the decreases in Mo uptake in subterranean clover pasture caused by S application (Table 14.19). Gupta and Cutcliffe (1968) showed that P applied as ordinary superphosphate decreased Mo uptake by Brussels sprouts. Gupta and Munro (1969) showed that under conditions where Mo uptake was decreased by application of S, as ammonium sulfate, the application of P as dicalcium phosphate increased Mo uptake by Brussels sprouts (Table 14.20). Feely (1990) found that increasing the rate of P application from 60 to 180 kg ha^{-1} reduced the Mo content of herbage grown on peatland from 44.4 to 25.8 μg g^{-1} when calcium ammonium nitrate or urea was used as the nitrogen (N) source, but there was no effect of P application on herbage Mo when the N source was ammonium sulfate or ammonium nitrate plus gypsum.

Table 14.20. *Effects of S and P application on Mo content of Brussels sprouts*

	Mo content ($\mu g\,g^{-1}$)	
Treatment	Tops	Roots
S applied to soil		
None	6.98	9.22
$50\,\mu g\,g^{-1}$	1.79	2.13
P applied to soil		
None	3.30	3.92
$33\,\mu g\,g^{-1}$	4.56	6.18
$66\,\mu g\,g^{-1}$	5.29	6.92

Table 14.21. *Effects of Cu and S on yield and Mo of groundnut seeds*

S added to soil ($\mu g\,g^{-1}$)	Cu added to soil ($\mu g\,g^{-1}$)	Groundnut seeds	
		Yield per pot (g)	Mo per pot (μg)
0	0	6.9	14.7
35	0	2.3	1.5
0	5	4.0	7.9
35	5	6.8	4.1

Table 14.22. *Effects of Fe and S on Fe and Mo contents of sorghum on an alkaline soil*

	Nutrient concentration ($\mu g\,g^{-1}$)	
Treatment	Mo	Fe
Check	1.53	62
S	0.83	71
Fe	0.77	64
Fe + S	0.53	72

Application of copper (Cu) can influence the interaction between S and Mo. Singh et al. (1977) found that application of Cu overcame the effect of S to depress the yield of groundnut seeds and also reduced the magnitude of the decrease in Mo content due to S application

(Table 14.21). Olsen and Watanabe (1979) found that on alkaline soils, iron (Fe) uptake increased as Mo concentration decreased with gypsum application (Table 14.22).

Molybdenum–Sulfur Relationships in Relation to Crop Quality

Forage Crops

Jones and Ruckman (1973) found that in subterranean clover, S application decreased the Mo concentration, and Mo application decreased the concentration of S (Table 14.23). They pointed out the forage-quality implications of that relationship. Application of Mo alone or in combination with P resulted in Mo concentrations in forage that were high enough to cause concern for animal health. When S was applied in combination with Mo, forage Mo concentrations were below the point at which Mo concentration becomes a health concern for animals ($10\,\mu g$ of Mo per gram of forage). Barshad (1948) indicated that concerns about animal health because of forages with high Mo concentrations are greater for legumes than for grasses, and they increase further with increasing maturity of the forage.

Because S is required for protein synthesis, and because Mo is a component of nitrate reductase, a balance between Mo and S is required to maintain normal protein metabolism, and that balance must be considered in the production of high-protein forage.

Gupta and MacLeod (1975) showed that Mo can more easily accumulate to toxic amounts in forage legumes than in grasses at Mo application rates of $1-2\,\mu g\,g^{-1}$ (Table 14.24). Although they found no effect of Mo or S on forage Cu concentration, they indicated the importance of considering Mo content in relation to Cu content in forage, because the animal health problems seen with high-Mo forages can be due to Mo-induced Cu deficiency (see Chapter 15).

Cereal Crops

The requirements for Mo in cereal crops are low, and the Mo concentrations in the seeds of wheat, oats, and barley are low (see Chapter 5). Hence there is little danger of accumulation of Mo in cereal grains to an extent that would cause concern for the health of livestock. Consequently, S–Mo relationships in cereal crops are unlikely to cause concern for feed quality. Seed legumes such as soybeans, however, accu-

Table 14.23. *Effects of P, S, and Mo on Mo and SO$_4^{2-}$-S contents of subclover*

Treatment	Mo content ($\mu g\,g^{-1}$)	SO$_4^{2-}$-S content ($\mu g\,g^{-1}$)
Check	0.6	70
P	0.5	120
S	0.1	330
Mo	6.7	180
P-S	0.5	240
S-Mo	5.4	120
P-Mo	4.8	130
P-S-Mo	6.6	160

Table 14.24. *Effects of S and Mo on Mo content in forages*

S added to soil ($\mu g\,g^{-1}$)	Mo content in dry matter ($\mu g\,g^{-1}$)		
	Mo added to soil: none	Mo added to soil: $1\,\mu g\,g^{-1}$	Mo added to soil: $2\,\mu g\,g^{-1}$
Timothy			
0	0.22	6.98	14.82
100	0.15	1.80	3.09
Alfalfa			
0	0.85	17.43	41.14
100	0.28	8.09	13.43
Red clover			
0	0.25	34.52	78.24
100	0.20	7.92	17.97

mulate higher amounts of Mo in their seeds (Table 14.13). Sulfur application can be useful in preventing accumulation of Mo to toxic amounts in soybeans grown on soils with high amounts of available Mo.

Vegetable Crops

Schnug (1990) summarized the prior studies on the effects of S fertilization on vegetable crops. Such effects include suppression of Mo accumulation in vegetables grown on high-Mo soils. Reisenauer (1963) reported that S application decreased the N content of peas grown on low-Mo soils. Although Mo is required for nitrate reduction, Pal et al. (1976) were unable to detect any decrease in nitrate reductase activity when S was applied to solution cultures with low concentrations of Mo.

Plant (1956) described the development of "whiptail" in cauliflower, which was severe when S was applied to low-Mo soils. He also described Mo deficiency symptoms in other brassicas and lettuce that increased in severity when S was applied to low-Mo soils.

Summary

Application of S, either as gypsum or from other S sources, generally decreases the uptake of Mo by crops. Studies of soybeans have shown that decreases in plant Mo are not limited only to Mo–S interactions in the soil, because foliar-applied Mo has also been shown to decrease Mo concentrations in soybean seeds and leaves in the presence of soil-added S. On low-Mo soils this can lead to induced Mo deficiency, which can decrease yields and crop quality. On soils with high amounts of Mo, application of S can prevent the accumulation of high Mo concentrations and decrease the potential for Mo toxicity to livestock. Competition between sulfate and molybdate anions, competition between bicarbonate and molybdate, and root-zone pH changes have been suggested as explanations for the action of S to reduce Mo uptake.

References

Barshad, I. (1948). Molybdenum content of pasture plants in relation to toxicity to cattle. *Soil Sci.* 66:187–95.

Barshad, I. (1951). Factors affecting the molybdenum content of pasture plants. II. Effect of soluble phosphates, available nitrogen, and soluble sulfates. *Soil Sci.* 71:387–98.

Bush, L. P., Leggett, J. E., King, M. J., and Vincent, J. E. (1981). Sulfur and molybdenum nutrition effects in tall fescue. *Can. J. Bot.* 59:536–41.

Chaphale, S. D., Naphade, P. S., and Kene, D. R. (1991). Effect of molybdenum and sulphur application on performance of mung (*Phaseolus aureus* L.) grown in black calcareous soil. *PKV Res. J.* 15:176–8.

Dhankar, J. S., Kumar, V., Sangwan, P. S., and Karwasra, S. P. S. (1993). Effect of sulphur and molybdenum application on dry matter yield, uptake and utilization of soil and fertilizer-sulphur in raya crop (*Brassica juncea* Coss). *Agrochimica* 37:4–5, 316–29.

Feely, L. (1990). Agronomic effectiveness of nitrogen, sulfur and phosphorus for reducing molybdenum uptake by herbage grown on peatland. *Irish J. Agric. Res.* 29:129–39.

Gupta, J. K., and Gupta, Y. P. (1972). Note on effect of sulphur and molybdenum on quality of pea. *Indian J. Agron.* 17:245–7.

Gupta, U. C. (1969). S × Mo interaction in plant nutrition. *Sulphur Inst. J.* 5:4–6.

Gupta, U. C., and Cutcliffe, J. A. (1968). Influence of phosphorus on molybdenum content of Brussels sprouts under field and greenhouse conditions and on recovery of added molybdenum in soil. *Can. J. Soil Sci.* 48:117–23.

Gupta, U. C., and MacLeod, L. B. (1975). Effects of sulfur and molybdenum on the molybdenum, copper and sulfur concentrations of forage crops. *Soil Sci.* 119:441–7.

Gupta, V. K., and Mehla, D. (1979). Copper, manganese and iron concentration in berseem (*Trifolium alexandrinum*) and copper-molybdenum ratio as affected by molybdenum in two types of soil. *Plant Soil* 51:587–602.

Gupta, V. K., and Mehla D. S. (1980). Influence of sulphur on the yield and concentration of copper, manganese, iron and molybdenum in berseem (*Trifolium alexandrinum*) grown on two different soils. *Plant Soil* 56:229–34.

Gupta, U. C., and Munro, D. C. (1969). Influence of sulfur, molybdenum and phosphorus on chemical composition and yields of Brussels sprouts and of molybdenum of sulfur contents of several plant species grown in the greenhouse. *Soil Sci.* 107:114–18.

Jones, M. B., and Ruckman, J. E. (1973). Long-term effects of phosphorus, sulfur, and molybdenum on a subterranean clover pasture. *Soil Sci.* 115:343–8.

Kumar, V., and Singh, M. (1980). Sulfur, phosphorus, and molybdenum interactions in relation to growth, uptake, and utilization of sulfur in soybean. *Soil Sci.* 129:297–304.

MacLeod, J. A., and Gupta, U. C. (1994). Effect of molybdenum and sulfur on grain yield and nutrient concentration in soybeans. *Trends Agric. Sci.: Soil Sci.* 12:9–19.

Mulder, E. G. (1954). Molybdenum in relation to growth of higher plants and micro-organisms. *Plant Soil.* 5:386–415.

Neenan, M., and Walsh, T. (1956). Some soil and herbage factors associated with the incidence and treatment of molybdenum conditional hypocuprosis on Irish pastures. In *Proceedings of the 7th International Grasslands Congress, Wellington, N.Z.*, pp. 345–56.

Olsen, S. R. (1972). Micronutrient interactions. In *Micronutrients in Agriculture*, ed. J. J. Mortvedt, P. M. Giordano, and W. L. Lindsay, pp. 243–64. Madison, WI: Soil Science Society of America.

Olsen, S. R., and Watanabe, F. S. (1979). Interaction of added gypsum in alkaline soils with uptake of iron, molybdenum, manganese, and zinc by sorghum. *Soil Sci. Soc. Am. J.* 43:125–30.

Pal, U. R., Gossett, D. R., Sims, J. L., and Leggett, J. E. (1976). Molybdenum and sulfur nutrition effects on nitrate reduction in burley tobacco. *Can. J. Bot.* 54:2014–22.

Pasricha, N. S., Nayyar, V. K., Randhawa, N. S., and Sinha, M. K. (1977). Influence of sulfur fertilization on suppression of molybdenum uptake by berseem (*Trifolium alexandrinum* L.) and oats (*Avena sativa* L.) grown on molybdenum deficient soil. *Plant Soil* 46:245–50.

Pasricha, N. S., and Randhawa, N. S. (1971). Available molybdenum status of some recently reclaimed saline-sodic soils and its effect on concentration of molybdenum, copper, sulphur and nitrogen in berseem (*Trifolium alexandrinum*) grown on these soils. In *Proceedings of the International Symposium on Soil Fertility Evaluation*, ed. J. S. Kanwar, pp. 1017–25. New Delhi.

Pasricha, N. S., and Randhawa, N. S. (1972). Interaction effect of sulfur and molybdenum on the uptake and utilization of these elements by raya (*Brassica juncea* L.). *Plant Soil* 37:215–20.

Phillips, R. L., and Meyer, R. D. (1993). Molybdenum concentration of alfalfa in Kern County, California: 1950 versus 1985. *Commun. Soil Sci. Plant Anal.* 24:2725–31.

Plant, W. (1956). The effects of molybdenum deficiency and mineral toxicities on crops in acid soils. *J. Hort. Sci.* 31:163–76.

Reisenauer, H. M. (1963). The effect of sulfur on the absorption and utilization of molybdenum by peas. *Soil Sci. Soc. Am. Proc.* 27:553–6.

Schnug. E. (1990). Sulphur nutrition and quality of vegetables. *Sulphur in Agric.* 14:3–7.

Shukla, P., and Pathak, A. N. (1973). Effect of molybdenum, phosphorus and sulphur on the yield and composition of berseem in acid soil. *J. Indian Soc. Soil Sci.* 21:187–92.

Sims, J. L., and Atkinson, W. O. (1974). Soil and plant factors influencing accumulation of dry matter in burley tobacco growing in soil made acid by fertilizer. *Agronomy J.* 66:775–8.

Sims, J. L., Leggett, J. E., and Pal, U. R. (1979). Molybdenum and sulfur interaction effects on growth, yield, and selected chemical constituents of burley tobacco. *Agronomy J.* 71:75–8.

Singh, A. P., Prasad, B., and Prasad, R. N. (1977). Sulfur, copper and molybdenum interrelationship in groundnut. *Fert. Technol.* 14:214–17.

Singh, M., and Kumar, V. (1979). Sulfur, phosphorus, and molybdenum interactions on the concentration and uptake of molybdenum in soybean plants. *Soil Sci.* 127:307–12.

Sisodia, A. K., Sawarkar, N. J., and Rai, M. M. (1975). Effect of sulphur and molybdenum on yield and nutrient uptake in berseem (*Trifolium alexandrinum*). *J. Indian Soc. Soil Sci.* 23:96–102.

Stout, P. R., and Meagher, W. R. (1948). Studies of the molybdenum nutrition of plants with radioactive molybdenum. *Science* 108:471–3.

Stout, P. R., Meagher, W. R., Pearson, G. A., and Johnson, C. M. (1951). Molybdenum nutrition of crop plants. I. The influence of phosphate and sulfate on the absorption of molybdenum from soils and solution cultures. *Plant Soil.* 3:51–87.

Widdowson, J. P. (1966). Molybenum uptake by French beans on two recent soils. *N.Z. J. Agric. Res.* 9:59–67.

15

Molybdenum in the Tropics

N. S. PASRICHA, V. K. NAYYAR, and R. SINGH

Introduction

Molybdenum (Mo) is an essential plant nutrient. It acts as a metallic cofactor in plant and animal enzymes. At high concentrations in forages, it can be toxic to ruminants by interfering with assimilation of copper (Cu). The range between toxicity and deficiency in animals is narrow, and therefore careful control of Mo in animal diets is essential.

Bear (1956) reviewed the early literature dealing with Mo in soils and plants and in animal nutrition in a special issue of *Soil Science*. The agricultural importance of Mo has been discussed (Mortvedt, Giordano, and Lindsay, 1972), and various aspects of the presence of Mo in the environment have been examined (Chappell and Petersen, 1977). Underwood (1977) and Beeson and Matrone (1976) have reviewed the biochemical importance of Mo in animal and human nutrition.

Molybdenum deficiency in tropical and subtropical soils is more widespread than elsewhere because soils in the tropics are highly weathered. In fact, many soils in the tropics are so highly weathered that little remains except sesquioxides and some 1:1 layer silicates (Pasricha and Fox, 1993). The acidic nature of many of these soils with a dominant content of sesquioxides is the major cause of Mo deficiency. In some cases there can be ample amounts of MoO_4^{2-} present in the adsorbed form in these sesquioxide-rich low-pH soils, but the solubility of Mo and its concentration in the soil solution are the important factors from the point of view of the availability of Mo to crop plants and natural herbage growing on such soils, and those conditions of high sesquioxide content and low soil pH result in a low concentration of Mo in the soil solution.

Multiple nutrient deficiencies tend to be the rule rather than the exception in the leached and weathered soils in the humid tropics. Heavy

clays account for a significant fraction of the acidic soils in many cultivated lands in tropical and subtropical areas (Munns and Franco, 1981).
These soils are characterized by high degrees of acidity, high concentrations of aluminum (Al) and manganese (Mn), very high percentages of
clays, and low concentrations of calcium (Ca). It is noteworthy that the
relatively poor performances of some legumes grown without added lime
in those soils can be largely attributed to problems of Ca deficiency and
decreased Mo uptake, as reported by Clark (1976) and Folscher and
Barnard (1985) in South Africa. Liming obviously would have extremely
beneficial effects on both of these elements, as was observed by Barnard
and Folscher (1988), although a nutritional factor commonly associated
with excess acidity is Mo deficiency, and it may not always be alleviated
by application of lime only (Coventry, Hirth, and Fung, 1987). In the acid
soils in southeastern Australia, addition of Mo has frequently been required for growth of subterranean clover (*Trifolium subterraneum* L.),
along with liming (Anderson and Moye, 1952). As a result of legume-
based farming systems, those soils became more acidic (William, 1980),
and the productivity of pastures declined (Bromfield et al., 1983). In this
chapter we attempt to consolidate the relevant published information on
the role of Mo in the soil, plant, and animal interactions in tropical
regions.

Soil Molybdenum Content

Total Molybdenum

The total Mo in the majority of surface soils can vary from 1 to $2\mu g\,g^{-1}$.
However, concentrations up to $18.1\mu g\,g^{-1}$ have been reported in a few
soils in Bihar, India (Verma and Jha, 1970), and concentrations between
3.0 and $8.0\mu g\,g^{-1}$ have been recorded in some black, red, and alluvial soils
(Misra and Misra, 1972; Gupta and Dabas, 1980) (Table 15.1). The
lowest reported Mo content, less than $0.1\mu g\,g^{-1}$, was found for certain
Bundelkhand, Bhabar, and Vindhyan soils and alluvial soils in Uttar
Pradesh, India (Kanwar and Randhawa, 1974), and $0.2\mu g\,g^{-1}$ was reported for some alfisols, spodosols, and inceptisols of Himachal Pradesh
(Sharma, Minhas, and Masand, 1988). Alluvial soils more recently derived from granites and metamorphic crystalline basalts contained
higher total amounts of Mo, 1.54–$3.01\mu g\,g^{-1}$, as compared with 1.50–
$1.84\mu g\,g^{-1}$ in black soils formed from traprock (basalt) and limestone.
The acid soils of Himachal Pradesh contained less Mo than the neutral

Table 15.1. *Total Mo and available Mo in different soils of India*

Soil	Classification	Total Mo ($\mu g\,g^{-1}$)	Available Mo ($\mu g\,g^{-1}$)	Reference
Alluvial	Ustipsamment, Ustifluvents, Ustochrepts, Hapludulfs, Haplustalfs, Ochraqualfs Natrustalfs, Natraqualf, Ustorthents, Calciorthents, Natrustalfs, Natraqualf, Fluvaquent	0.44–14.5	0.01–2.71	Durate et al. (1961), Pathak et al. (1968), Verma and Jha (1970), Pasricha and Randhawa (1972), Misra and Misra (1972), Nayyar et al. (1977a), Gupta and Dabas (1980), Chakraborty et al. (1982)
Black	Chromusterts, Pellusterts	0.87–7.50	traces–2.02	Pathak et al. (1968), Rai et al. (1972)
Red and laterite	Haplustalf, Rodustalf, Ustorthents	1.12–5.25	0.10–0.65	Pathak et al. (1968), Misra and Misra (1972)
Desert	Torripsamments	1.37–1.82	0.17–0.23	
Others	Hapludalfs, Dystrochrepts, Argiudolls, Haplustalfs, Palendalfs, Haplorthoid, Eutrochrepts	0.20–3.80	0.02–0.81	Pathak et al. (1968), Shukla and Pathak (1973), Rawat and Mathpal (1981), Bhandari and Randhawa (1985), Sharma and Minhas (1987), Sharma et al. (1988)

and alkaline soils of Punjab (Takkar, 1978). Vertisols in Tamil Nadu contained more Mo than did alfisols (Balaguru and Mosi, 1973). In the soils of Gujarat, a range of Mo content from 0.5 to $4.1 \mu g \, Mo \, g^{-1}$, with an average value of $1.8 \mu g \, g^{-1}$, has been reported.

There is little information on the distribution of Mo in soil profiles. The Mo content was generally found to be higher in surface soils as compared with subsurface soils in some alluvial, black, red, and hill soils in Uttar Pradesh (Pathak, Shanker, and Misra, 1968; Misra and Misra, 1972). In the salt-affected soils of Haryana, the total Mo content decreased with depth (Gupta and Dabas, 1980), whereas in the soils of Assam and Himachal Pradesh no regular pattern of its distribution was found (Chakraborty, Sinha, and Prasad, 1982; Sharma et al., 1988).

Factors Affecting Total Molybdenum

The Mo content of a soil is dependent on the nature of its parent material. An alfisol derived from mixed shale revealed a total Mo content of $0.90 \mu g \, g^{-1}$ as compared with $0.20 \mu g \, g^{-1}$ in a spodosol formed from mixed schist under similar climate conditions (Sharma et al., 1988). An average Mo content in alluvial soils derived in part from Okchon uraniferous black shale was $136 \, ng \, g^{-1}$, and in alluvial soils derived in part from those black shales it was reported to be $20 \, ng \, g^{-1}$ (Kim and Thornton, 1993). Total Mo tends to be higher in neutral and alkaline soils (Verma and Jha, 1970; Takkar, 1978). Positive and significant correlations between total Mo and pH and the electrical conductivity (EC) of salt-affected soils were observed by Gupta and Dabas (1980). The higher content of total Mo in vertisols can be related to their high clay contents (Balaguru and Mosi, 1973). However, a negative relationship between total Mo and clay content has been reported (Gupta and Dabas, 1980; Chakraborty et al., 1982). The influences of soil characteristics on the total Mo content of soils cannot be ascertained, because of our limited information. Verma and Jha (1970), however, observed that 53% of the variation in total Mo could be accounted for on the basis of soil texture, pH, and organic carbon.

Available Molybdenum

Acid ammonium oxalate, buffered to pH 3.3, has commonly been used to extract plant-available Mo from a soil. Its content in a majority of Indian soils has ranged between 0.10 and $0.8 \mu g \, g^{-1}$. However, a few alluvial and black soils have contained fairly high amounts (1.32–2.71 µg g^{-1}) of available Mo (Rai, Pal, and Shitoley, 1972; Gupta and Dabas, 1980). The amount of available Mo was found to be greater in alluvial soils than in alfisols and vertisols in Tamil Nadu (Balaguru and Mosi, 1973). Generally, alkaline alluvial soils have had higher amounts of available Mo, as compared with soils that have been acidic in reaction or sedentary in origin (Pathak et al., 1968; Verma and Jha, 1970). In the acid soils of Assam, India, the amount of available Mo was generally less than 0.05 µg g^{-1} (Chakraborty et al., 1982). In salt-affected and floodplain soils, its content was higher than in normal soils. The available Mo constituted 29–65% of total Mo in alkali soils, as compared with 17–50% in the adjoining normal soils, in Uttar Pradesh. While compiling their findings from micronutrient research in India, Takkar, Chhibba, and Mehta (1989) reported that the red and yellow soils of Madhya Pradesh contained an average of 0.08 µg of available Mo per gram of soil, as compared with 0.16–0.38 µg g^{-1} in shallow and deep black soils. In the soils of Gujarat, its average content varied from 0.07 to 0.23 µg g^{-1}. They also reported that 30% of the 544 soil samples from Madhya Pradesh and 6% of the 401 soil samples from Gujarat contained available Mo in the deficient range. The responses of crops to Mo application in Gujarat confirmed that finding (Reddy, 1964). In the soils of Himachal Pradesh, 70% of the 47 surface soil samples collected from representative areas had available Mo at less than the critical concentration of 0.15 µg g^{-1} (Sharma et al., 1988). In greenhouse experiments with added Mo, soybean yields in such soils increased by 20.7–96.1% (Sharma and Minhas, 1987).

Little information is available on the distribution of Mo in the soil profile. In alkaline soils, Mo is more mobile, and if it is not leached from the profile it can accumulate in plants. Pasricha and Randhawa (1971) reported that the available Mo in the surface horizons of recently reclaimed sodic soils in Punjab varied from 0.012 to 0.449 µg g^{-1}, with an average value of 0.112 µg g^{-1}, as compared with subsurface samples in which the values of available Mo ranged from 0.065 to 0.720 µg g^{-1}, with a mean of 0.207 µg g^{-1}. Ahmad, Khathak, and Perveen (1991) reported that the available Mo in the Dir district of Pakistan varied from 0.10 to 1.39 µg g^{-1} in the topsoil and from 0.12 to 1.31 µg g^{-1} in the subsoil. Misra and Misra (1972) observed high amounts of available Mo in surface soils and a tendency for it to decrease with depth in an alkali soil profile, but not so in the black and red soils, in Uttar Pradesh. The available Mo content decreased with depth in salt-affected soils in Haryana and

Table 15.2. *Mo uptake by several leguminous species as influenced by liming of acid soils*

Legume species	Mo uptake per pot (mg)		
	No lime	Lime (6 t ha⁻¹)	Factor of increase
Lespedeza	0.020	0.023	1.15
Caucasian clover	0.0096	0.140	14.58
Crown vetch	0.015	0.120	8.00
Kenian white clover	0.0099	0.160	16.16
Palestinian strawberry clover	0.0036	0.130	36.00
Lucern	0.0085	0.068	8.00
Birdsfoot trefoil	0.023	0.074	3.21
Greater lotus	0.029	0.104	3.59
White clover	0.05	0.26	5.20

Source: Adapted from Barnard and Folscher (1988).

Uttar Pradesh (Gupta and Dabas, 1980), but an irregular trend for its distribution was reported for soils in Tamil Nadu, Assam, and Himachal Pradesh (Balaguru and Mosi, 1973; Chakraborty et al., 1982; Sharma et al., 1988).

Factors Affecting Molybdenum Availability

The total content of a micronutrient is generally a poor indicator of its phytoavailability. However, significant correlations between total Mo and available Mo in the soils of Uttar Pradesh, Haryana, and Himachal Pradesh speak of the dependence, to some extent, of the available Mo on the total Mo content in soils (Misra and Misra, 1972; Gupta and Dabas, 1980; Sharma et al., 1988). Low pH values and larger amounts of sesquioxides tend to reduce the availability of Mo to plants. The available Mo content has been shown to increase with increases in pH and/or exchangeable sodium percentage (ESP) (Pathak et al., 1968; Grewal, Bhumbla, and Randhawa, 1969; Nayyar, 1972; Gupta and Dabas, 1980). Such relationships were not observed in vertisols in Madhya Pradesh and hill soils in Himachal Pradesh (Rai et al., 1972; Sharma et al., 1988). Liming of acid soils will increase their available Mo content (Verma and Jha, 1970; Nayyar, 1972). Barnard and Folscher (1988) reported 3-fold to 36-fold increases in Mo uptake by several legume crops as a result of application of $CaCO_3$ at 6 t ha⁻¹ to an acidic, brown, sandy clay loam oxisol in the Piet Ritief area of South Africa (Table 15.2). The decrease in Mo uptake at higher rates of liming has been attributed to Mo adsorption by the $CaCO_3$ (Nayyar, 1972).

The available Mo was found to increase significantly with increases in soil organic matter in Vidhurbha soils, black soils in Madhya Pradesh, and salt-affected soils in Haryana (Rai et al., 1972; Gupta and Dabas, 1980), but it decreased with increases in organic matter in red, black, and alluvial soils in Tamil Nadu and hill soils in Uttar Pradesh (Pathak et al., 1968; Balaguru and Mosi, 1973; Rawat and Mathpal, 1981). Whereas the available Mo significantly decreased with increases in the clay content in the soils of Assam (Chakraborty et al., 1982) and the salt-affected soils of Haryana (Gupta and Dabas, 1980), it increased with increases in the silt-plus-clay content in the alkali soils of Uttar Pradesh. Nayyar (1972) reported a maximum increase in available Mo at field-capacity moisture conditions, followed by a rate of 50% field capacity and saturation moisture conditions.

Availability of Soil Molybdenum

The availability of soil Mo to plants varies with soil characteristics as well as with crop species. The ideal Mo concentration in soil solution should range between 10^{-6} and 10^{-7} M, the threshold value being 10^{-8} M, below which crop plants show Mo deficiency. Molybdenum concentrations in soil solution of about 10^{-5} M are toxic to plants (Kubota, Lemon, and Allaway, 1963).

Among the soil factors that can profoundly affect the availability of Mo are the soil reactions; the other factors are mineralogy, degree of Mo saturation, and the presence of other competing anions. The usual concentration of Mo in legumes is several times higher than the general value for most other plants of $2\,mg\,kg^{-1}$ (dry matter). This is because of its involvement in nitrogen (N) fixation. Thus the amounts of Mo needed for plant growth are much lower than the amounts needed to supply the root nodule with sufficient Mo for N fixation. Hence legumes are the first plants to show Mo deficiencies and responses to addition of Mo fertilizer. *Brassica* crop species are known to accumulate excessive amounts of Mo when grown in soils with high amounts of available Mo (Nayyar, Randhawa, and Pasricha, 1977b).

Some 40% of 1,550 soils sampled throughout Poland were estimated to be deficient in Mo, and 20–30% were deficient in Cu (Gembarzewski and Stanislawska, 1987). The critical concentrations of Cu for wheat

(*Triticum aestivum* L.) and oats (*Avena sativa* L.) were 1.9 mg kg⁻¹ and 1.0 mg kg⁻¹. The contents of cadmium (Cd), cobalt (Co), and Mo in various Chilean soils and crops were determined by Gonzalez, Raez, and Lachica (1988); normal amounts of Mo were found in plants grown on desert soils, alluvial soils, and volcanic ash. Acidic peat soils are widely distributed in Southeast Asia. More than 20 million hectares of peat soils are located in Indonesia, Malaysia, and Thailand alone (Dent, 1986). Molybdenum availability in those soils was shown to increase with increases in pH up to 4.7, but it decreased with increases in pH above 4.7 as a result of excessive lime application to those soils.

Movement of Molybdenum in Soil

The reactions of molybdate with soils and clays have been widely studied (Barshad, 1951; Jones, 1957; Pasricha and Randhawa, 1977). Molybdate retention in soil occurs through anion exchange, primarily with surface OH groups in soil materials. Pasricha and Randhawa (1977) investigated the cationic effects on Mo adsorption for four soils representing different agroclimatic regions of Punjab, India. The relative capacities of the four soils for adsorbing molybdate followed this order: gray-brown podzolic > reddish chestnut > sierozem > arid brown soil. The magnitude of MoO_4^{2-} adsorption to soils saturated with different cations followed the order of their chemical valences: Fe-soil > Al-soil > Ca-soil > original soil > K-soil > NH₄-soil (Table 15.3). Soils saturated with different cations exhibited appreciable differences in pH values. Fe-soil and Al-soil showed a lowering of pH, whereas K-soil and NH₄-soil showed a slight increase in the pH of the equilibrium suspension (Table 15.4). The complementary ions on the soil exchange complex were shown to influence the Mo availability. In clay (humic gleysol) and loam (eutric cambisod) samples, the sorbed phosphorus (P) decreased the rate of Mo sorption, especially with high amounts of added P (Xie and Mackenzie, 1991). The presence of P reduced the intercept values of the Temkin equation, indicating reduced affinity of surfaces for Mo.

The plant availability of Mo applied to soil will decline with time because of reactions between Mo and soil constituents, leaching of Mo to regions below the rooting zone of plants, and removal of Mo taken up by plants. The relative importance of these processes will depend on the pH of the soil, the sesquioxide content of the soil, and the texture of the soil, as well as on the amount of rainfall. Jones and Belling (1967) found that 60–95% of added Mo was leached from 16-cm columns of calcareous

Table 15.3. *Molybdate adsorption from Mo solution ($10 \mu g\, mL^{-1}$) by four soils saturated with different cations (soil:solution ratio = 1:5)*

Soil	Mo adsorbed per kilogram of soil (mg)					
	Original soil	K-soil	NH_4-soil	Ca-soil	Al-soil	Fe-soil
Gray brown						
podzolic	42.9	25.0	19.0	44.3	49.8	49.9
Reddish chestnut	29.0	24.8	20.0	29.1	48.8	49.8
Arid brown	11.0	12.0	12.0	23.0	35.0	49.6
Sierozem	16.0	18.2	12.6	17.7	31.6	47.0

Source: Adapted from Pasricha and Randhawa (1977).

Table 15.4. *pH values in the equilibrium soil suspension for the four soils saturated with different cations when equilibrated with $10 \mu g$ of Mo per $1\, mL$ of solution*

Soil	Mo adsorbed per kilogram of soil (mg)					
	Original soil	K-soil	NH_4-soil	Ca-soil	Al-soil	Fe-soil
Gray brown						
podzolic	5.9	6.5	6.6	6.0	3.8	4.7
Reddish chestnut	6.7	7.5	7.0	6.9	4.2	5.1
Arid brown	8.1	8.4	8.3	8.4	4.6	5.6
Sierozem	8.8	9.0	8.6	8.7	8.2	8.4

Source: Adapted from Pasricha and Randhawa (1977).

sands in Western Australia with the equivalent of 450 mm of water. However, less Mo is likely to be lost by leaching from acid soils, as adsorption of MoO_4^{2-} is greater in such soils. Less than 4% of added Mo was removed by the equivalent of 250 mm of water from 3.75-cm columns of seven acidic soils (pH 5.5–6.5) (Smith and Leeper, 1969). The major cause of the decreasing Mo concentrations in a soil solution, and consequently the cause of the decline in the plant availability of Mo applied to the soil, appears to be irreversible fixation of Mo on the surfaces of particles (Barrow and Shaw, 1975). Riley et al. (1987) studied the extent of leaching of Mo from acidic sandy soils in Western Australia and found that approximately 10% of added Mo was removed by leaching from two gray sands (pH 5.8–6.1), whereas negligible quantities were removed from three other acidic sandy soils (pH 5.0–5.4). Leaching does not appear to be an important factor in the occurrence of Mo deficiency on yellow-brown acidic sandplain soils of the Western Australian wheat belt, where cereals and pasture legumes are grown and where the availability of applied Mo declines rapidly with time. Perhaps losses associated with reactions between Mo and soil constituents are more important than its leaching losses.

The influences of competing ions in the soil solution on the relationship between sorption of Mo by soils and Mo uptake by grass were investigated by Smith, Brown, and Devel (1987). Plants did not show any toxicity symptoms with addition of high amounts of Mo (up to 100 mg L^{-1}). Sorption data fit the Freundlich isotherm (Pasricha and Randhawa, 1977). Soils with the highest amounts of $CaCO_3$ sorbed the least amount of Mo; Cl^- in the soil increased Mo sorption, whereas SO_4^{2-} decreased it.

Molybdenum Accumulation in Plants

Excessive amounts of Mo seldom retard plant growth (Nayyar et al., 1977b), but they can be toxic to ruminants feeding on such plants. Molybdenum toxicity is an endemic nutritional problem in ruminants, primarily on wet, poorly drained, neutral or alkaline soils. Nayyar et al. (1977a), in survey of floodplain soils in Punjab, India, observed that the amounts of extractable Mo ranged from 0.025 to 0.618 $\mu g\,g^{-1}$. More than 26% of plant samples collected from 105 sites representing the calcareous floodplains contained toxic amounts of Mo ($>10\,\mu g\,g^{-1}$). On the basis of the interrelations of Mo with Cu and S in animal nutrition, 27% of the sites on the calcareous floodplains were found to have Mo accumulations in forage in clearly toxic amounts (Table 15.5).

Excessive Mo uptake in the floodplain regions of Punjab is attributed to alkaline reactions and moist soil conditions. Floodplain soils are periodically flooded, and they are spread to both sides of the rivers of Punjab. The water table is very shallow, and the soils are stratified. Forage grown on those soils transpires relatively more water while growing, owing to the greater availability of moisture in such wet soils. The Mo-enriched soil solution is absorbed into the plants at an increased rate of transpiration under wet soil conditions, resulting in an increase in passive accumulation of Mo in the plant system. That process operates more vigorously in the tropics and subtropics, where the prevailing high temperatures accentuate the transpiration process and continue to increase the rate of Mo uptake.

Table 15.5. *Available Mo content for calcareous floodplain sites and Mo concentrations in Egyptian clover grown on those soils*

District	No. of locations	Extractable Mo (μg g^{-1})		Mo content in plants (μg g^{-1})		Percentage toxic[a] samples
		Range	Mean	Range	Mean	
Ludhiana	35	0.025–0.233	0.151	0.50–14.20	6.56	17.2
Ropar	14	0.045–0.328	0.156	1.25–20.70	6.11	21.4
Kapurthala	13	0.080–0.268	0.133	2.10–13.00	5.91	15.4
Hoshiarpur	7	0.100–0.185	0.131	4.30–16.00	7.10	42.9
Ferozepur	24	0.030–0.398	0.171	2.90–19.50	7.44	20.8
Amritsar	12	0.168–0.618	0.410	3.42–12.90	7.68	41.7
Total	105	0.025–0.618	0.192	0.50–20.70	6.80	26.6

[a] Mo concentration greater than 10μg g^{-1}.
Source: Adapted from Nayyar et al. (1977a).

Molybdenum accumulation by forage plants to amounts that are toxic to ruminants is a soil-borne nutritional problem. Considerable areas of wet, saline, alkali soils (Pasricha and Randhawa, 1971) and alkaline calcareous floodplains (Nayyar et al., 1977a) in Punjab, India, are recognized as areas where leguminous forages have been found to contain toxic amounts of Mo. Plant species differ markedly with respect to their capacities to accumulate Mo. The average Mo content of a plant indicates its relative ability to accumulate Mo, rather than the concentration that would be expected to be found on a soil where high concentrations of Mo in forages are produced. Nayyar et al. (1977b) grouped forages into the three classes of high, moderate, and low accumulators of Mo, as shown in Table 15.6. The wide variation among plant species allows growers to select an adapted species that accumulates the least amount of Mo. Molybdenum accumulation varies significantly among the forage species and varies with the rate of its application to the soil. Leguminous forages accumulate more Mo than nonlegumes. Forages from *Brassica* species have been found to have the highest Mo concentrations (538μg g^{-1}), and cowpeas [*Vigna unguiculata* (L.) Walp.] rank next in accumulation (410μg g^{-1}). Maize (*Zea mays* L.), sorghum [*Sorghum bicolor* (L.) Moench], and teosinte (*Euchlaena mexicana* Schrad) are classified as low accumulators, with Mo concentrations of 117, 110, and 132μg g^{-1}, respectively. Accumulations of Mo in Japan rape (*Brassica*

Table 15.6. *Relative abilities of forage crops to accumulate Mo under the influence of varying amounts of applied Mo*

Forage species	No added Mo	Mo added at 0.5 µg g⁻¹	Mo added at 1.0 µg g⁻¹	Mo added at 1.5 µg g⁻¹	Mo added at 2.0 µg g⁻¹	Mean
High						
Japan rape (*Brassica campestris* L.)	2.2	100.3	248.3	402.0	538.0	258.2
Cowpea [*Vigna unguiculata* (L.) Welp.]	1.5	99.7	195.7	297.3	410.0	200.8
Oats (*Avena sativa* L.)	1.0	72.2	130.0	236.0	288.0	145.4
Moderate						
Indian clover (*Melilotus parviflora* Desv.)	1.3	81.0	164.0	198.3	241.3	137.2
Egyptian clover (*Trifolium alexandrinum* L.), first cut	1.8	72.5	154.0	198.0	288.0	142.9
Egyptian clover (*Trifolium alexandrinum* L.), second cut	1.4	61.5	118.0	182.0	248.0	122.2
Club beans (*Cyamopsis tetragonoloba* Tenb.)	1.2	54.0	123.0	178.0	262.0	123.6
Alfalfa (*Medicago sativa* L.), first cut	1.9	80.0	171.0	234.0	280.0	155.4
Alfalfa (*Medicago sativa* L.), second cut	1.1	34.0	63.0	116.0	137.0	70.7
Low						
Teosinte (*Euchlaena mexicana* Schrad)	1.2	37.8	60.8	113.3	132.3	69.1
Sorghum [*Sorghum bicolor* (L.) Moench]	1.1	23.8	55.3	85.3	109.8	56.1
Maize (*Zea mays* L.)	2.2	23.8	53.0	76.5	116.7	54.8

Note: Mo content in dry matter (mg kg⁻¹)

Source: Adapted from Nayyar et al. (1977b).

campestris L.), cowpeas, and maize increased by 275, 203, and $56 \mu g g^{-1}$, respectively, for each $1 \mu g$ of Mo added per gram of soil.

Molybdenum Nutrition and Animal Health

In ruminants, Mo toxicity is often associated with Cu deficiency, and the malady called molybdenosis can be corrected by supplementing the mineral composition of the feed with $CuSO_4$ or by giving Cu glycinate injections (Clawson, 1972). In general, tolerance to excessive amounts of Mo is to some extent determined by the concentration of SO_4^{2-}-S in the feed. However, plants can accumulate excessive amounts of Mo without any apparent signs of toxicity (Nayyar et al., 1977b). Thus the concentration at which Mo in forage plants becomes toxic to animals is higher in the presence of adequate amounts of Cu and S than at lower concentrations of these elements.

In a report on the use of municipal sewage sludge for agricultural land, the Council for Agricultural Science and Technology (1976) identified Mo as a potentially hazardous element. There have been reports of increases in the Mo content in plants grown on soils fertilized with sludge (Williams and Gogna, 1981). Plants growing on restored oil-shale-disposal sites have Mo concentrations high enough ($>8 mg kg^{-1}$) and Cu:Mo ratios low enough (<2.0) to cause molybdenosis in grazing ruminants (Stark and Redente, 1990). Such studies indicate that although Cu applications can be used for short-term improvements in Cu:Mo ratios, such applications cannot be relied on as the sole technique for preventing molybdenosis on restored shale-disposal sites. Murray and Baker (1990) studied forage and soil Mo values in the vicinity of a plant for processing Mo and reported that Mo concentrations in red clover (*Trifolium pratense* L.) measured at 23 sites ranged from 1.1 to $56.6 mg kg^{-1}$. The red clover at 14 of 23 sites sampled had Mo concentrations above the threshold value of $6 mg kg^{-1}$. The red clover Cu:Mo ratios were below the recommended ratio of 4:1 at 19 of the sites, showing a potential risk for molybdenosis in ruminants. They developed regression equations to predict red clover uptake of soil Mo based on soil extracts and soil pH data.

In a reconnaissance survey of the minerals in soil and plant samples in relation to adequate concentrations of the same elements in blood samples from animals in Lake Nakura National Park, Kenya, Maskall and Thornton (1989) showed that Mo concentrations in all plants were relatively high, and the availability of the element appeared to be increased in the wetter soils of high pH near the lake shore. More than 30% of the impalas sampled had blood Cu concentrations below the limit considered normal for domestic animals. The relatively high Mo concentrations in the grasses and forage plants were believed to have contributed to the possible Cu deficiencies among the impala and waterbuck in the park.

Interactions with Other Nutrients

Plants show a tendency to accumulate Mo at concentrations several times higher than those required for their normal growth and development, without showing any visible signs of toxicity (Nayyar et al., 1977b). However, such herbage, when fed to ruminants, appears to cause Mo toxicity, commonly called molybdenosis, but it is in fact Mo-induced Cu deficiency. Thus there is a strong relationship between Mo and Cu as far as ruminant nutrition is concerned. Similarly, Mo and S and Mo and P strongly interact in determining the uptake of Mo by herbage and the concentration at which it will be toxic to animals. The concentration at which Mo becomes toxic is higher in the presence of adequate S nutrition in the animal ration. Moreover, S is known to suppress, and P to synergize, Mo uptake by plants (Pasricha et al., 1977).

Molybdenum and Copper

Increasing the amount of Cu in forage plants through Cu fertilization can help to counteract the toxic effect of Mo. Nayyar et al. (1980) reported a significant interaction between Mo and Cu that affected the yield of Egyptian clover (*Trifolium alexandrinum*). The Mo concentrations in the crop tissues, at both cuttings, increased almost linearly with an increase in the amount of added Mo, and the Mo content ranged from $1.1 \mu g\,g^{-1}$ to as high as $276.5 \mu g\,g^{-1}$. Even at such high Mo concentrations, plant growth appeared normal, and there were no signs of Mo toxicity (Table 15.7).

Excessive Mo in herbage causes a decrease in its Cu content, and such a decrease is more conspicuous when no Cu fertilizer has been applied. Copper deficiency was reported among penned sheep in eastern Saudi Arabia (Ali and Al-Noaim, 1989), but there was no such problem among grazing sheep in the same area. The mean serum Cu concentration for grazing rams and ewes was higher than that for penned sheep (Table 15.8). Ali and Al-Noaim (1992) further reported that the Cu content of

Table 15.7. *Interaction effects of Mo and Cu on their contents in Egyptian clover*

Mo applied ($\mu g\,g^{-1}$)	Cu applied ($\mu g\,g^{-1}$)					
	0.0	2.5	5.0	10.0	20.0	Mean
Mo content ($\mu g\,g^{-1}$)						
0	2.4	1.9	2.6	0.7	0.8	1.5
0.25	40.5	40.7	41.5	38.5	34.7	39.2
0.50	110.8	113.0	104.0	90.5	78.7	99.4
1.00	195.0	192.0	166.5	146.0	135.7	167.0
2.00	337.5	335.5	330.3	312.2	303.6	323.8
Mean	137.2	136.6	128.8	117.6	110.7	
Cu content ($\mu g\,g^{-1}$)						
0	15.4	20.3	24.3	32.3	47.1	27.9
0.25	16.5	22.0	25.2	34.2	47.7	29.1
0.50	12.4	19.6	24.1	33.6	45.7	27.3
1.00	11.1	17.4	20.0	29.2	45.0	24.5
2.00	9.7	15.5	18.7	28.0	42.2	22.6
Mean	13.0	18.8	22.5	31.4	45.7	
Cu:Mo ratio						
0	6.42	10.68	9.35	46.14	58.88	18.60
0.25	0.41	0.54	0.61	0.89	1.37	0.74
0.50	0.11	0.15	0.14	0.37	0.58	0.27
1.00	0.06	0.09	0.12	0.20	0.33	0.15
2.00	0.03	0.05	0.06	0.25	0.14	0.11
Mean	0.09	0.14	0.17	0.27	0.41	

Source: Adapted from Nayyar et al. (1980).

the forage given to the penned sheep was low, whereas the Mo content was high, giving a low Cu:Mo ratio. As for the grazing sheep, the Cu content in their pasture plants was higher. Although the Mo content was also high, the Cu:Mo ratio was higher than that for the penned sheep, which is favorable for Cu assimilation in animals (McDowell, Conrad, and Ellis, 1983).

A Cu:Mo ratio of 4:1 in an animal diet is considered ideal; however, that ratio can easily be decreased severalfold if there is increased availability of Mo to plants (Table 15.7). Herbage plants utilize Mo from applied sources in a highly efficient manner. Therefore, even in situations where the Cu content in plants would ordinarily be adequate, Cu

Table 15.8. *Concentrations of serum Cu in*
penned and grazing Najdi sheep

| | Serum Cu (μmol L^{-1}) | | | |
| | Penned | | Grazing | |
Category	Site I	Site II	Site I	Site II
Rams				
Mean	7.74	3.74	14.44	14.26
SD	2.86	1.51	5.31	3.88
n	5	42	5	5
Ewes				
Mean	7.80	—	13.99	11.96
SD	3.56	—	4.80	2.72
n	45	—	20	20

Source: Adapted from Ali and Al-Noaim (1992).

Table 15.9. *Average plant Cu, Mo, and S contents of Echium*
plantagineum in comparison with control diet

Plant	Cu (μg g^{-1})	Mo (μg g^{-1})	S (%)	Cu:Mo
E. plantagineum	14	0.36	0.20	47
Control diet	8	0.44	0.33	18

Source: Adapted from Seaman and Dixon (1989).

availability can be disturbed by excessive plant intake of Mo, thereby
causing a drastic decrease in the Cu:Mo ratio. The Cu:Mo ratio in the
plants in the study by Nayyar et al. (1980) was above the threshold value
(>4:1) in the absence of applied Mo, but it decreased drastically with
application of even the lowest amount of fertilizer Mo (0.25 μg g^{-1}). At
that rate of Mo application, the ratio could not be brought to the normal
value even with application of the highest amount (20 μg g^{-1}) of Cu
fertilizer tried. On the other hand, a very high ratio is equally harmful. In
an investigation into the toxicity of *Echium plantagineum* in sheep, Seaman and Dixon (1989) found that sheep on an *Echium* diet lost weight,
and deaths occurred, with histological evidence of excessive Cu accumulation, usually accompanied by pyrrolidine alkaloid damage in the liver

and biochemical evidence of liver toxicity due to a high Cu:Mo ratio (Table 15.9).

The critical range for the Cu:Mo ratio in animal feeds is 2:1 to 4:1, and feeds and pastures with lower ratios can be expected to result in conditioned Cu deficiency (Miltimore and Mason, 1971). In the presence of a high ratio, toxicity due to excessive Cu intake may cause metabolic problems in animals. Haque, Aduayi, and Sibanda (1993) reviewed the available information from sub-Saharan Africa on the roles of Cu in soil, plant, and animal systems. They emphasized the need to maintain adequate amounts of Cu in the soil and in forage crops for ruminant consumption. Copper deficiencies and toxicities have been diagnosed in many sub-Saharan African countries. Although there are several ways of correcting Cu imbalances, success depends on the ability to identify the underlying causes. Interactions of Cu with other elements, particularly Mo, are important factors in determining whether the deficiency or toxicity is primary or secondary. Balbuena and Mastandrea (1992) observed clinical signs of diarrhea, anorexia, and lameness in steers and cows grazing on *Melilotus alba* pastures with high concentrations of Mo growing in the slightly alkaline soils of the central Chaco area in Argentina. Parenteral administration of Cu led to remission of those systems.

Sulfur and Molybdenum, Phosphorus and Molybdenum

The availability of soil Mo or fertilizer Mo to plants will vary significantly with the presence of competing ions, such as PO_4^{2-} and SO_4^{2-}. Stout et al. (1951) found that adding SO_4^{2-} depressed Mo uptake, possibly because SO_4^{2-} and MoO_4^{2-} compete for the same absorption sites on roots. Singh, Mehta, and Singh (1986) observed that Mo concentrations in plants decreased with S application, but increased with P application (Table 15.10). Addition of phosphate fertilizer perhaps can stimulate Mo uptake from soil solution.

In a comparison of how N fertilizers that also contained S affected the elemental composition of celery (*Apium graveolens*) grown on polluted marsh soil, Schnug and Schnier (1986) observed that with the use of ammonium sulfate (as compared with urea or calcium ammonium nitrate) there was a conspicuous increase in total S, accompanied by a significant decrease in Mo concentration of up to 80% (Figure 15.1). In soils that are inherently poor in available Mo, or where its availability is limited by acid conditions, addition of S fertilizer may further aggravate

Table 15.10. *Mo uptake by* Brassica campestris *as influenced by applied S and P*

Sulfur applied ($mg\,kg^{-1}$)	Mo uptake per pot (mg)			
	No added P	P added at $40\,mg\,kg^{-1}$	P added at $60\,mg\,kg^{-1}$	Mean
0	3.00	3.53	4.5	3.68
25	3.27	3.67	4.6	3.85
50	2.93	3.47	3.5	3.29
100	2.77	3.13	3.0	2.97
Mean	2.99	3.45	3.9	

Source: Adapted from Singh et al. (1986).

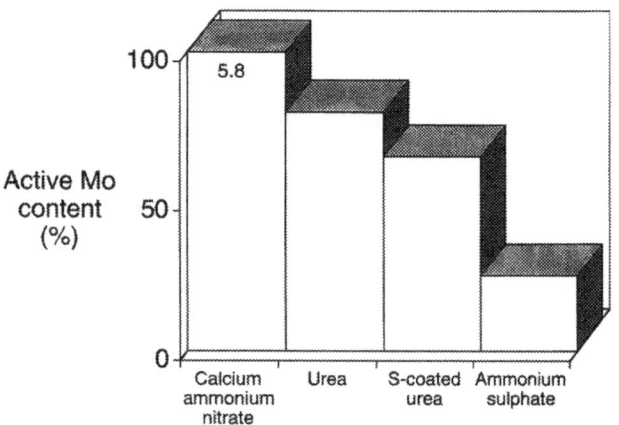

Figure 15.1. Relative Mo contents of young celery leaves 160 days after fertilization with different nitrogen sources on contaminated soil.

the problem. Fodder plants have often been reported to contain deficient amounts of Mo, but S is often in excess in ruminant feeds (Lamand, 1989). When the minimum concentration of Mo that would trigger the Cu-S-Mo interference in sheep was measured, with the S-Mo-Cu interference quantified on the basis of plasma Cu fractions insoluble in 5% trichloroacetic acid, it was found that interference occurred at Mo concentrations higher than $2.4\,mg\,kg^{-1}$ (dry matter), irrespective of the source of S (elemental S or SO_4^{2-}-S).

Table 15.11. *Effects of fertilizers on the yield and yield components of*
V. faba *in Xichang, China*

Treatment	Fruiting branches per plant	No. of pods per plant	No. of seeds per plant	Weight of 100 seeds (g)	Seed yield (kg m²)
Control	2.32	10.23	12.34	112.7	0.304
$(NH_4)_2MoO_4$	2.84	12.33	15.10	125.6	0.421
KH_2PO_4	2.81	11.75	14.75	128.3	0.470
$(NH_4)_2MoO_4$ + KH_2PO_4	3.71	12.95	17.31	131.5	0.506
SE (15 df)	0.341	0.318	0.581	1.841	0.0197

Source: Adapted from Xia and Xiong (1991).

Uptake of Mo by tomato (*Lycopersicon esculentum* Mill.) plants grown in a leached cinnamon meadow soil was reduced by long-term treatment with mineral fertilizer that decreased its pH. However, application of P fertilizer at high rates increased Mo uptake, as did the use of lime and organic fertilizers (Gezenchova, Mirchev, and Rankov, 1987). Many fertilizer trials with *Vicia faba* have been reported from China. Xia, Tang, and Bai (1984) reported that plant dry matter, chlorophyll content, and photosynthesis rate for *V. faba* were increased significantly by soaking the seed in ammonium molybdate. Xia and Xiong (1991) studied how the interaction of ammonium molybdate and KH_2PO_4 affected the growth of root nodules, total N, soluble sugar, chlorophyll content, and photosynthesis rate for *V. faba*. Treatment with $(NH_4)_2MoO_4$ + KH_2PO_4 increased the seed yield by 66% over the control value in field experiments (Table 15.11).

Molybdenum applied alone or in combination with S was readily available to *Brassica juncea* grown on loam soil with a moderate amount of available Mo (0.122 $\mu g\,g^{-1}$). Application of S at 50 $\mu g\,g^{-1}$ (as sulfate) decreased the Mo content to one-third, and the effect of applied S to decrease the Mo content of plants was more conspicuous when Mo was applied with S. With Mo addition at a rate of 0.5 $\mu g\,g^{-1}$, the recovery of applied Mo was reduced by 7.1% when S was added at 12.5 $\mu g\,g^{-1}$, and by 76.2% when S was added at 50 $\mu g\,g^{-1}$, with the same rate of Mo application (Table 15.12). In another investigation of the suppression of Mo uptake in fodder crops, Pasricha et al. (1977) observed that the Mo concentrations in plant tissues of berseem (*Trifolium alexandrinum* L.)

Table 15.12. *Percentage recovery of applied Mo as affected by S application*

S treatment (ppm)	Recovery (%)				
	Mo at $0.05\mu g\,g^{-1}$	Mo at $1.0\mu g\,g^{-1}$	Mo at $1.5\mu g\,g^{-1}$	Mo at $3.0\mu g\,g^{-1}$	Mean
0	3.82	3.68	2.97	2.49	3.24
12.5	3.54	3.20	2.50	2.37	2.90
25.0	3.10	2.16	1.59	1.64	2.12
37.5	2.01	1.77	1.38	1.51	1.67
50.0	0.91	1.34	1.09	1.18	1.13
Mean	2.68	2.43	1.91	1.84	

Source: Adapted from Pasricha and Randhawa (1972).

and oats were enhanced by application of P, whereas application of S reduced the Mo concentrations of plants. The effect of P to enhance the Mo content of the plants was stronger than the suppressing effect of S. The decrease in Mo content with S was more pronounced in the oat crop. An application of S (as gypsum) at $100\mu g\,g^{-1}$ caused a 57.6% decrease in the concentration of Mo, which was below the toxic threshold. But in berseem, even an application of S at $100\mu g\,g^{-1}$ could not lower the Mo to a safe concentration. Legume fodder can be more toxic than nonlegume fodder when grown on soils with sufficient Mo to produce toxicity, and under such conditions, application of S to decrease Mo content is more effective for nonlegumes than for legumes.

Crop Responses to Molybdenum Application

The primary role of Mo in plant nutrition is associated with N metabolism, where it acts as an enzyme activator. Under most soil conditions, the source of N is NO_3^-, and Mo is required by all crop species when N is provided as NO_3^-. Molybdenum is a critical constituent of the enzyme nitrate reductase. Different crop species have different requirements for Mo. In general, legumes have high requirements, but Mo accumulations in some of the *Brassica* crop species can be several times higher than those in legumes (Nayyar et al., 1977b).

It is suggested that Mo in *Vigna mungo* is phloem-immobile at a low Mo supply, but is phloem-mobile in all plant parts (with the possible exception of stem segments) at an adequate Mo supply (Jongruaysup,

Dell, and Bell, 1994). In a field study of Mo uptake by common beans (*Phaseolus vulgaris* L.), Nicoloso, Santos, and Camargo (1990) found that the stems contained the highest concentrations of Mo at all ages. Nodulation, plant dry weight, and seed yields for chickpeas (*Cicer arietinum* L.), lentils (*Lens culinaris* Medic), and *Lupinus albus* grown in pots were increased by seed inoculation. Seed yields generally were highest in plants given Mo (Yanni, 1992). In acid soils, yields of annual ryegrass (*Lolium multiflorum* Lam) increased with application of Mo in unlimed soils, but no such effect was observed in limed plots (Hillard et al., 1993). Vargas and Ramirez (1989), in field trials at Canas, Costa Rica, concluded that seed inoculation with low amounts of N and adequate P and Mo was necessary to increase soybean [*Glycine max* (L.) Merr.] and groundnut (*Arachis hypogaea*) yields.

Summary

In the tropics and subtropics, the soils are highly weathered and rich in sesquioxides, and the soil reaction is acidic. Under such conditions, MoO_4^{2-} is adsorbed by soils, and its concentration in the soil solution will determine its availability. The approach in establishing guidelines for the available Mo status of such soils must be based on adsorption–desorption relationships for Mo. Such a relationship can be described by plotting the amount of Mo adsorbed per unit mass of the soil against the Mo concentration in the equilibrium solution. In alkaline calcareous and alkali soils, excessive Mo in the soil solution can be encountered, especially in low-lying floodplains, where soils remain wet because of their topographic position, or where the water table is quite shallow. Molybdenum becomes more concentrated in the soil solution as water is lost by evapotranspiration under arid and semiarid tropical and subtropical conditions. On the basis of the existing information, it seems quite apparent that toxicity of Mo to ruminants because of its excessive accumulation in forage species is a greater problem than is its deficiency for plant growth. Liming of acid soils has been found to be helpful for increasing Mo availability for normal plant growth, but that may not be the case everywhere. Soils that are inherently poor in extractable Mo will require Mo application along with liming.

References

Ahmad, R. S., Khathak, J. K., and Perveen, S. (1991). Acid ammonium oxalate extractable molybdenum status of DIR soils and its relationship with soil properties. *Sashad J. Agric.* 7:105–13.

Ali, K. E., and Al-Noaim, A. A. (1989). Mineral status of indigenous sheep of
    Saudi Arabia under intensive system. In *Abstracts of the Saudi Biological
    Society 12th Symposium on the Biological Aspects of Saudi Arabia. King
    Saud University, Riyadh*, p. 139.
Ali, K. E., and Al-Noaim, A. A. (1992). Copper status of Najdi sheep in
    eastern Saudi Arabia under penned and grazing conditions. *Trop. Anim.
    Health Prod.* 24:121–4.
Anderson, A. J., and Moye, D. V. (1952). Lime and molybdenum in clover
    development in acid soils. *Aust. J. Agric. Res.* 3:95–110.
Balaguru, T., and Mosi, A. D. (1973). Studies on soil molybdenum in Tamil
    Nadu soils. *Madras Agric. J.* 60:147–51.
Balbuena, O., and Mastandrea, O. (1992). Clinical symptoms of cattle grazing
    on *Melilotus alba* with high molybdenum concentration in central-Chaco
    area. *Veterinaria-Argentina* 89:612–15.
Barnard, R. O., and Folscher, W. J. (1988). Growth of legumes at different
    levels of liming. *Trop. Agric. (Trinidad)* 65:113–16.
Barrow, N. J., and Shaw, T. C. (1975). The slow reaction between soil and
    anions. 4. Effect of time and temperature on contact between soil and
    molybdate on the uptake of molybdenum by plants and on the molybdate
    concentration in the soil solution. *Soil Sci.* 119:301–10.
Barshad, I. (1951). Factors affecting the molybdenum content of pasture
    plants. I. Nature of soil molybdenum, growth of plants and soil pH. *Soil
    Sci.* 71:297–13.
Bear, F. E. (1956). Molybdenum in plant and animal nutrition. *Soil Sci.* 81:159.
Beeson, K. C., and Matrone, G. (1976). *The Soil Factor in Nutrition: Animal
    and Human.* New York: Marcel Dekker.
Bhandari, A. R., and Randhawa, N. S. (1985). Distribution of available
    micronutrients in soils of apple orchards in Himachal Pradesh. *J. Indian
    Soc. Soil Sci.* 33:171–4.
Bromfield, S. M., Cumming, R. W., David, D. J., and Williams, C. H. (1983).
    The assessment of available manganese and aluminium status in acid soils
    under subterranean clover pastures of various ages. *Aust. J. Exp. Agric.
    Anim. Husb.* 23:192–200.
Chakraborty, S. K., Sinha, H., and Prasad, R. (1982). Distribution of boron
    and molybdenum and their relationship with certain properties of soils
    from Assam. *J. Indian Soc. Soil Sci.* 30:92–3.
Chappell, W. B., and Peterson, K. K. (eds.) (1977). *Molybdenum in the
    Environment.* New York: Marcel Dekker.
Clark, R. B. (1976). Plant efficiencies in use of calcium, magnesium and
    molybdenum. In *Plant Adaption to Mineral Stress in Problem Soils*, ed.
    M. J. Wright, pp. 175–91. Ithaca, NY: Cornell University Press.
Clawson, W. J. (1972). Copper, molybdenum and selenium in the west. In
    Proceedings of the Nevada-California Beef Conference. Reno: University
    of Nevada.
Council for Agricultural Science and Technology (1976). *Application of
    Sewage Sludge to Cropland: Appraisal of Potential Hazards of the Heavy
    Metals to Plants and Animals.* Report 64. Ames, IA: CAST.
Coventry, D. R., Hirth, J. R., and Fung, K. K. H. (1987). Nutritional restraints
    on subterranean clover grown on acid soils used for crop–pasture
    rotation. *Aust. J. Agric. Res.* 38:163–7.
Dent, F. J. (1986). Southeast Asian coastal peats and their use. An overview.
    In *Proceedings of the Second International Soil Management Workshop.*

Thailand/Malaysia Classification, Characterization and Utilization of Peat Land, ed. H. Eswaran et al., pp. 27–49.

Durate, V. M., Leley, V. K., and Narayana, N. (1961). Micronutrient status of the Bombay state soils. *J. Indian Soc. Soil Sci.* 9:41–53.

Folscher, W. J., and Barnard, R. O. (1985). Groei in chemiose samestelling van verskillende penlge-wasse onder suurgrondtoes tande. *South Afr. J. Plant Soil* 2:93–7.

Gembarzewski, H., and Stanislawska, E. (1987). Molybdenum and copper fertilization requirements of soils in Poland estimated on the basis of investigations of control farms of Institute of Soil and cultivation of plants. *Roczniki-Gleboznawcze* 38:161–74.

Gezenchova, L., Mirchev, S., and Rankov, V. (1987). Effect of prolonged intensive fertilizer application and liming on Mo uptake by tomato plants. *Rasteniev "D.-Nauki"* 24:98–104.

Gonzalez, C., Raez, M., and Lachica, M. (1988). A survey of contents of Cd, Co and Mo in some crops and soils of Chile. *Agrochimica* 32:90–3.

Grewal, J. S., Bhumbla, D. R., and Randhawa, N. S. (1969). Available micronutrient status of Punjab, Haryana and Himachal soils. J. Indian Soc. Soil Sci. 17:27–31.

Gupta, V. K., and Dabas, D. S. (1980). Distribution of molybdenum in some saline-sodic soils from Haryana. *J. Indian Soc. Soil Sci.* 28:28–30.

Haque, I., Aduayi E. A., and Sibanda, S. (1993). Copper in soils, plants and ruminant animal nutrition with special reference to sub-Saharan Africa. *J. Plant Nutr.* 16:2149–212.

Hillard, J. B., Haby, V. A., Hons, F. M., Hussey, M. A., and Gates, C. E. (1993). Factors associated with annual ryegrass yield response to limestone. *Commun. Soil Sci. Plant Anal.* 24:9–10.

Jones, G. B., and Belling, G. B. (1967). The movement of copper, molybdenum and selenium in soils as indicated by radioactive isotopes. *Aust. J. Agric. Res.* 18:733–40.

Jones, L. H. P. (1957). The solubility of molybdenum in simplified systems and aqueous suspensions. *J. Soil Sci.* 5:313–27.

Jongruaysup, S., Dell, B., and Bell, R. W. (1994). Distribution and redistribution of molybdenum in black gram (*Vigna mungo* L. Hepper) in relation to molybdenum supply. *Ann. Bot.* 73:161–7.

Kanwar, J. S., and Randhawa, N. S. (1974). *Micronutrient Research in Soils and Plants in India (A Review)*. ICAR technical bulletin (Agric.) 50.

Kim, K. W., and Thornton, I. (1993). Influence of uriniferous black shales on cadmium, molybdenum and selenium in soils and crop plants in the Deog-Pyoung area of Korea. *Environ. Geochem. Health* 15:119–33.

Kubota, J., Lemon, E. R., and Allaway, W. H. (1963). The effect of soil moisture content upon the uptake of molybdenum, copper and cobalt by alsike clover. *Soil Sci. Soc. Am. Proc.* 27:679–83.

Lamand, M. (1989). Influence of molybdenum and sulphur on copper metabolism in sheep: comparison of elemental S and SO_4. *Ann. Rech. Vet.* 20:103–6.

McDowell, L. R., Conrad, L. R., and Ellis, G. L. (1983). Mineral deficiencies and imbalances and their diagnosis. Presented at the Symposium on Herbivora Nutrition in Sub-Tropics and Tropics, Pretoria, South Africa.

Maskall, J. E., and Thornton, I. (1989). The mineral status of lake Nakuru National Park, Kenya: a reconnaissance survey. *Afr. J. Ecol.* 27:191–200.

Miltimore, J. E., and Mason, J. L. (1971). Cu to Mo ratio and Cu concentrations in ruminant foods. *Can. J. Anim. Sci.* 51:193–200.

Misra, S. G., and Misra, K. C. (1972). Distribution of total and available molybdenum in soils of U. P. *J. Indian Soc. Soil. Sci.* 20:193–6.

Mitra, G. N., Sahu, S. K., and Das, B. (1993). Available molybdenum status of the soils of Orissa. *J. Indian Soc. Soil Sci.* 41:168–9.

Mortvedt, J. J., Giordano, P. M., and Lindsay, W. L. (eds.) (1972). *Micronutrients in Agriculture.* Madison, WI: Soil Science Society of America.

Munns, D. N., and Franco, A. A. (1981). Soil constraints to legume production. In *Biological Nitrogen Fixation Technology for Tropical Agriculture*, ed. P. H. Graham and S. C. Harris. pp. 133–52. Cali, Colombia: Centro International de Agricultura Tropical.

Murray, M. R., and Baker, D. E. (1990). Monitoring and assessment of soil and forage molybdenum near an atmospheric source. *Environmental Monitoring and Assessment* 15:25–33.

Nayyar, V. K. (1972). Studies on molybdenum and copper in calcareous flood plain soils of Punjab. Ph.D. dissertation, PAU, Ludhiana.

Nayyar, V. K., Randhawa, N. S., and Pasricha, N. S. (1977a). Molybdenum accumulation in forage crops. I. Distribution of Mo, Cu, S and N in berseem grown on calcareous flood plains. *J. Res. Punjab Agric. Univ.* 14:245–51.

Nayyar, V. K., Randhawa, N. S., and Pasricha, N. S. (1977b). Molybdenum accumulation in forage crops. III. Screening of forage species for their capacity to accumulate molybdenum as Mo-toxic soils. *J. Res. Punjab Agric. Univ.* 14:406–10.

Nayyar, V. K., Randhawa, N. S., and Pasricha, N. S. (1980). Effect of interaction between molybdenum and copper on the concentration of these nutrients in berseem and its yield. *Indian J. Agric. Sci.* 50:434–40.

Nicoloso, F. T., Santos, O. S., and Camargo, R. P. (1990). Molybdenum uptake by common beans. *Revista do Centro de Ciencias Rurais, Universidate Federal de Santa Maria* 20:37–49.

Pasricha, N. S., and Fox, R. L. (1993). Plant nutrient sulphur in the tropics and the sub-tropics. *Adv. Agron.* 50:209–69.

Pasricha, N. S., Nayyar, V. K., Randhawa, N. S., and Sinha, M. K. (1977). Molybdenum accumulation in forage crops. II. Influence of sulphur fertilization on suppression of molybdenum uptake by berseem (*Trifolium alexandrinum* L.) and oat (*Avena sativa* L.) grown on Mo-toxic soil. *Plant Soil* 46:245–50.

Pasricha, N. S., and Randhawa, N. S. (1971). Available Mo status of some reclaimed saline-sodic soils and its effect on the concentration of Mo, Cu, S, and N in berseem (*Trifolium alexandrinum*) grown on these soils. In *Proceedings of the International Symposium on Soil Fertility Evaluation*, ed. J. S. Kanwar, N. P. Datta, S. S. Bains, D. R. Bhumbla and T. D. Biswas. New Delhi: Indian Society of Soil Science.

Pasricha, N. S., and Randhawa, N. S. (1972). Interaction effect of sulphur and molybdenum on the uptake and utilization of these elements by raya (*Brassica juncea* L.). *Plant Soil* 37:215–20.

Pasricha, N. S., and Randhawa, N. S. (1977). Molybdenum adsorption by soils as affected by different cations. *Agrochimica* 21:105–13.

Pathak, A. N., Shanker, H., and Misra, R. V. (1968). Molybdenum status of certain Uttar Pradesh soils. *J. Indian Soc.Soil Sci.* 16:400–4.

Rai, M. M., Pal, A. R., and Shitoley, D. B. (1972). Available molybdenum status of deep black soils of Madhya Pradesh. *J. Indian Soc. Soil Sci.* 20:53–8.

Rawat, P. S., and Mathpal, K. N. (1981). Micronutrient status of some soils of U.P. hills. *J. Indian Soc. Soil Sci.* 29:208–14.

Reddy, G. R. (1964). Molybdenum status of western India soils. *Indian J. Agric. Sci.* 34:219–33.

Riley, M. M., Robson, A. D., Gartrell, J. W., and Jeffery, R. C. (1987). The absence of leaching of molybdenum in acidic soils from Western Australia. *Aust. J. Soil Res.* 25:179–84.

Schnug, E., and Schnier, C. (1986). The effect of sulfur-containing nitrogen fertilizers on the elemental composition of celery (*Apium graveolens*) grown on polluted marsh soil. *Plant Soil* 91:273–8.

Seaman, J. T., and Dixon, R. J. (1989). Investigations into the toxicity of *Echium plantagineum* in sheep. 2. Pen feeding experiments. *Aust. Vet. J.* 66:286–8.

Sharma, C. M., and Minhas, R. S. (1987). Evaluation of extractants for available molybdenum in some acid soils of Himachal Pradesh. *J. Indian Soc. Soil Sci.* 35:329–30.

Sharma, C. M., Minhas, R. S., and Masand, S. S. (1988). Molybdenum in surface soils and its vertical distribution in profiles of some acid soils. *J. Indian Soc. Soil Sci.* 36:252–6.

Shukla, P., and Pathak, A. N. (1973). Effect of molybdenum, phosphorus and sulphur on the yield and composition of berseem in acid soil. *J. Indian Soc. Soil Sci.* 21:187–92.

Singh, V., Mehta, V. S., and Singh, B. (1986). Individual and interaction effect of sulphur, phosphorus and molybdenum in mustard. *J. Indian Soc. Soil Sci.* 34:535–8.

Smith, B. H., and Leeper, G. W. (1969). The fate of applied molybdate in acidic soils. *J. Soil Sci.* 20:246–54.

Smith, C., Brown, K. W., and Devel, L. E., Jr. (1987). Plant availability and uptake of molybdenum as influenced by soil type and competing ions. *J. Environ. Qual.* 16:377–82.

Stark, J. W., and Redente, E. E. (1990). Copper fertilization to prevent molybdenosis on restored oil-shale deposit piles. *J. Environ. Qual.* 19:502–4.

Stout, P. R., Meagher, W. R., Peasson, G. A., and Johnson, C. M. (1951). Molybdenum nutrition of crop plants. I. The influence of phosphate and sulphate on the absorption of molybdenum from soils and solution cultures. *Plant Soil* 3:51–87.

Takkar, P. N. (1987). Micronutrients: their forms, distribution, areas of deficiency and toxicity and factors affecting availability in the Punjab, Haryana and Himachal Pradesh. In *Proceedings of the National Symposium on Land and Water Management in the Indus Basin (India)*, ed. A. S. Atwal, S. S. Prihar, A. K. Srivastava, Sucha Singh, V. V. N. Murty, Balraj Singh, M. K. Sinha and Y. P. S. Bajaj vol. 1, pp. 348–80. Ludhiana, Punjab, India: Indian Ecological Society.

Takkar, P. N., Chhibba, I. M., and Mehta, S. K. (1989). *Twenty Years of Coordinated Research on Micronutrients in Soils and Plants.* Bulletin I, Indian Institute of Soil Science, Bhopal.

Underwood, E. J. (1977). *Trace Elements in Human and Animal Nutrition,* 4th ed. New York: Academic Press.

Vargas, R., and Ramirez, C. (1989). Response of soybean and groundnuts to *Rhizobium* and to N, P and Mo fertilizer on a Typic Pellustert from Canas, Guanacaste. *Agronomica-Costarricense* 13:175–81.

Verma, K. P., and Jha, K. K. (1970). Studies on soil molybdenum of Bihar. *J. Indian Soc. Soil Sci.* 18:37.

William, C. H. (1980). Soil acidification under clover pasture. *Aust. J. Exp. Agric. Anim. Husb.* 20:561–7.

Williams, J. H., and Gogna, J. C. (1981). Molybdenum uptake from sewage sludge treated soil. In *Heavy Metals in the Environment*, pp. 189–92. Edinburgh, UK: CEP Consultants Ltd.

Xia, M. Z., Tang, Y., and Bai, H. Y. (1984). Effect of trace elements on physiological function and yield of *Vicia faba. Plant Physiol. Commun.* 6:28–30.

Xia, M. Z., and Xiong, F. Q. (1991). Interaction of molybdenum, phosphorus and potassium on yield of *Vicia faba, J. Agric. Sci. (Cambridge)* 117:85–9.

Xie, R. J., and Mackenzie, A. F. (1991). Molybdate sorption–desorption in soils treated with phosphate. *Geoderma* 48:321–33.

Yanni, Y. G. (1992). Performance of chickpea, lentil and lupin nodulated with indigenous or inoculated rhizobia, boron, cobalt and molybdenum fertilization schedules. *World J. Microbiol. Biotech.* 8:607–13.

Index

For EU product safety concerns, contact us at Calle de José Abascal, 56–1°, 28003 Madrid, Spain or eugpsr@cambridge.org.

www.ingramcontent.com/pod-product-compliance
Ingram Content Group UK Ltd.
Pitfield, Milton Keynes, MK11 3LW, UK
UKHW010854090126
466816UK00011B/224